LONDON MATHEMATICAL SOCIETY LECTURE NOTE SERIES

Managing Editor: Professor I.M.James,
Mathematical Institute, 24-29 St Giles, Oxford

1. General cohomology theory and K-theory, P.HILTON
4. Algebraic topology: a student's guide, J.F.ADAMS
5. Commutative algebra, J.T.KNIGHT
8. Integration and harmonic analysis on compact groups, R.E.EDWARDS
9. Elliptic functions and elliptic curves, P.DU VAL
10. Numerical ranges II, F.F.BONSALL & J.DUNCAN
11. New developments in topology, G.SEGAL (ed.)
12. Symposium on complex analysis, Canterbury, 1973, J.CLUNIE & W.K.HAYMAN (eds.)
13. Combinatorics: Proceedings of the British combinatorial conference 1973, T.P.McDONOUGH & V.C.MAVRON (eds.)
14. Analytic theory of abelian varieties, H.P.F.SWINNERTON-DYER
15. An introduction to topological groups, P.J.HIGGINS
16. Topics in finite groups, T.M.GAGEN
17. Differentiable germs and catastrophes, Th.BROCKER & L.LANDER
18. A geometric approach to homology theory, S.BUONCRISTIANO, C.P.ROURKE & B.J.SANDERSON
20. Sheaf theory, B.R.TENNISON
21. Automatic continuity of linear operators, A.M.SINCLAIR
23. Parallelisms of complete designs, P.J.CAMERON
24. The topology of Stiefel manifolds, I.M.JAMES
25. Lie groups and compact groups, J.F.PRICE
26. Transformation groups: Proceedings of the conference in the University of Newcastle upon Tyne, August 1976, C.KOSNIOWSKI
27. Skew field constructions, P.M.COHN
28. Brownian motion, Hardy spaces and bounded mean oscillation, K.E.PETERSEN
29. Pontryagin duality and the structure of locally compact abelian groups, S.A.MORRIS
30. Interaction models, N.L.BIGGS
31. Continuous crossed products and type III von Neumann algebras, A.VAN DAELE
32. Uniform algebras and Jensen measures, T.W.GAMELIN
33. Permutation groups and combinatorial structures, N.L.BIGGS & A.T.WHITE
34. Representation theory of Lie groups, M.F.ATIYAH et al.
35. Trace ideals and their applications, B.SIMON
36. Homological group theory, C.T.C.WALL (ed.)
37. Partially ordered rings and semi-algebraic geometry, G.W.BRUMFIEL
38. Surveys in combinatorics, B.BOLLOBAS (ed.)
39. Affine sets and affine groups, D.G.NORTHCOTT
40. Introduction to H_p spaces, P.J.KOOSIS
41. Theory and applications of Hopf bifurcation, B.D.HASSARD, N.D.KAZARINOFF & Y-H.WAN
42. Topics in the theory of group presentations, D.L.JOHNSON
43. Graphs, codes and designs, P.J.CAMERON & J.H.VAN LINT
44. Z/2-homotopy theory, M.C.CRABB
45. Recursion theory: its generalisations and applications, F.R.DRAKE & S.S.WAINER (eds.)
46. p-adic analysis: a short course on recent work, N.KOBLITZ
47. Coding the Universe, A. BELLER, R. JENSEN & P. WELCH
48. Low-dimensional topology, R. BROWN & T.L. THICKSTUN (eds.)
49. Finite geometries and designs, P. CAMERON, J.W.P. HIRSCHFELD & D.R. HUGHES (eds.)

London Mathematical Society Lecture Note Series. 49

Finite Geometries and Designs

Proceedings of the Second Isle of Thorns Conference 1980

Edited by

P.J.CAMERON
Fellow and Tutor at Merton College, Oxford

J.W.P.HIRSCHFELD
Lecturer in Mathematics, University of Sussex

D.R.HUGHES
Professor of Mathematics, Westfield College, London

CAMBRIDGE UNIVERSITY PRESS

CAMBRIDGE

LONDON NEW YORK NEW ROCHELLE

MELBOURNE SYDNEY

CAMBRIDGE UNIVERSITY PRESS
Cambridge, New York, Melbourne, Madrid, Cape Town,
Singapore, São Paulo, Delhi, Tokyo, Mexico City

Cambridge University Press
The Edinburgh Building, Cambridge CB2 8RU, UK

Published in the United States of America by Cambridge University Press, New York

www.cambridge.org
Information on this title: www.cambridge.org/9780521283786

First published 1981
Re-issued 2011

A catalogue record for this publication is available from the British Library

ISBN 978-0-521-28378-6 Paperback

CONTENTS

PREFACE

This book contains the proceedings of a conference that took place from 15th to 19th June, 1980, at the White House, Isle of Thorns, Chelwood Gate, England, which is a conference centre run by the University of Sussex.

There are 36 articles in all. The Introduction provides a background for non-specialists and places in context the 35 papers submitted. An asterisk following an author's name in some multi-authored papers indicates the presenter of the paper. At the back of the book are a list of talks given for which there is no paper and a list of participants.

We are indebted to the British Council for supporting some of the participants and to the publishers John Wiley, McGraw Hill, Oxford University Press, Pitman and Springer for supporting a book exhibition.

Above all, we are profoundly grateful to Mrs. Jill Foster of the University of Sussex for the excellent typing of the camera-ready copy from which this book was produced.

P.J.C.
J.W.P.H.
D.R.H.

INTRODUCTION

One of the features of geometry, and of finite geometry in particular, is the difficulty of giving a concise definition of the subject. As well as the wide variety of structures that are studied and techniques that are used, an important factor contributing to this intractibility is the way in which different parts of the subject link up with and influence one another. This is part of the excitement of the subject for its practitioners, but may be off-putting for outsiders who see a confused tangle rather than an elegant network. The purpose of this introduction is to attempt to trace some of the main threads of finite geometry, and to locate the papers of this collection in the warp and weft of its fabric.

The structure of the subject militates against a linear tour of its highlights; but, of course, there is no other way to write an introduction! To simplify the task, we regard finite projective geometries (Galois spaces) as the central concept.

Let n be a positive integer, and q a prime power; let $GF(q)$ denote the Galois field with q elements. The elements of the n-dimensional *projective geometry* $PG(n,q)$ or $S_{n,q}$ are the subspaces of an $(n+1)$-dimensional vector space V over $GF(q)$; each has a geometric dimension which is one less than its vector space dimension. Thus the basic objects, the points, are the 1-dimensional subspaces of V . It is common to identify an arbitrary subspace with the set of points it contains.

We may loosely divide the study of finite projective geometry into two parts, whose extensions and relations cover a great part of combinatorics : characterisations, and the study of subsets. We deal with these in turn.

Axioms for projective spaces are "classical", and can be found in Veblen and Young's "Projective Geometry". In terms of points and lines, they may be stated as follows:

(i) any line has at least three points;

(ii) any two points lie on a unique line;

(iii) a transversal to two sides of a triangle meets the third side also.

A subspace can be defined as a set of points containing the line through any two of its points; and the dimension of the geometry is the number of subspaces in a maximal chain (excluding the empty set and the whole space).

The characterisation has a very important feature. Any finite structure satisfying the axioms and having dimension at least 3 is isomorphic to PG(n,q) for some n and q ; but this is not so for dimension 2 . Here, *projective planes* not isomorphic to the "classical" PG(2,q) exist. They are charcterised by the first two of the above axioms together with the requirement that any two lines are concurrent. The planes PG(2,q) are called *Desarguesian,* since they are characterised by the additional requirement that the theorem of Desargues is valid.

A finite projective plane has an *order* n , with the property that any line has $n+1$ points, and there are n^2+n+1 points altogether. The order of PG(2,q) is q .

If a line of a projective plane and all its points are deleted, the resulting structure is called an *affine plane*. Its lines fall into parallel classes in such a way that Euclid's parallel postulate holds. The projective plane can be reconstructed from the affine plane by adjoining a "point at infinity" to the lines of each parallel class, the new points lying on a "line at infinity". In a similar way, the *affine space* AG(n,q) is obtained from PG(n,q) by deleting a hyperplane (a subspace of codimension 1) together with all the non-empty subspaces it contains.

The study of finite projective and affine planes is of enormous importance in finite geometry; much of Dembowski's compendious "Finite Geometries" is devoted to this topic. The principal tool has been the use of collineation groups, and the interplay between groups and planes has provided a lot of information. Four papers in the present collection, those by Cohen *et al.*, Foulser, Hering and Kallaher, develop this theme.

Prominent in such studies is the notion of a *central collineation,* one which fixes all the lines through a point (called its *centre*). Such a collineation also has the property that it fixes all the points on a line (called the *axis*); it is called an *elation* or a *homology* according as the centre lies on the axis or not. If we choose the axis as line at infinity, then elations and homologies are translations and dilatations respectively of the affine plane, fixing every parallel class.

A typical example of the role of central collineations in characterisation theorems is the following well-known result. It is easy to see that, in a plane of order n, there are at most n elations with given centre and axis. If every incident point and line are centre and axis for n elations, then the plane is necessarily Desarguesian.

Hering shows that, if an affine plane possesses central collineations of certain types, then its full collineation group has a unique minimal normal subgroup, which is a simple group. Thus it is to be expected that the effect of recent work on the classification of finite simple groups will be felt in this part of geometry: we should examine the known simple groups to see how they can act on planes. Both Hering and Kallaher take up this theme.

In a projective plane, an involution (a collineation of order 2) which is not central is called a *Baer involution,* and has the property that its fixed points and lines form a subplane whose order is the square root of the order of the plane (a *Baer subplane*). (In general, the order of a subplane of a projective plane of order n cannot exceed $\sqrt{n}$. However, there can be affine subplanes of larger order: for example, both PG(2,4) and PG(2,7) "contain" AG(2,3). This situation is studied by Vedder.)

A feature of affine spaces is the existence of a group of translations, acting trasitively on the points but fixing every parallel class. The most important affine planes, the translation planes (studied by Cohen *et al.*, Foulser and Kallaher) have the same property. Cohen, Ganley and Jha consider translation planes admitting a collineation group fixing a subplane and acting transitively on the parallel classes outside this subplane; Kallaher deletes the transitivity condition but assumes that the group is of known type; while Foulser takes the specific case where the plane has order 81 but assumes only the existence of two collineations of order 3 whose fixed point sets are overlapping Baer subplanes.

Other classes of structures admitting translations are considered by Jungnickel and Marchi. Among Jungnickel's structures are the class-regular Hjelmslev planes, studied further by Sane.

A large area of combinatorics, the theory of designs, involves a generalisation of the first two axioms for projective geometries. If t is a positive integer, a t-*design* consists of a set of points equipped with a collection of proper subsets called "blocks", the blocks having a constant size k, and any t points lying in a non-zero constant number λ of blocks. If $\lambda = 1$, the design is called a *Steiner system*. Thus, the subspaces of fixed dimension in a projective space are the blocks of a 2-design; the lines form a Steiner system.

As yet, no non-trivial t-design with $t \geq 6$ is known; but there are two remarkable 5-designs, both Steiner systems, having 12 and 24 points, which have been known since the 1930s. These designs are connected with Mathieu's simple groups and Golay's perfect error-correcting codes, and have been intensively studied; yet this remarkable seam has further wealth to yield. Beth gives a new construction and uniqueness proof for the first of them.

The design formed by the points and hyperplanes of $PG(n,q)$ has the property that the number of points and blocks are equal. More generally, any 2-design with this property is called *symmetric*. The symmetric designs are extremal with respect to *Fisher's inequality*, which asserts that a 2-design has at least as many blocks as points.

Biggs and Ito describe another situation in which extremal configurations are often provided by symmetric designs. An ordinary (undirected) regular graph with girth 6 and valency k has at least $2(k^2 - k + 1)$ vertices. Equality is attained if and only if the graph can be constructed as follows: the vertices are the points and lines of a projective plane of order $k - 1$, two vertices being adjacent exactly when they correspond to a point and a line which are incident in the plane. A similar construction, applied to a symmetric design with $\lambda > 1$, would yield a graph of girth 4; yet this graph may have a λ-fold covering graph whose girth is 6. If such a covering graph exists, the number of its vertices exceeds the bound by just $2(\lambda - 1)$. Several examples exist.

A recent approach to design theory has made use of error-correcting codes. The setting for these is the set S^n of *words* or n-tuples of elements from an *alphabet* S of q symbols. The *distance* between two words is the number of coordinates in which they differ. A *code* is simply a subset of S^n. If it is to correct d errors, we must ensure that there is at most one codeword distant d or less from any given word; the triangle inequality shows that this will be achieved if the shortest distance between two codewords is at least $2d + 1$.

An important special case is that where S is a finite field $GF(q)$, and the code C is *linear* (that is, a subspace of S^n). In this case, the distance between two codewords is just the *weight* of their difference (the number of non-zero coordinates). Thus it is important to know the *weight distribution* of a code, the number of codewords of each weight. The *MacWilliams identities* show that the weight distribution of a code C determines that of its *dual code* $C^\perp$ (with respect to the standard inner product on S^n). Using them, together with classical invariant theory, Gleason found the general form of the weight distribution of a self-dual code over GF(2). For details and generalisations, we refer to the book "Error-correcting Codes" by MacWilliams and Sloane.

Given a design on v points, each block can be represented by a v-tuple, having ones in the positions corresponding to the points of the block and zeros elsewhere. The subspace spanned by all such v-tuples over GF(q) (for some q) is a linear code. Under suitable conditions, this code, or a modification of it, is self-dual. Information about the weight distribution of the code interacts with structural information about the design. Furthermore, the permutation representation of any automorphism group of the design admits the code as an invariant subspace. Hall gives a survey of the connection between codes and designs. Highlights include the proof that a projective plane of order 10 (if any exists) can possess no collineation of order 5, and the construction of a new symmetric design on 41 points.

Two other papers also study planes of order 10. A *k-arc* in a projective plane is a set of k points no three of which are collinear; a k-arc is *complete* if it is not contained in a larger arc. Bruen relates the existence of a complete 6-arc to the existence of a set of complete 9-arcs and a set of complete 10-arcs in a plane of order 10. Coding theory methods enter into this paper, and are central to the paper of Assmus and Novillo Sardi. By considering the geometric configurations (on 16 or 20 points) defined by codewords of weight 16 or 20, they are led to consider a generalization of Steiner systems, in which blocks have 4 or 6 points and any 3 points lie in a unique block.

Finally in this area, Liebler turns his attention to projective geometry codes, where the structures in question are defined by subspaces of various dimensions in PG(3,q), and proves a new inclusion relation among the codes, using the representation theory of a suitable cyclic collineation group.

Tallini's paper is related in a different way to the problem of characterising projective geometry. So far, we have considered the points of a projective space as its basic objects. It is possible to take the set of i-dimensional subspaces for any value of i, and assign structure to it. (These sets are sometimes known as Grassman manifolds.) The Grassman manifold whose "points" are the lines of projective space has certain subsets called "lines", a Grassman "line" consisting of all the projective lines lying in a plane and passing through a point. Tallini gives axioms characterising this geometry.

We turn now to the other aspect of projective geometry, the study of properties of subsets of the point set. This can be subdivided, somewhat arbitrarily, into two parts: the geometric structure of a subset (for example, the configuration formed by the lines it contains), and the cardinalities of the intersections of lines (or other subspaces) with the subset.

Some of the most familiar subsets of projective spaces are quadrics and Hermitian varieties (the sets of zeros of non-singular quadratic or Hermitian forms). Each of these, equipped with the subspaces it contains, forms a geometry known as a (classical) *polar space*. Further polar spaces are defined by non-singular alternating bilinear forms: in this case the points of the polar space are all the points of the projective space, but only those subspaces which are totally isotropic (that is, on which the form is identically zero) belong to the polar space. These polar spaces are sometimes known by the same names as the classical groups associated with them, namely *orthogonal*, *unitary* and *symplectic*.

Classical polar spaces are given to us embedded in projective spaces. One may ask, supposing the polar spaces known, just how can they be embedded in projective spaces? Lefèvre-Percsy in her paper, surveys results on this question, and proves some new ones.

The problem of axiomatising polar spaces was solved by Tits in 1974, in his lecture notes on "Buildings of Spherical Type and Finite BN-Pairs". Tits defines an (abstract) polar space of rank n to be a set equipped with a collection of distinguished subsets called subspaces, having the following properties:

1. Any subspace, together with the subspaces it contains, is a projective space of dimension at most $n-1$.

2. Any intersection of subspaces is a subspace.
3. If M is an $(n-1)$-dimensional subspace and x a point, then the union of the set of lines joining x to points of M is an $(n-1)$-dimensional subspace.
4. There exist two disjoint $(n-1)$-dimensional subspaces.

(Tits uses the term "projective space" in a more general sense than the one we have defined. In place of our axiom 1, it is required only that each line contains at least two points. However, it is straightforward to show that the point set of such a generalised projective space is a disjoint union of point sets of restricted projective spaces, the lines being all those of the constituent spaces together with all pairs of points in different constituents.)

The result of Tits relevant to us is that a finite abstract polar space of rank at least 3, in which all lines have at least three points, is a classical polar space of one of the types described earlier.

Subsequently, Buekenhout and Shult gave a simpler axiom scheme, involving only the points and lines. Their axioms were as follows:

1. Any line has at least three points.
2. If a point x is not incident with a line L, then x is collinear with one or all points of L.
3. There exists a line; and no point is collinear with all others.

They showed that, starting from these hypotheses, it is possible to reconstruct all the subspaces and verify Tits' axioms.

It should be noticed that, as in the axiomatisation of projective spaces, the theorem only applies provided the dimension is sufficiently large: polar spaces of rank 2 are not covered by Tits' theorem. For these, the maximal subspaces are lines, and a stronger version of Buekenhout and Shult's second axiom holds: if x is not incident with L, then x is collinear with just one point of L. Such a geometry is called a *generalised quadrangle*. Thus, generalised quadrangles stand in the same relation to polar spaces as projective planes do to projective spaces. It might then be expected that the theory of generalised quadrangles would parallel that of projective planes, with the place of PG(2,q) being taken by the classical quadrangles (the classical polar spaces of rank 2). This is indeed the case. A number of characterisations of classical quadrangles by configuration theorems as properties of automorphism groups have been given.

One of the most important of these is Tits' theorem on Moufang quadrangles. These are quadrangles admitting sufficiently many "root automorphisms" (analogous to central collineations of projective planes), and Tits has shown, in particular, that finite Moufang quadrangles are classical. (This should be seen as the analogue of the characterisation of Desarguesian projective planes mentioned earlier.) Payne and Thas, in their paper in this volume, give an alternative, more elementary, approach to this important theorem. The basic idea is to establish relations between root automorphisms and structural properties, similar to the well-known relation between central collineations and Desargues' theorem for planes.

These classes of geometries have been generalised further. It is possible to define *generalised n-gons* for any $n \geq 2$. The precise definition does not concern us here; it suffices to say that projective planes are generalised 3-gons, and just as we saw that a projective plane gives rise to an extremal graph of girth 6, so a generalised n-gon gives an extremal graph of girth 2n. Generalised polygons are the objects from which *buildings* are constructed, in much the same way as polytopes (or Coxeter complexes) are built from ordinary polygons. The class of buildings includes projective and polar spaces. Buildings are associated with the finite simple groups of "Lie type".

A more general class of *diagram geometries* has been defined by Buekenhout. He allows a wider class of "building blocks", and a more general "construction method". As a result, he finds geometries associated with many of the recently-discovered "sporadic" simple groups. A diagram geometry (in Buekenhout's sense) which is built from generalised polygons is called a *Tits geometry* by Ott and Ronan, who continue the programme (initiated by Tits) of studying such geometries. Much of this study has a topological flavour. Any "geometry" which consists of subspaces of various dimensions or "types" can be described as a simplicial complex, where a simplex is a flag or set of mutually incident subspaces. (Our graph of girth 6 constructed from a projective plane earlier is an example.) Ott and Ronan study the universal covers of Tits geometries, showing that under certain conditions they are necessarily buildings. Ronan examines similar problems of universal covers for some sporadic Buekenhout geometries.

Another method, that of Hecke algebras or incidence algebras, underlies the celebrated result of Feit and Higman : this asserts that a finite generalised n-gon with $s+1$ points on each line and $t+1$ lines through each point $(s, t > 1)$ can exist only if $n = 2,3,4, 6$ or 8. Ott extends this method to wider classes of diagram geometries.

The article by Hall and Shult may be fitted loosely into this framework. Consider a polar space with three points on every line. (Such a space is necessarily of symplectic or orthogonal type over $GF(2)$: there are no non-classical quadrangles with $s = 2$.) Form a graph, whose vertices are the points of the polar space, two vertices adjacent whenever they are collinear. Examining the Buekenhout-Shult axioms, we see that such graphs are characterised by the following *triangle property :*

(*) any edge xy lies in a triangle xyz having the property that any further vertex is joined to one or all of x, y and z ;

together with the nondegeneracy conditions that there is at least one edge and that no vertex is joined to all others. Indeed, this characterisation, given by Shult and Seidel, preceded and motivated the Buekenhout-Shult theorem. The same authors considered a variation, the *cotriangle property,* obtained from (*) by replacing "edge" and "triangle" by "non-edge" and "cotriangle" (a cotriangle being a set of three pairwise nonadjacent vertices).

Graphs with the cotriangle property include the symplectic geometries and the complements of quadrics over GF(2) : in these cases a cotriangle is a hyperbolic line. Details of the characterisation are given by Hall and Shult, who proceed to determine the *locally cotriangular* graphs, those in which the induced subgraph on the set of neighbours of any vertex has the cotriangle property.

The diagram geometries of Buekenhout are not the only extension of generalised polygons. Consider the following system of axioms : any point lies on $t+1$ lines, any line has $s+1$ points, (with at most one line through any two points), and if a point x does not lie on a line L then x is collinear with exactly α points of L. Such a structure is called a *partial geometry*. For $\alpha = s+1$, it is a Steiner system (and, in particular, a projective plane if $\alpha = s+1 - t+1$), while for $\alpha = 1$ it is a generalised quadrangle. Examples of partial geometries with $\alpha = t$ (called nets) are obtained by taking the points of an affine plane of order $s+1$ together with the lines in $t+1$ distinguished parallel classes. These types and their duals include most of the known partial geometries. Haemers describes an unusual new example, not of any of these types.

Neumaier generalises still further, requiring that any two points lie on 0 or λ "lines" (so that partial geometries are the case $\lambda = 1$). He observes that many structures studied previously in their own right are special cases of such $1\frac{1}{2}$-*designs*, and that they arise naturally in the context of strongly regular graphs. Wolff's *point stable designs* are still more general, and indicate the important role of linear algebra in this subject.

Yet another extension is given by the *near n-gons* of Shult and Yanushka. These are interesting because, as well as the generalised n-gons, they include the dual polar spaces. These are just polar spaces with the inclusion relations reversed, so that the points of a dual polar space are the maximal subspaces of the corresponding polar space. Axioms for dual polar spaces (including the near n-gon condition) were given by Shult.

Cameron's article describes this, an embedding of certain dual polar spaces in projective spaces derived from spinors, and some characterisations. Shult and Yanushka had given some examples of sporadic near 6-gons; Cohen, in his article, constructs the first known sporadic near 8-gon.

All this has taken us some way from our original motivating example, the study of quadrics in projective space. Of course, another generalisation would involve varieties of higher degree. The most familiar example is the cubic surface in 3-space with its 27 lines. This configuration is realisable over various finite fields. Hirschfeld determines all such realisations in which every point of the variety lies on one of its 27 lines.

Percsy considers a much more general situation. Given an arbitrary set S of points in $PG(n,q)$, the subspaces spanned by subsets of S form a lattice, an example of a geometric lattice. Conversely, which geometric lattices can be embedded in projective space in this way? A great deal of work has been done on this. Percsy provides a survey, together with substantial improvements to some known results.

On the other hand, we may have much weaker information about S, for example, the cardinalities of its intersections with subspaces (perhaps just with lines). Under special conditions, it is possible to characterise S as a variety of known type from this information. This subject too has connections with coding theory. These are discussed by Ceccherini and Tallini, but we give an outline here of the connection.

Let S be a spanning subset of $PG(n,q)$ of cardinality N. For each point of S, choose a spanning vector of the corresponding 1-dimensional subspace $V(n+1, q)$; write these column vectors as an $(n+1) \times N$ matrix A. Then A has rank $n+1$ and no two of its columns are linearly dependent. Multiplying A on the left by a non-singular matrix corresponds to applying a linear collineation to S, giving an "equivalent" set; while multiplying on the right by a monomial matrix simply re-orders S and changes the choice of spanning vectors.

Now let C be a linear code in $GF(q)^N$ of dimension $n+1$ and having the property that $C^\perp$ is 1-error-correcting. (Such a code is called *projective*.) Choose a basis for C, and write the basis vectors as rows of a matrix A. Again A has rank $n+1$ and no two dependent columns. Multiplying A on the right by a monomial matrix permutes the coordinates of C and multiplies them by nonzero scalars, giving an "equivalent" code; while multiplying on the left by a non-singular matrix simply changes the chosen basis.

So we have a bijection between spanning subsets of $PG(n,q)$ of size N (up to linear equivalence) and projective codes of length N and dimension $n+1$ (up to monomial equivalence). Of course, much information can be transferred back and forth. For example, the positions of zeros in words of C give the hyperplane sections of S, while the supports of words of weight 3 in $C^\perp$ give the collinear triples in S. Greene has remarked that the weight distribution of C is determined by the Tutte polynomial of the geometric lattice on S. And so on.

It is suprising how little is known about subsets of $PG(n,q)$ subjected to even very simple intersection conditions with subspaces. A subset K is of type $(m_1, m_2, \dots, m_r)_d$ with $m_1 < m_2 < \dots < m_r$ if $|\Pi_d \cap K| \in \{m_1, \dots, m_r\}$ for every d-dimensional subspace Π_d. Hermitian varieties have been characterized as subsets K of type $(1, m, q+1)$ with $m \neq \frac{1}{2}q + 1$ by Tallini Scafati : no restriction is put on the size of K. If a subset K has the same number of points as a quadric and is of type $(0, 1, 2, q+1)_1$, then Tallini has shown that K is a quadric. If K is of type $(1, m, q+1)_1$ with $m \neq \sqrt{q} + 1$, then Hirschfeld and Thas have shown that K is the projection of a quadric with q even. For each of these results, there is some further minor condition. Lefèvre-Percsy has determined all sets of type $(0, 1, 2, q+1)_1$ in $PG(3,q)$ and $PG(4,q)$ as well as showing the difficulty of a complete determination in higher dimensions.

Here, Bichara determines all sets of type $(0, 1, 2, m)_2$ in $PG(n,q)$ with $m \geq q + 1$ and $n \geq 3$. de Finis gives some examples of sets of type $(m_1, m_2)_1$ in $PG(2,q)$ constructed from disjoint subplanes $PG(2, \sqrt{q})$ and from Hermitian curves. de Resmini considers an analogous problem in Steiner systems.

One elusive but much studied problem is to determine the maximum size of a set of type $(0, 1, 2)_1$ in $PG(n,q)$. The answer is known when $q = 2$ for all n, when $q = 3$ for $n = 5$, and when $q \geq 4$ for $n \leq 3$. A $(k; n)$-*arc* in $PG(2,q)$ is a k-set K such that $|\Pi_1 \cap K| \leq n$ for all lines Π_1 and that $|\Pi_1 \cap K| = n$ for some line Π_1. Trivially $k \leq (n-1)q + n$. If the upper bound is achieved, K is *maximal* and necessarily n divides q. The existence of maximal arcs has been established when $q = 2^h$ and $n = 2^g$, and their non-existence when $q = 3^h$ and $n = 3$ or $q/3$; in all other cases the answer is still unknown. When n does not divide q, it had previously been conjectured that $k \leq (n-1)q + 1$. Hill and Mason show that this is false, and determine the upper bound in almost all cases with $n \leq 8$ and $q \leq 9$.

The two papers in the collection which we have not yet discussed treat topics which have some roots in projective geometry but are not so closely connected. A *Latin square* is an $n \times n$ array with entries $\{1,\ldots,n\}$, in which each entry occurs exactly once in each row and once in each column. Suppose we are given an affine plane of order n, and select two parallel classes, the "horizontal" lines $H_1,\ldots,H_n$ and the "vertical" lines $V_1,\ldots,V_n$. Any further parallel class $L_1,\ldots,L_n$ can be described by a Latin square as follows: the (i,j) entry of the square is k if the lines H_i, V_j and L_k are concurrent. (Of course, not every Latin square arises in this way.) Another source of Latin squares is as multiplication tables of groups. A Latin square is called *row-complete* if each ordered pair of distinct entries occurs exactly once in the $n(n-1)$ pairs of positions which are consecutive in some row. Column-completeness is defined similarly. Keedwell gives a survey of these concepts. Most of the examples are multiplication tables of groups.

We saw, in discussing collineations of projective planes, that the present situation in the study of finite simple groups is bringing a new flavour to permutation group theory and related parts of finite geometry, that of studying representations of known groups. Saxl's contribution concerns a class of permutation representations which is smaller than the very general class of transitive representations but yet wider than that of multiply-transitive representations, the so-called "multiplicity-free representations". He determines all such representations of symmetric groups, and obtains some striking results on the general linear groups.

Of the 72 mathematicians who attended the conference, 54 gave talks on their research. A considerable core has been attending a peripatetic, almost annual conference of this size for some years:

1974/1976/1978	"Finite Geometries" at Oberwolfach, West Germany;
1976/1979	"Finite Geometries and Groups" at Han-sur-Lesse, Belgium;
1975/1980	"Finite Geometries and Designs" at the Isle of Thorns, Chelwood Gate, England.

This is the first of these conferences to be published as a single entity. It is expected that these or similar conferences will continue on an annual basis.

GENERALIZED STEINER SYSTEMS OF TYPE 3-(v, {4,6}, 1)

E.F. Assmus, Jr.* and J.E. Novillo Sardi

Let S be the set of entry positions of a 4×4 square. To each point of S associate the 6-subset of S consisting of those entry positions in its row and column but not including it. It is well-known that these sixteen 6-subsets form one of the three biplanes of order four - in fact, the "best" of the three. And this fact is easy to verify [3, 7].

What is not so well-known is that this biplane's ovals also have an easy description in terms of the 4×4 square. An oval is either the four corners of one of the square's thirty-six subrectangles or the four entry positions for the 1's of any of the twenty-four 4×4 permutation matrices.

Only a few minutes reflection are needed to convince oneself that these sixty 4-subsets are, indeed, the biplane's ovals. Moreover, the sixteen 6-subsets that constitute the biplane and the sixty 4-subsets that are its ovals do, when taken together, form a generalized Steiner system of type 3-(16, {4,6}, 1). That is, every 3-subset of S is contained in a unique member of this collection. Again, a minute's reflection is all that is needed - given this geometric description.

The description does have a disadvantage : it suggests that there are two kinds of ovals when indeed there are not. The automorphism groups of the biplane is the automorphism group of the generalized Steiner system and it acts doubly-transitively on S, [3] .

The structure just described occupies a rather remarkably unique niche in the museum of combinatorial curiosities and one purpose of this note is to prove a theorem further justifying its place there.

Before proceeding to a statement of the theorem we make the necessary definitions.

A *generalized Steiner system of type* $3-(v, \{4,6\}, 1)$ is a collection, Γ, of subsets of a v-set, S, satisfying the following two conditions:

(i) Every subset in Γ is of cardinality 4 or 6.

(ii) Every 3-subset of S is contained in a unique member of Γ.

Such systems have been investigated by Hanani [8], Mills [9], and van Buggenhaut [5]. The main result is that they exist whenever v is even. From our point of view this result is disappointing for it makes virtually impossible any classification of the incredibly large number of generalized Steiner systems of the stated type. For example, all quadruple systems occur as generalized Steiner systems of type $3-(v, \{4,6\}, 1)$ and thus, even for $v = 16$, we have at least thirty-one thousand such systems. We eliminate those by insisting that there be 6-subsets of S in Γ, and further restrict the systems under discussion by pretending that there is a transitive automorphism group at hand. If there were such an automorphism group and if there were 6-subsets in Γ - but not too many - system of type $3-(v, \{4,6\}, 1)$ would satisfy the following further conditions.

(iii) The collection of 4-subsets in Γ form a 1-design;

(iv) there are precisely v 6-subsets in Γ and they form a 1-design.

From now on we restrict ourselves to systems satisfying (i), (ii), (iii), and (iv) and refer to such a system as a *homogeneous* generalized Steiner system of type $3-(v, \{4,6\}, 1)$.

Here, then, is the result which we hope will help justify our remark about the remarkable uniqueness of the system on sixteen points which we described at the outset.

THEOREM: *If a homogeneous, generalized Steiner system of type* $3\text{-}(v, \{4,6\}, 1)$ *exists, then* v *is congruent to* 2 *or* 4 *modulo* 6 *and at least* 16. *Moreover, there is a unique such system with* $v = 16$ *consisting of the best biplane on sixteen points and its sixty ovals.* □

REMARKS: (1) The necessary congruence condition on v, $v \equiv 2,4 \pmod 6$, is precisely the condition for the existence of a quadruple system. The lower bound, $v \geq 16$, results from the fact that we are insisting on the presence of v 6-subsets.

(2) Except for the uniqueness portion of the Theorem, the proof is only slightly more difficult than the usual proofs of necessary congruence conditions, [11].

(3) Hanani's recursive techniques allow one to construct infinitely many homogeneous generalized Steiner systems of type $3\text{-}(v, \{4,6\}, 1)$.

(4) The smallest such system constructible by Hanani's recursive techniques has $v = 26$ and it results from a quadruple system on fourteen points. One would guess that there are many homogeneous generalized Steiner systems of type $3\text{-}(26, \{4,6\}, 1)$.

(5) The smallest open case has $v = 20$; we discuss this case below.

(6) The geometric description of the unique system of type $3\text{-}(16, \{4,6\}, 1)$ has given rise to a game similar to Battleship (dubbed "Battlestar"). The defense picks a 3-subset and the offense picks one 6-subset of Γ and one 4-subset (in either order) attempting to cover the (unknown to offense) 3-subset chosen by the defense. There are ten "frames" and scoring is as in bowling.

(7) There is also a graph-theoretic description of the homogeneous $3\text{-}(16, \{4,6\}, 1)$: The sixteen points consist of the fifteen edges of K_6, the complete graph on six vertices together with another point, ∞. A 6-subset of Γ is either ∞ together with five edges of K_6 emanating from a point (a claw) or the six edges of two disjoint triangles in K_6. A 4-subset of Γ is

either the four edges of a square of K_6 or ∞ together with three "parallel" edges of K_6. Again, it is easy to see that conditions (i) - (iv) are satisfied.

(8) Perhaps the most sophisticated description of the homogeneous 3-(16, {4,6}, 1) is obtained from the extension of the projective plane of order four, i.e., a 3-(22, 6, 1) design. For S one takes the complement of a block and for Γ the intersections of the remaining seventy-six blocks with S. One can, in fact, reverse the procedure, constructing the 3-(22, 6, 1) from the 3-(16, {4,6}, 1), but since this is not the most elegant road to this extended plane we will not describe this construction. See instead [2, 4].

We wish, finally, to describe the motivation for this theorem and its application to the question of the existence or non-existence of an extended projective plane of order ten, i.e. a design of type 3-(112, 12, 1).

It has been known for well over a decade now [1] that should a design of type 3-(112, 12, 1) exist, then the span of the rows of its incidence matrix over the field with two elements forms a (112, 56) self-dual, doubly-even code with minimum weight 12. Moreover, the vectors of this minimum weight are precisely the rows of the incidence matrix.

Early on, Denniston [6] and, independently, MacWilliams, Sloane, and Thompson [10] carried out a computation that showed, among other things, that this code did not contain vectors of weight sixteen.

Although there is no doubt about the validity of this fact, the Theorem above allows of an independent proof that can be carried out by hand. We sketch that proof: Given a weight - 16 vector with support S one shows by simple counting techniques that there are precisely sixteen blocks of the design meeting S six times and precisely sixty meeting S four times and that, moreover, S together with these intersections form a homogeneous generalized Steiner system of type 3-(16, {4,6}, 1). Thus, by the Theorem we know precisely the structure of this weight - 16 vector and one can show (and it is quite instructive to do so) that one cannot expand it to a design of type 3-(112, 12, 1), [11].

Now, Gleason's Theorem allows one to compute the precise weight distribution of the code. In particular, one finds that there are precisely

868,560

weight - 20 vectors. Only

341,880

of these arise in an obvious way (as the sum of two blocks that meet twice). Fixing one of the other, exotic, weight - 20 vectors with support S, say, and making the same easy counting arguments as were made for the weight - 16 vector, we have at hand a homogeneous generalized Steiner system of type

3-(20, {4,6}, 1).

We are not able to find, nor to disprove the existence of such a system. Preliminary electronic calculation seems to suggest non-existence. Chester Salwach is now engaged in an effort to exhaust the problem and we are hoping for an answer shortly.

The most likely result is non-existence and this would prove (albeit with electronic computation) the non-existence of an extendable projective plane of order 10. Should such a design be uncovered, however, one would have to classify them up to isomorphism and then investigate each to see if it "expands" to a

3-(112, 12, 1),

a task that would undoubtedly best be carried out electronically.

POSTSCRIPT: During the Conference, Andries Brouwer produced a homogeneous Steiner system of type 3-(20, {4,6}, 1). We have not yet been able to decide whether or not it expands to a 3-(112, 12, 1). The estimates of the computer time necessary to decide the question seem, at this time, too great to warrant going ahead since one expects a negative answer and since the classification of all homogeneous 3-(20, {4,6}, 1) designs seems to be of even greater electronic complexity.

BIBLIOGRAPHY

1. E.F. Assmus, Jr., "The projective plane of order 10", *Tagungsbericht* 10 (Math. Forsch. Inst. Oberwolfach, 1970).
2. E.F. Assmus, Jr., Joseph A. Mezzaroba, and Chester J. Salwach, "Planes and biplanes", *Higher Combinatorics,* Berlin 1976 (ed. M. Aigner, Reidel, Dordrecht, 1977), pp. 205-212.
3. E.F. Assmus, Jr. and Chester J. Salwach, "The (16,6,2) designs", *J. Math. Math. Sci.,* 2 (1979), 261-281.
4. E.F. Assmus, Jr. and J.H. van Lint, "Ovals in projective designs", *J. Combin. Theory Ser. A,* 27 (1979), 307-324.
5. J. van Buggenhaut, "On some Hanani's generalized Steiner systems", *Bull. Soc. Math. Belg.,* 23 (1971), 500-505.
6. R.H.F. Denniston, "Nonexistence of a certain projective plane", *J. Austral. Math. Soc.,* 10 (1969), 214-218.
7. C. Jordan, "Sur une équation du 16 ème degré", *J. Reine Angew. Math.,* 70 (1869), 182-184.
8. H. Hanani, "On truncated finite planes", *Combinatorics, Proc. Symp. Pure Math.* 19, Univ. California, 1968 (Amer. Math. Soc., 1971), pp. 115-120.
9. W.H. Mills, "On the covering of triples by quadruples", *Proc. Fifth Southeastern Conf. on Combinatorics, Graph Theory and Computing,* Florida Atlantic Univ., 1974, (Utilitas Math., Winnipeg, 1974), pp. 563-581.
10. F.J. MacWilliams, N.J.A. Sloane, and J.G. Thompson, "On the existence of a projective plane of order 10", *J. Combin. Theory Ser. A,* 14 (1973), 66-78.
11. J.E. Novillo Sardi, "On generalized Steiner systems and the extension of a projective plane of order 10", Ph.D. Thesis, Lehigh University, 1980.

Department of Mathematics
Lehigh University
Bethlehem
Pennsylvania 18015
U.S.A.

SOME REMARKS ON D.R. HUGHES' CONSTRUCTION OF M_{12} AND ITS ASSOCIATED DESIGNS

Thomas Beth

1. SUMMARY

The aim of this note is to present a simple way of constructing the Mathieu group M_{12} and the Witt design $S_1(5,6;12)$ from the Hadamard design $S_2(2,5;11)$. The method presented here is based on the ideas of Hughes [1]. While Hughes' approach yields an excellent way of constructing $S_1(5,6;12)$ and M_{12} at the same time, our approach is even simpler and, what is more, gives the uniqueness of $S_1(5,6;12)$ as well.

2. INTRODUCTION

Let H_{11} be the unique Paley-Hadamard design $S_2(2,5;11)$ which over $GF(11)^+$ is given by the difference set of quadratic residues $S = \{1, 3, 4, 5, 9\}$ (mod 11).

There is a unique extension $\overline{H}_{11}$ of H_{11} by complementation of the blocks of H_{11}, thus yielding an $S_2(3,6;12)$ consisting of 11 pairs of disjoint blocks of size 6. Hughes [1] pointed out that the new incidence structure, in which points are these pairs of disjoint blocks, lines are pairs of points in $\overline{H}_{11}$ and incidence is given if a pair of points lies completely in a block of a pair of disjoint blocks, is an $S_1(4,5;11)$ admitting a sharply 4-transitive group of automorphisms of degree 11. Standard complementation yields an $S_1(5,6;12)$ and a sharply 5-transitive-group of automorphisms. Hughes' construction does not automatically give the uniqueness of $S_1(5,6;12)$. In the sequel we want to show that a slight modification of this idea gives a simpler way of constructing $S_1(5,6;12)$ at the same time proving the uniqueness of this design.

3. EXTENSIONS OF PALEY-HADAMARD DESIGNS

For a prime power $q \equiv 3 \pmod 4$ the Paley-Hadamard design H_q is given by $(GF(q), \mathcal{O})$, where

$\mathcal{O} = \{ S + g \mid g \in GF(q) \}$ and

$S = \{ a^2 \mid a \in GF(q)^* \}$ is the set of non-zero squares in $GF(q)$. Clearly H_q admits the 2-homogeneous group $ASL(1,q)$ of transformations

$$x \longmapsto a^2 x + b, \quad a \in GF(q)^*, \quad b \in GF(q),$$

as an automorphism group.

As H_q has a unique extension $\overline{H}_q$ [1], [4] and $PSL(2,q)$ is a transitive extension of $ASL(1,q)$, the question arises whether $PSL(2,q)$ in its 3-homogeneous representation of degree $q+1$ is an automorphism group of $\overline{H}_{11}$. This question is even more natural, when recalling that $PSL(2,q)$ is an automorphism group of the extended quadratic residue codes, cf. [2][3][4]. The following lemma answers this question.

LEMMA 3.1: $PSL(2,q)$ *in its* 3-*homogeneous representation of degree* $q+1$ *is an automorphism group of a non-trivial* $\overline{H}_q$ *iff* $q = 7$.

PROOF: If $q = 7$ there is nothing to prove. If $PSL(2,q)$ acts in the required way then the setwise stabilizer H in $PSL(2,q)$ of a block B in $\overline{H}_q$ has to have order $\frac{1}{2}(q+1) \cdot \frac{1}{2}(q-1)$. The permutation group H acts transitively on B; the stabilizer H_x of an element $x \in B$ is cyclic of order $\frac{1}{2}(q-1)$. Thus H is a sharply 2-transitive Frobenius group. Thus the Frobenius kernel of order $\frac{1}{2}(q+1)$ is elementary abelian with $\frac{1}{2}(q+1) = 2^e$ for some $e \in \mathbb{N}$, cf. [5]. As the subgroups of $PSL(2,q)$ are known [5], this is only possible if $e = 1$ or 2. As $e = 1$ is excluded for non-triviality reasons, the case $e = 2$ implies $q = 7$. □

REMARK 3.2 : This lemma explains why, when proving that the extended quadratic residue codes admit PSL(2,q), the generating matrix is not left invariant by PSL(2,q).

In the next section we will study the action of PSL(2, 11) more closely.

4. CONSTRUCTING S(5,6; 12) VIA PSL(2, 11).

In the preceding section we have shown that the set $E = \{\infty, 1, 3, 4, 5, 9\}$ under the action of PSL(2, 11) generates a 3-design $S_\lambda(3,6;12)$ with $\lambda = 2 \cdot \lambda'$ where $\lambda' \neq 1$ and $\lambda' \mid 6$. We will show that $\lambda' = 6$ and the $S_\lambda(3,6;12)$ given by $M = (GF(11) \cup \{\infty\}, \mathcal{E})$ with $\mathcal{E} = \{E^\pi \mid \pi \in PSL(2,11)\}$ is in fact a 5-design $S_1(5,6;12)$.

NOTATION 4.1 : Let $\tau \in PSL(2,11)$ be given by the function $\tau : x \longmapsto -1/x$. Then $E \cap E^\tau = \emptyset$. Define $F := E^\tau$.

LEMMA 4.2 : *Let* $\overline{H_{11}} = (GF(11) \cup \{\infty\}, \mathcal{B})$. *Then for any two distinct blocks* $B, B' \in \mathcal{B}$ *with* $B \cap B' \neq \emptyset$ *the symmetric difference* $B \,\Delta\, B'$ *is a block of* M, *i.e.* $B \,\Delta\, B' \in \mathcal{E}$.

PROOF : It is sufficient to prove this for any pair $F, E + j \in \mathcal{B}$. For this we make the following table:

$$\left.\begin{aligned} F \,\Delta\, (E+1) &= (E-1)^\tau \\ F \,\Delta\, (E+2) &= (F+5)^\tau \\ F \,\Delta\, (E+3) &= (D-4)^\tau \\ F \,\Delta\, (E+4) &= (D-3)^\tau \\ F \,\Delta\, (E+5) &= (D+2)^\tau \end{aligned}\right\} \qquad (4.2.a)$$

Observing that ASL(1,q) and τ generate the group PSL(2, 11), the proof is complete. □

LEMMA 4.3 : $\mathcal{E} = \mathcal{B} \cup \{B \,\Delta\, B' \mid B, B' \in \mathcal{B},\ B \neq B',\ B \cap B \neq \emptyset\}$.

PROOF : Observing that in the 3-design $\overline{H}_{11}$ for $B, B' \in \mathcal{B}$ we either have $|B \cap B'| = 0$ or $|B \cap B'| = 3$ it suffices to observe that

$$|\{ B \Delta B' \mid B, B' \in \mathcal{B}, || B \cap B'| = 3 \}| = 110 .$$

By lemma 4.2, the assertion is proved. □

COROLLARY 4.4 : M *is an* $S_{12}(3,6;12)$. □

In order to show that M is an $S_1(5,6;12)$, it is necessary to study the derived incidence structures $M_{x,y,z}$ for any 3-subset $\{x,y,z\}$ of $GF(11) \cup \{\infty\}$. As $PSL(2,11)$ is a 3-homogeneous automorphism group of M, it suffices to show that for some 3-subset $\{x,y,z\}$ the derived structure $M_{x,y,z}$ is an $S_1(2,3;9)$.

OBSERVATION 4.5 : Let $x = \infty$, $y = 1$, $z = 4$. Then the two blocks in $\mathcal{B}$ containing $\{x,y,z\}$ are $E = \{\infty,1,3,4,5,9\}$ and $E+3 = \{\infty,1,4,6,7,8\}$. Then $F \Delta (E+3) = \{\infty,1,4,0,2,10\} \in \mathcal{E}$.

Thus the following situation holds

$A = \{\infty, 1, 4\}$	$B = \{6, 7, 8\}$
$C = \{3, 5, 9\}$	$D = \{0, 2, 10\}$

FIGURE 1 : The 4 triads generated by E and $E+3$

Then $A \cup B = E+3$, $A \cup C = E$ and $A \cup D = F \Delta (E+3)$.

Observing that the setwise stabilizer N of A in $PSL(2,11)$ has order 3 and is generated by $\nu : x \mapsto (x-2)/(x-4)$, we get that

$B^{\nu} = \{2, 9, 7\}$ $\quad$ $B^{\nu^2} = \{0, 8, 9\}$

$C^{\nu} = \{10, 3, 8\}$ $\quad$ $C^{\nu^2} = \{5, 10, 7\}$

$D^{\nu} = \{6, 0, 5\}$ $\quad$ $D^{\nu^2} = \{2, 6, 3\}$

After verifying that

$$(\{\infty, 1, 4, 2, 5, 8\} + 2)^{\tau} = F + 1$$

$$(\{\infty, 1, 4, 6, 9, 10\} + 2)^{\tau} = E - 5$$

$$(\{\infty, 1, 4, 0, 3, 7\} + 3)^{\tau} = F + 1$$

it is clear that $M_{\infty,1,4}$ is an $S_1\{2,3;9\}$. □

As a corollary we formulate the following theorem.

THEOREM 4.6 : $M = (GF(11) \cup \{\infty\}, E)$ *is an* $S_1(5,6;12)$ *admitting an automorphism group* G *which is* 5-*transitive on the points.* □

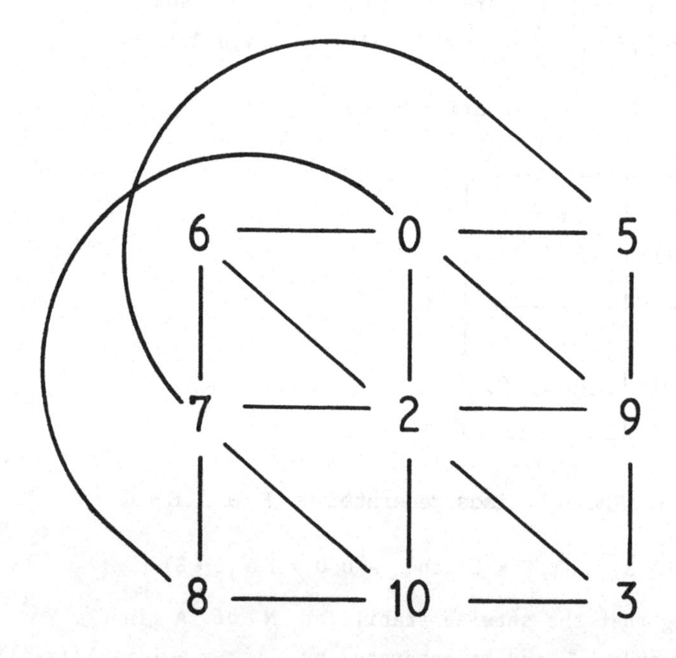

FIGURE 2 : The structure given by B, C, D and ν .

REMARK 4.7 : It can be concluded as usual that G is sharply 5-fold transitive and simple. The present construction allows us to show the uniqueness of $S_1(5,6;12)$. This is carried out in the next section.

5. THE UNIQUENESS OF $S_1(5,6;12)$

The construction of $S_1(5,6;12)$ presented in the preceding section can be used to show that the constructed $S_1(5,6;12)$ is the only one possible. First we recall a few facts about any $S_1(5,6;12)$.

OBSERVATION 5.1 : Let $\mathcal{W} = (GF(11) \cup \{\infty\}, \mathcal{C})$ be any $S_1(5,6;12)$. Then the following equations for the parameters of $\mathcal{W}$ hold:

$$\begin{aligned} b_0 &= b = |\mathcal{C}| = 132 \\ b_1 &= 66 \\ b_2 &= 30 \\ b_3 &= 12 \\ b_4 &= 4 \\ b_5 &= \lambda = 1 \end{aligned} \tag{5.1.a}$$

From these parameters it is easily computed that for any block $B \in \mathcal{C}$ and any i-subset $T \subseteq B$ the numbers B_i of blocks meeting B in exactly the i-subset T, i.e. $B_i = |\{C \in \mathcal{C} \mid C \cap B = T\}|$ for $T \subseteq B$, $|T| = i$ are given by

$$\begin{aligned} B_4 &= 3 \\ B_3 &= 2 \\ B_2 &= 3 \\ B_1 &= 0 \\ B_0 &= 1 \end{aligned} \tag{5.1.b}$$

With this observation we readily obtain the following propositions.

LEMMA 5.2 : *Let $\mathcal{W}$ be given as above. Let (G,G') be any pair of disjoint blocks of $\mathcal{C}$, i.e. $G,G' \in \mathcal{C}$, $G \cap G' = \emptyset$. Let $T \subseteq G$ be a 3-subset and let L and K be the two other blocks of $\mathcal{C}$ meeting G in the set T.*

With $A = G \cap L$, $B = G' \cap L$, $C = G \cap L'$ *and* $G' \cap L'$, *we then have*

$$L = A \cup B$$
$$K = A \cup D\ . \qquad (5.2.a)$$

For any 4-*subset* Q *with the property* $|Q \cap A| = 2 = |Q \cap C|$ *there exists exactly one block* H_Q *with* $Q \subset H$ *and* $|H \cap B| = 1 = |H \cap D|$.

PROOF : Recalling Figure 1, the sets A, B, C, D form the 4 triads

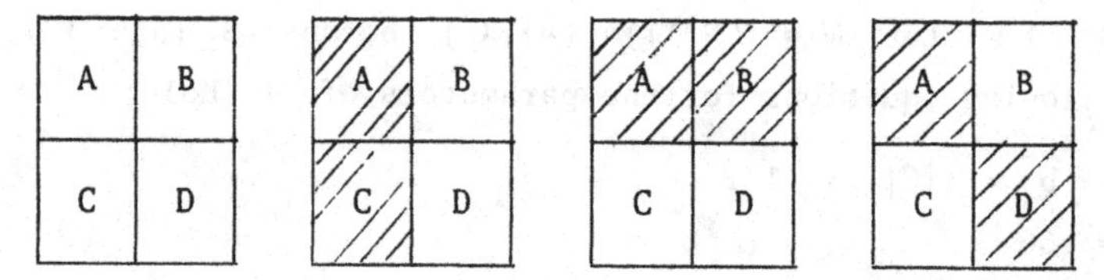

The four triads The block G The block L The block K

FIGURE 3

Since $\mathcal{W}$ is an S_1 (5,6; 12), the blocks containing D must form an S_1 (2,3,9) on $A \cup B \cup C$. Thus there is a block R with $D \subset R$ meeting G in $G \cap Q$ and B in one point. This implies that there is exactly one block H_Q with the required property. □

NOTATION 5.3 : Let $\mathcal{V}$ $\{ |Q \subset G|\, |Q \cap A| = 2 = |Q \cap C| \}$. For $Q \in \mathcal{V}$ let H_Q be the unique block in $\mathcal{C}$ with $|H_Q \cap B| = 1 = |H_Q \cap D|$. For any $R \in \mathcal{C}$ let R' be its complement.

Then the following theorem holds.

THEOREM 5.4 : *Let* $\mathcal{G} = \{ L, L', K, K' \} \cup \{ H_Q, H_Q' \mid Q \in \mathcal{V} \}$. *Then* $(GF(11) \cup \{\infty\}, \mathcal{G})$ *is an* S (3,6; 12).

PROOF : Clearly $\mathcal{G}$ has 22 blocks. In order to show that $\mathcal{G}$ forms a 3-design with $\lambda = 2$ it is sufficient to prove, for any two blocks $R, S \in G$, that $|R \cap S| = 0$ or 3.

This can easily be checked after proving the following statement:

With the notation of Lemma 5.2, take any 2-subset $T \subseteq G'$ with $|T \cap B| = 1 = |T \cap D|$. Then there exists a 4-subset $Q \in V$ with $Q \cup T = H_Q$. This is clear when observing that

(i) the three blocks M_1, M_2, M_3 meeting G' in exactly T are of the form $M_1 \Delta M_2 = M_3$;

(ii) $T \cup A$ must be on a block $M \in C$ such that $|M \cap C| = 1$.

Thus there is one-to-one correspondence between such sets T and the 4-subsets $Q \in V$. This implies that for $Q, Q' \in V$, $Q \neq Q'$, the block H_Q and $H_{Q'}$ meet in exactly 3 points. □

COROLLARY 5.5 : For G, let $G^* = G \cup \{R \Delta S \mid R, S \in G, R \cap S \neq \emptyset\}$. Then $G^* = C$. □

THEOREM 5.6 : *The* 5*-design* M *(constructed in* §4*) is unique.*

PROOF : This follows from Corollary 5.5 when we observe that any $S_2(3, 6; 12)$ is uniquely determined by its parameters. □

ACKNOWLEDGEMENT : The author is indebted to D.R. Hughes, who in the stimulating atmosphere of Westfield College (University of London) encouraged him to pursue these investigations.

BIBLIOGRAPHY

1. D.R. Hughes, "A combinatorial construction of the small Mathieu designs and groups", *Ann. Discrete Math.*, to appear.
2. F.J. MacWilliams and N.J.A. Sloane, *The Theory of Error-correcting Codes* (North-Holland, Amsterdam, 1977).

3. T. Beth and V. Strehl, *Materialien zur Codierungstheorie* (Universität Erlangen, 1978).
4. H. Lenz, T. Beth and D. Jungnickel, *Design Theory,* to appear.
5. B. Huppert, *Endliche Gruppen* (Springer-Verlag, Berlin, 1979).

Institut für Mathematische Maschinen
und Datenverarbeitung I
Universität Erlangen
Martensstrasse 3
8520 Erlangen
Federal Republic of Germany.

ON k-SETS OF CLASS $[0,1,2,n]_2$ IN PG(r,q)

Alessandro Bichara

1. INTRODUCTION

Characters, class and type of a k-set in PG(r,q), $r \geq 3$, $q = p^h$, p a prime, h a non-negative integer, were defined in [4].

Let K be a proper k-set in PG(r,q) and $x_1, x_2, \ldots, x_g$ non-negative integers such that $x_1 < x_2 < \ldots < x_g$; K will be said to be of *class* $[x_1, x_2, \ldots, x_g]_d$ *with respect to dimension d* if any d-dimensional subspace (d-space) in PG(r,q) meets K either in x_1, or $x_2, \ldots,$ or x_g points. K will be called of *type* $(x_1, x_2, \ldots, x_g)_d$ *with respect to dimension d* if K is of class $[x_1, x_2, \ldots, x_g]_d$ and for all $i \in \{1,2,\ldots,g\}$ there exists a d-space meeting K in exactly x_i points.

When K is of class $[x_1, x_2, \ldots, x_g]_d$, a d-space P_d is an x_i - secant d-space of K provided that $x_i = |P_d \cap K|$; a 0-secant, respectively 1-secant, d-space will be called an external, respectively tangent, d-space.

A k-set of class $[0,1,2,\ldots,d+1]_d$ will be called a *k-cap of kind at least d+1;* therefore, a *cap* is a cap of kind at least two [3].

A k-set K of class $[0,1,2,n]_2$, $3 \leq n \leq q^2+q+1$, with respect to planes in PG(r,q) will be denoted by k(n).

If K is a k(n) in PG(r,q) whose points span a d-space P_d, $d \geq 3$, then K is a k(n) in P_d. Therefore, k(n)'s, K, in PG(r,q) satisfying

$$K \text{ contains } r+1 \text{ independent points,} \tag{1.1}$$

will be considered.

$k(n)$'s in $PG(r,q)$, $r \geq 3$, such that (1.1) holds do exist; e.g.

(i) The complement of a hyperplane in $PG(r,q)$ is a $k(q^2)$, $k = q^r$.

(ii) A $(q^2 + 1)$-cap in $PG(3,q)$ is a $k(q+1)$, $k = q^2+1$.

(iii) A pair of skew lines in $PG(3,q)$ is a $k(q+2)$, $k = 2q+2$.

(iv) A k-cap of kind at least three not contained in any hyperplane in $PG(r,q)$, $r \geq 3$, is a $k(3)$.

$k(n)$'s satisfying (1.1) are studied and the following theorems proved.

THEOREM I : *In* $PG(r,q)$, $r \geq 3$, *no* $k(n)$ *of type* $(0,1,n)_2$ *satisfying (1.1) exists. Furthermore, if* K *is a k-set of type* $(0,1,n)_d$ *in* $PG(r,q)$, $r \geq 3$, *for which (1.1) holds, then* $d = 1$.

THEOREM II : *If* $n \geq q+1$ *and* K *is a* $k(n)$ *in* $PG(r,q)$, *satisfying (1.1), then one of the following is true:*

(i) K *is a cap of kind at least three in* $PG(r,2)$, *and* K *is not contained in any hyperplane.*

(ii) K *is the complement of a hyperplane.*

(iii) $r = 3$ *and* K *is either a* (q^2+1)*-cap or the set of points on two skew lines in* $PG(3,q)$.

2. SOME PROPERTIES OF $k(n)$'s IN $PG(r,q)$

In order to study $k(n)$'s, the case $r = 3$ will be considered first.

THEOREM III : *If* K *is a* $k(n)$ *in* $PG(3,q)$, *for which (1.1) holds, then there exists a line external to* K.

PROOF : Since not all planes are n-secant planes of K (see [4], Proposition I), there exists a plane π meeting K in at most two points; thus π contains some external line.

THEOREM IV : *Under the hypothesis of Theorem III, if* s *is a 2-secant line of* K *such that there are exactly* x *planes n-secant to* K *through it, then*

$$x = \frac{k-2}{n-2} \leq q+1 . \qquad (2.1)$$

PROOF : Obviously, $x \leq q+1$. The n-secant planes through s partition the $k-2$ points in $K \setminus s$ into x sets of size $n-2$; therefore, $k-2 = x(n-2)$.

THEOREM V : *Under the hypothesis of Theorem III, if there exists an m-secant line,* $m \geq 3$, *of* K *in* $PG(3,q)$, *then* $k(n)$ *is of class* $[0,1,2,m]_1$ *with respect to lines and*

$$k = (n-m)(q+1) + m, \quad 3 \leq m \leq n-1 . \qquad (2.2)$$

PROOF : Any plane through s containing $m \geq 3$ points of K meets K in exactly n points, hence $n \geq m$; if $n = m$, then K consists of m collinear points which contradicts (1.1); therefore, $m \leq n-1$.

The planes through s partition the $k-m$ points in $K \setminus s$ into $q+1$ sets of size $n-m$; thus $k-m = (n-m)(q+1)$ and (2.2) follows.

Now let s' be an m'-secant line, $m' \geq 3$. By the previous argument, writing s' and m' instead of s and m, we have $k = (n-m')(q+1) - m'$; thus, by (2.2), $m = m'$ and the statement is proved.

When $r \geq 3$, a similar argument proves:

THEOREM VI : *Let* K *be a* $k(n)$ *in* $PG(r,q)$, $r \geq 3$ *satisfying (1.1). If there exists a line* s *meeting* K *at exactly* m *points,* $m \geq 3$, *then* $k(n)$ *is of class* $[0,1,2,m]_1$ *with respect to lines in* $PG(r,q)$.

THEOREM VII : *Let* K *be a* $k(n)$ *in* $PG(r,q)$, $r \geq 3$, *satisfying (1.1). If* K *is of type* $(0,1,2,m)_1$ *with respect to lines* $(3 \leq m \leq q+1)$, *then* $n-2$ *and* q *are not coprime.*

PROOF : There exist two lines, s and s', in $PG(r,q)$, which are a 2-secant and an m-secant of K, respectively. Let P_3 be the 3-space in $PG(r,q)$ through s and s', containing four independent points of K. Then $K' = K \cap P_3$ is of class $[0,1,2,n]_2$ and (by Theorem V) of class $[0,1,2,m]_1$ in P_3. Substituting in (2.1) for k the size $|K'|$ given by (2.2), one gets

$$x = q + 1 - (m-2)q/(n-2) ; \qquad (2.3)$$

as x is an integer, so $(n-2, q) = 1$ would imply $(n-2) | (m-2)$, whence $m \geq n$, a contradiction.

3 : PROOF OF THEOREM I

Theorem I is partly a consequence of

THEOREM VIII : *In* $PG(3,q)$ *no* k-*set of type* $(0,1,n)_2$, *satisfying (1.1) exists.*

PROOF : Suppose the contrary and let K be a k-set of type $(0,1,n)_2$ in $PG(3,q)$, satisfying (1.1). The set K is of type $(0,1,m)_1$ with respect to lines and $2 \leq m \leq q$, [2]. Moreover, there exists a positive integer e such that

$$n - 1 = (m-1)p^e, \quad 1 < p^e < p^h = q ; \qquad (3.1)$$

$$k = (n-m)(q+1) + m . \qquad (3.2)$$

Let E, T, and N be the numbers of external, tangent and n-secant planes of K; then, from [4],

$$\left.\begin{aligned} E + T + N &= (q+1)(q^2+1) \\ T + nN &= k(q^2+q+1) \\ n(n-1)N &= k(k-1)(q+1) . \end{aligned}\right\} \qquad (3.3)$$

The simultaneous equations (3.3) have exactly one solution and, taking into account (3.2), we have

$$E = [\, n\,(q^2+1)(q+1) - \{(n-m)(q+1) + m\}\,] \times$$
$$[\,(mq+1)(q+1) - qn\,]/\,n\,. \tag{3.4}$$

With the help of (3.1), it is proved in [1] that $E < 0$; on the other hand, K being of type $(0,1,n)_2$ implies $E > 0$, a contradiction.

Esposito [2] proved that, if K is a k-set of type $(0,1,n)_d$ in $PG(r,q)$, $r \geq 3$, $2 \leq d \leq r-1$, and (1.1) holds, then $r = 3$ and $d = 2$.

From this result and Theorem VIII, Theorem I follows.

4 . PROOF OF THEOREM II

First we remark (see [4], Proposition XIV) :

THEOREM IX : *If* K *is a* $k(n)$ *of type* $(0,n)_2$ *in* $PG(r,q)$, $r \geq 3$, *satisfying (1.1), then* K *is the complement of a hyperplane and* $n = q^2$.

Now we prove

THEOREM X : *Let* K *be a* $k(n)$ *in* $PG(r,q)$, $r \geq 3$, *satisfying (1.1). If* K *is not the complement of a hyperplane, then there exist two lines,* t *and* s *, which are a tangent and a 2-secant, respectively; moreover,* $n \leq q+2$ *and when* $n = q+2$ *and* $r = 3$ *there is no plane tangent to* $k(n)$.

PROOF : By Theorem IX, the set K cannot be of type $(0,n)_2$; by (1.1) and Theorem I, the set K is not of type $(0,1)_2$, $(0,2)_2$, $(1,2)_2$, $(0,1,2)_2$, and $(0,1,n)_2$. If K is of type $(1,n)_2$, then K is a cap [8], and there are tangent and 2-secant lines. In all other cases (i.e. types: $(0,1,2,n)_2$, $(1,2,n)_2$, $(0,2,n)_2$, and $(2,n)_2$) , there exists a 2-secant plane on which $2q$ tangent and one 2-secant lines lie.

Now, let P_3 be the 3-space in $PG(r,q)$, through s and t, and containing four independent points of K. Clearly, $K' = K \cap P$ is a $k'(n)$, $k' = |K'|$, in P_3, satisfying (1.1). Let

$$y = |\{\text{planes in } P_3 \text{ through } t \text{ and 2-secant of } K'\}| ,$$
$$z = |\{\text{planes in } P_3 \text{ through } t \text{ and n-secant of } K'\}| ;$$

then

$$k' = 1 + y + z(n-1) , \quad q+1 \geq y+z . \tag{4.1}$$

Since $k' > n$ (recall (1.1)), (4.1) implies $1+y+z(n-1) > n$; therefore

$$y + z \geq 2 . \tag{4.2}$$

The line s is a 2-secant of K', and, by Theorem IV, we have $k' = x(n-2) + 2$; comparing this value with (4.1), we get

$$y + z = 1 + (x-z)(n-2) . \tag{4.3}$$

(4.2) and (4.3) imply $x - z \geq 1$; thus $y + z \geq 1 + (n-2) = n - 1$; by (1.1), as $y + z \geq n-1$,

$$q + 1 \geq y + z \geq n-1 , \tag{4.4}$$

from which $q + 1 \geq n - 1$ and $q + 2 \geq n$ follow.

When $n = q + 2$ and $r = 3$, (4.4) implies $y + z = q + 1$ and a plane through t meets K either at 2 or at $q+2$ points and no tangent line belongs to a tangent plane. Since on a possible tangent plane there would lie $q+1$ tangent lines, the statement is proved.

THEOREM XI : *If* K *is a* $k(q+1)$ *in* $PG(r,q)$, *satisfying (1.1), then either* K *is the set in Theorem II(i), or* $r = 3$ *and* K *is a* (q^2+1)*-cap in* $PG(3,q)$.

PROOF : If $q = 2$, then K is of class $[0,1,2,3]_2$; not being contained in any hyperplane, K is the set in Theorem II(i).

Suppose $q \geq 3$. Since $q^2 > q + 1$, by Theorems III, IX, and X, there exist tangent, 2-secant and external lines. Moreover,

$(q-1, q) = 1$ and, by VI and VII, the set $k(q+1)$ is of type $(0,1,2)_1$ with respect to lines; that is, K is a cap.

If, from a point Q in K, the points in $K \setminus \{Q\}$ are projected onto a hyperplane $P_{r-1} \not\ni Q$, a $(k-1)$-set K' is obtained. The join of any line s meeting K' in at least two points with Q is a plane $\pi = Qs$ meeting K in at least three, and hence $q+1$ points. Therefore, s is a q-secant of K' and this set is of class $[0,1,q]_1$. Thus K' consists of the points in a subspace P_d in P_{r-1}, not belonging to some P_{d-1} in P_d [7]. Clearly, K belongs to the $(d+1)$-space joining P_d and Q; from (1.1) it follows that $d+1 = r$ and K' is the complement, in P_{r-1}, of a P_{r-2}. Then $k - 1 = |K'| = q^{r-1}$, whence $k = q^{r-1} + 1$. Thus, K is a $(q^{r-1}+1)$-cap in PG(r,q). Furthermore, the lines tangent to K at Q form the hyperplane $\{Q\} \cup P_{r-2}$. By [5], Proposition I, we have $r = 3$, and $k = q^2 + 1$. The statement follows.

The complement of a hyperplane in PG(r,2) is a k(4), so a $k(q+2)$. When $n = q + 2$, we prove:

THEOREM XII : *Let* K *be a* $k(q+2)$ *in* PG(r,q), *satisfying (1.1). If* K *is not the complement of any hyperplane in* PG(r,2), *then* $r = 3$ *and* K *consists of the points belonging to two skew lines in* PG(3,q).

PROOF : Firstly, we will prove that

$$\text{if } r = 3, \text{ then } K \text{ is of type } (0,1,2,m)_1 \text{ and } 3 \le m \le q+1. \tag{4.5}$$

By Theorems III, IX, X, as K is not the complement of a hyperplane, there exist external, tangent and 2-secant lines. If K were of type $(0,1,2)_1$, then K would be a cap such that any plane meeting it in at least three points meets it in a $(q+2)$-arc; but such a cap is the complement of a hyperplane in PG(r,2) [3], a contradiction. Therefore, K is not of type $(0,1,2)_1$ and there exists an m-secant line, $3 \le m$. From Theorem IV, the assertion (4.5) follows.

By Theorem XI, the set K is of class $[0,2,q+2]_2$; by (4.5) and (2.2), we have $k = (q+2-m)(q+1) + m = (q+3-m)q+2$. Let E, D, and N be the numbers of planes in PG(3,q) which are external, 2-secant and $(q+2)$-secant to K; then, from [4],

$$\left.\begin{aligned} E + D + N &= q^3 + q^2 + q + 1 \\ 2D + (q+2)N &= ((q+3-m)q+2)(q^2+q+1) \\ 2D + (q+1)(q+2)N &= ((q+3-m)q+2)((q+3-m)q+1)(q+1) . \end{aligned}\right\} \quad (4.6)$$

These simultaneous equations have a unique solution; in particular,

$$E = q^3 + q^2 + q + 1 - [(q+3-m)q+2][q^2m + qm + 2]/[2(q+1)] . \quad (4.7)$$

Since $E \geq 0$, it follows from (4.7) that

either $m \leq 1 + 3/(q+1)$ *or* $m \geq q+1$. (4.8)

Now, $q \leq 2 \Longrightarrow m \geq q+1$ and

if $r = 3$, *then* K *is of type* $(0,1,2,q+1)_1$. (4.9)

Let s be a line belonging to K. Since $k = 2(q+1)$, there exist two distinct points Q and Q' in $K \setminus s$. The plane through Q and s meets $k(q+2)$ in the points of $s \cup \{Q\}$ and does not contain Q'. So the line $s' = QQ'$ is skew to s and meets K in at least two points. Any plane through s' meets K at $q+2$ points. If s' met K in exactly two points, by (2.1), $k = (q+1)q+2$, a contradiction. Therefore, s' belongs to K. The $2(q+1)$ points on s and s' are exactly the points in K. When $r = 3$, the statement is proved.

Suppose $r \geq 3$. The 3-space P_3 joining four independent points in K meets it in a set K' consisting of two skew lines s and s'. Let Q be a point on s' and let $\pi = Qs$. Any 3-space P_3' through π and a point in $K \setminus \pi$ meets K in two skew lines: one of them is s, the other one, s'', contains Q. If $s'' \neq s'$, the plane through s' and s'' would meet K in at least $2q+1$ points, a contradiction (as $2q+1 > q+2$); therefore, $s'' = s'$. Thus, $P_3' = P_3$ and all points in K belong to this space. By (1.1), $r = 3$ and the statement is proved.

From Theorems IX - XII, Theorem II follows.

BIBLIOGRAPHY

1. A. Bichara, "Sui k-insiemi di PG(r,q) di classe $[0,1,2,n]_2$" *Rend. Mat.*, 13 (1980), to appear.
2. R. Esposito, "Sui k-insiemi di tipo $(0,1,m_d)_d$, $2 \leq d < r$, di uno spazio di Galois $S_{r,q}$ $(r \geq 3)$", *Riv. Mat. Univ. Parma*, 3 (1977), 131-140.
3. B. Segre, "Le geometrie di Galois", *Ann. Mat. Pura Appl.*, 48 (1959), 1-97.
4. G. Tallini, "Graphic characterization of algebraic varieties in a Galois space", *Teorie Combinatorie*, Tomo II, Atti dei Convegni Lincei 17, (Accad. Naz. Lincei, Rome, 1976), pp. 153-165.
5. G. Tallini, "I k-insiemi di classe $[0,1,n,q+1]_1$ regolari di $S_{r,q}$", Atti del Convegno sulle Geometrie Combinatorie, 1975, Quaderno del G.N.S.A.G.A. (Consiglio Nazionale delle Richerche, 1976).
6. G. Tallini, "Spazi di rette e geometrie combinatorie", Seminario di Geometrie Combinatorie 3, (Università di Roma, 1977).
7. M. Tallini-Scafati, "Caratterizzazione grafica delle forme hermitiane di un $S_{r,q}$", *Rend. Mat. e Appl.*, 26 (1967), 273-303.
8. J.A. Thas, "A combinatorial problem", *Geom. Dedicata*, 1 (1973), 263-240.

Istituto Matematico
Università di Roma
00185 Roma
Italy

COVERING GRAPHS AND SYMMETRIC DESIGNS

N.L. Biggs* and T. Ito

1. INTRODUCTION

Let P and B be the sets of points and blocks, respectively, of a symmetric (v,k,λ)-design, so that $v = |P| = |B| = (k^2 - k + \lambda)/\lambda$. The *incidence graph* of the design has vertex-set $P \cup B$, and its edges are the pairs $\{p, b\}$, where p is a point incident with the block b. We shall denote such a graph by $D(k,\lambda)$. For given k and λ there may be no such graph, or there may be many graphs. A graph $D(k, \lambda)$ is bipartite and regular with valency k; it has order $2v$ and diameter 3.

A graph G is an r-*fold covering* of a graph H if there is a local isomorphism ϕ from G to H such that $|\phi^{-1}(h)| = r$ for each vertex h of H. (An example of a 2-fold covering is shown in Figure 1.) If H is regular then G is regular with the same valency, and $|VG| = r|VH|$.

In this paper we shall investigate λ-fold coverings of the graphs $D(k, \lambda)$. The motivation stems from a problem in graph theory, concerning the *girth* (length of a shortest cycle) of regular graphs: basically, we are seeking the smallest graph with a given girth and valency. Our present investigations are relevant to the case when the girth is 6.

The first remark is that the girth of a graph $D(k, \lambda)$ with $\lambda \geq 2$ is just 4, since any two points p and q of the design are incident with λ blocks $b,c,\ldots$, and we have a 4-cycle $pbqc$. On the other hand, the girth of $D(k, 1)$ is 6. Now the order of $D(k, 1)$ is

$$n_0(k) = 2(k^2 - k + 1),$$

and it is easy to see that this is the smallest possible order for a graph with girth 6 and valency k; furthermore, a graph of order

$n_0(k)$ which has these properties must be a $D(k, 1)$ graph [12]. Thus there is a graph with girth G, valency k, and order $n_0(k)$ if and only if there is a projective plane with k points on each line. Since we know that this is impossible for some values of k (for example, $k = 7$), the general question of finding a graph of smallest order is non-trivial.

It is here that the λ-fold coverings of $D(k, \lambda)$ are relevant, for $\lambda \geq 2$. Although $D(k, \lambda)$ itself has girth 4, a λ-fold covering of it may have girth 6, and its order is

$$2(k^2 - k + \lambda) \quad = \quad n_0(k) + 2(\lambda - 1) .$$

In other words, the *excess* $2(\lambda - 1)$ is relatively small, and it does not depend on k. In a recent paper [3] it is shown that if G is a graph with valency k and girth 6 then:

(i) if G has excess $e \leq k - 2$ then it is bipartite and e is even;

(ii) provided that an additional graph-theoretic condition is satisfied, any such graph is a λ-fold covering of $D(k, \lambda)$, with $e = 2(\lambda - 1)$;

(iii) the conclusion of (ii) holds without restriction if $e = 2$.

In view of these remarks we have a revised and restricted version of our original problem: given k, what is the smallest value of λ for which there is a λ-fold covering of $D(k, \lambda)$ with girth 6? If $k - 1$ is a prime power the answer is $\lambda = 1$, but for other values the answer is less clear.

In Section 2 of this paper we shall outline a method of construction for the graphs we are seeking, and investigate some of its theoretical consequences. In Section 3 we shall survey the current state of the art, in particular for small values of k.

2. MATRICES AND AUTOMORPHISMS

Suppose that a graph $D(k, \lambda)$ is given, corresponding to a known symmetric design, and we wish to describe a λ-fold covering of it. Each vertex v in $P \cup B$ will be covered by λ vertices $(v,1), (v,2), \ldots, (v,\lambda)$, where the numbering may be assigned arbitrarily. Each edge of $D(k, \lambda)$ is a pair $\{p,b\}$ and we have to define the λ covering edges of the form $\{(p,\alpha), (b,\beta)\}$, where α and β belong to the set $\Lambda = \{1,2,\ldots,\lambda\}$. We do this by prescribing a permutation x_{pb} of Λ such that

$$(p,\alpha) \text{ is joined to } (q,\beta) \iff x_{pb}(\alpha) = \beta .$$

A convenient way of describing this prescription is by modifying the incidence matrix of the design: we replace each 1 by the relevant x_{pb}. We obtain a matrix X, whose rows and columns correspond to the points and blocks of the design, and

$$(X)_{pb} = \begin{cases} x_{pb} & \text{if } p \text{ and } b \text{ are incident;} \\ 0 & \text{otherwise.} \end{cases}$$

For example, there is a unique symmetric (4,3,2)-design, and its incidence graph $D(3,2)$ is the ordinary cube. Let 1 (identity) and $\tau = (12)$ denote the permutations of $\{1,2\}$ and set

$$X = \begin{bmatrix} 0 & 1 & 1 & 1 \\ 1 & 0 & 1 & \tau \\ 1 & \tau & 0 & 1 \\ 1 & 1 & \tau & 0 \end{bmatrix}$$

In this case the 2-fold covering of $D(3,2)$ has girth 6. It is depicted in Figure 1.

If $D = D(k, \lambda)$ we shall use the symbol XD to denote a λ-fold covering defined by a matrix X as above. Given points p and q of the design, we let

$$B_{pq} = \{ b \in B \mid b \text{ is incident with } p \text{ and } q \},$$

so that $|B_{pq}| = \lambda$.

THEOREM 1: *Let* XD *and* B_{pq} *be as above, and let* y_b $(b \in B_{pq})$ *be the permutation of* Λ *defined by*

$$y_b = x_{qb}^{-1} x_{pb} .$$

The graph XD *has girth* 6 *if and only if, for each pair of points* $p, q,$ *and each* α *in* Λ, *the function* $B_{pq} \longrightarrow \Lambda$ *given by* $b \mapsto y_b(\alpha)$ *is a bijection.*

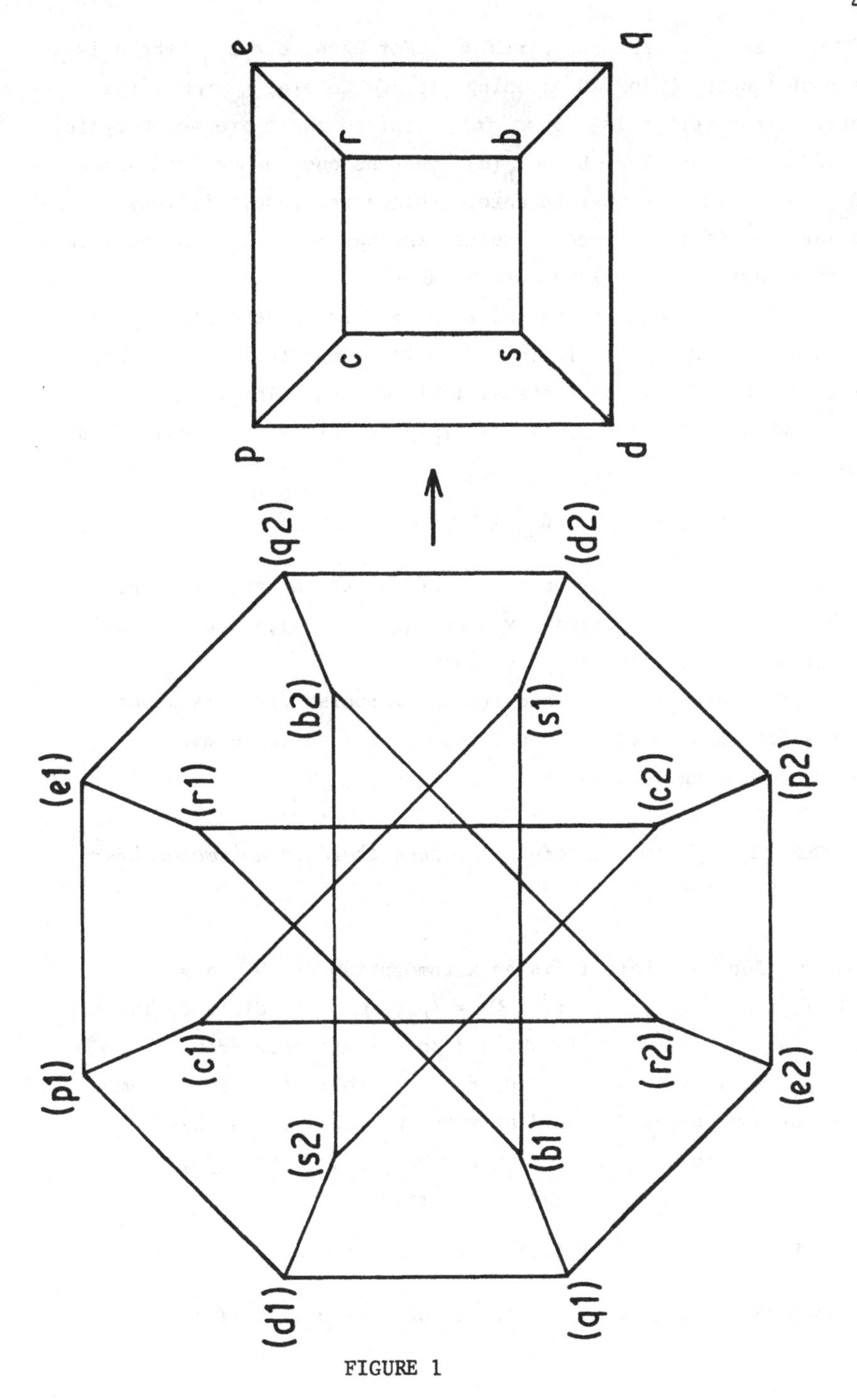

FIGURE 1

PROOF: Suppose XD has girth 6. For each $b \in B_{pq}$ there is a path of length 2 in XD joining (p, α) to $(q, y_b(\alpha))$ - the intermediate vertex is $(b, x_{pb}(\alpha))$. Since there are no 4-cycles in XD, the function $b \longmapsto y_b(\alpha)$ must be one-to-one, and since $|B_{pq}| = \lambda = |\Lambda|$ it must be onto. Hence the result follows. Conversely, if this property holds XD has no 4-cycles and (since it is bipartite) its girth must be 6. □

Let $\Gamma(X)$ denote the subgroup of $\mathrm{Sym}\,\Lambda$ generated by the permutations x_{pb}. It follows from Theorem 1 that if XD has girth 6 then $\Gamma(X)$ acts transitively on Λ. The extreme case occurs when $\Gamma(X)$ is regular (sharply transitive): we call this the *regular case*. We remark that

$$\Gamma(X) = \{ y_b \mid b \in B_{pq} \} \tag{1}$$

in the regular case, for each pair of points p and q. This is because the permutations y_b belong to $\Gamma(X)$, they are all different, and there are λ of them.

If we are given D, and its automorphism group is known, we may find a clue to the construction of a suitable matrix X in the following results.

THEOREM 2: *If* XD *has girth* 6 *then there is a homomorphism* $\pi : \mathrm{Aut}\,XD \to \mathrm{Aut}\,D$.

PROOF: Suppose that f is an automorphism of XD, $\alpha \neq \beta$, and $f(p, \alpha) = (p_1, \gamma)$, $f(p, \beta) = (p_2, \delta)$. The distance in XD from (p, α) to (p, β) is 4, and so the distance from (p_1, γ) to (p_2, δ) is also 4. If $p_1 \neq p_2$ it follows from Theorem 1 that for some block b incident with p_1 and p_2 we have $y_b(\gamma) = \delta$, so that the distance from (p_1, γ) to (p_2, δ) would be 2. Hence $p_1 = p_2$, and if we write

$$f(v, \alpha) = (v', f_v(\alpha)) \qquad (v \in P \cup B),$$

we have an automorphism πf of D defined by $(\pi f)(v) = v'$. □

THEOREM 3 : *Suppose* XD *has girth* 6, *and* K *is the kernel of the homomorphism* π. *Then*

(i) K *may be regarded as a subgroup of* Sym Λ, *and it acts semi-regularly;*

(ii) in the regular case, K *is the centralizer of* $\Gamma(X)$ *in* Sym Λ.

PROOF : For any automorphism f of XD we have a set of permutations $\{f_v\}$ of Λ defined as in the proof of the previous theorem. Since f is an automorphism, it follows that for all edges $\{p,b\}$ of D

$$f_b x_{pb} = x_{p'b'} f_p , \quad \text{where} \quad p' = (\pi f)(p), \quad b' = (\pi f)(b).$$

In particular, if h is in K,

$$h_b = x_{pb} h_p x_{pb}^{-1} . \tag{2}$$

Repeated application of (2) shows that, since D is connected, the whole set $\{h_v\}$ is determined by one of them, h_p say. Thus we may regard K as a subgroup of Sym Λ by means of the monomorphism $h \longmapsto h_p$ (p fixed).

Now fix p, c, and q, with c in B_{pq}, and let

$$\Delta = \{ y_b^{-1} y_c \mid b \in B_{pq} \} .$$

By Theorem 1, given α and β in Λ, there is some b in B_{pq} such that $y_b(\beta) = y_c(\alpha)$, and so the set Δ is transitive on Λ. Also, four applications of (2) show that h_p commutes with each member of Δ. Thus K centralizes Δ and acts semi-regularly.

(ii) In the regular case it follows from (1) that $\Delta = \Gamma(X)$, and so K is contained in the centralizer of $\Gamma(X)$. On the other hand, if we are given any permutation h_* which centralizes $\Gamma(X)$, and we define $h_v = h_*$ for all vertices v of D, then (2) is satisfied and the mapping h defined by $h(v, \alpha) = (v, h_v(\alpha))$ is an automorphism of XD belonging to K. □

If $D = D(k, \lambda)$ is given, let us write $XD \sim X'D$ to signify that XD and $X'D$ are isomorphic graphs. We might hope that if a graph XD with girth 6 does exist, then X' can be found with $XD \sim X'D$ and $\Gamma(X')$ regular. In other words, the permutations x'_{pb} generate a small subgroup of $\mathrm{Sym}\,\Lambda$. We shall see that in some very special cases our hope is fulfilled.

The fundamental observation is that we can always assign the permutations x'_{pb} arbitrarily on any acyclic subset of the edges of D, since this is simply equivalent to relabelling the vertices of the covering graph.

THEOREM 4: *Suppose that* XD *has girth* 6 *and* $\lambda \leq 3$. *Then there is a matrix* X' *such that* $\Gamma(X')$ *is regular and* $XD \sim X'D$.

PROOF: If $\lambda \leq 2$ there is nothing to prove. If $\lambda = 3$ we have to show that $\Gamma(X') = A_3 = \{1, (123), (132)\}$ for some X' such that $XD \sim X'D$.

Classify the edges of D as follows, where p is an arbitrarily chosen point:

Level 1: $\{p,b\}$ b incident with p;

Level 2: $\{q,b\}$ $q \neq p$, b incident with p;

Level 3: $\{q,a\}$ $q \neq p$, a not incident with p.

Order the blocks b incident with p in some way, and do the same for the points $q \neq p$. Then the following subset of edges forms a tree in D:

Level 1: all edges;

Level 2: for each q, the edge joining q to the first block b incident with q and p;

Level 3: for each a, the edge joining a to the first point $q \neq p$ which is incident with a.

Since this subset of edges contains no cycles, we may assign the entries of X' to be 1 on these edges and assign the remaining entries so that $X'D \sim XD$.

Given any points r, s, and a block c in B_{rs}, consider the set

$$\left\{ y_d'^{-1} y_c' \mid d \in B_{rs} \right\},$$

where $y_d' = x_{sd}'^{-1} x_{rd}'$. This set contains the identity $(d = c)$ and two other distinct fixed-point-free permutations (otherwise $X'D$ would contain a 4-cycle). Hence it is just A_3. Now the set

$$\left\{ y_d' \mid d \in B_{rs} \right\}$$

is a coset of A_3 - either A_3 itself, or the class of transpositions. The same holds for the analogous set with points and blocks interchanged.

We now show that all non-zero entries of X' must lie in A_3. This is immediately true of all Level 1 edges. Let $\{q,d\}$ be a Level 2 edge not in the basic tree, and let $\{q,b\}$ be the unique Level 2 edge at q which is in the tree. Taking $r = q$ and $s = p$ above, we see that $y_b' = 1$, so that the coset must be A_3 itself. It follows that $x_{qd}' = x_{pd}'^{-1} x_{qd}' = y_d'$ is in A_3. Thus all the entries of X' corresponding to Level 2 edges are in A_3. A similar argument establishes the same result for Level 3 edges, and so we have the result. □

Theorem 4 tells us that if we wish to construct a graph XD with girth 6, and $\lambda = 3$, then we can restrict the entries of X to the regular group A_3. It is possible that a similar result may hold in the case $\lambda = 4$, but unlikely for larger values of λ, since the combinatorial properties of the set $\{ y_d^{-1} y_c \}$ no longer force it to be a group.

In general it can be shown that if XD has girth 6 and the order of the kernel of π is λ, then we can find X' as in Theorem 4. However, the hypothesis here is not particularly useful in practice.

3. CONCLUSIONS AND QUESTIONS

We now return to our original problem, the construction of graphs with girth 6, valency k, and small excess. In the present context, we are seeking a λ-fold covering of $D(k, \lambda)$ with girth 6, and with λ as small as possible.

Let us begin by tabulating those values of k and λ in the range $3 \le k \le 15$ for which a symmetric (v, k, λ)-design may exist. With the exception of two elusive projective planes ($k = 11$ and 13), it is now known that every set of parameters in this range which satisfies the arithmetical conditions does in fact correspond to a design. Hall's book [7] is the basic reference, and this is updated in an article by the same author [8]. The only case not covered by these references is $k = 15$, $\lambda = 3$, and this has now been settled by Haemers (see [1,5]). In the following table a numerical entry v indicates the existence of at least one symmetric (v, k, λ)-design for that value of k and λ, and ?v indicates the possibility of such a design.

k \ λ	1	2	3	4	5	6	7	8	9	10	11	12	13	14
3	7✓	4✓												
4	13✓	7✓	5✓											
5	21✓	11×		6✓										
6	31✓	26×	11×		7									
7			15✓			8✓								
8	57✓			15			9✓							
9	73✓	37	25	19		13✓		10✓						
10	91✓				19	16			11					
11	?111	56	31		23					12✓				
12	133✓		45			23					13			
13	?157	79×		40		27						14✓		
14	183✓						27						15	
15			71			36	31							16

The symbol following v indicates a definite answer to the question of the existence of a λ-fold covering of $D(k, \lambda)$, with girth 6, in the few cases where such an answer is known.

The case $\lambda = 1$ is trivial. The case $\lambda = 2$ was first studied by Payne [11]. He showed that a necessary condition for a 2-fold covering of $D(k, 2)$ with girth 6 is that k or $k-2$ is a perfect square. This result was rediscovered in reference [2]; it justifies the entries 11× and 79× in the table. The entry 16× is a consequence of a calculation first done by Carter in 1974 (unpublished); an equivalent calculation may be found on pages 81-82 of reference [6]. The entry 4✓ is justified by the example in Figure 1, and the entry 7✓ by the following matrix:

$$\begin{bmatrix} 0 & 0 & \tau & 0 & 1 & 1 & \tau \\ 1 & 0 & 0 & \tau & 0 & 1 & 1 \\ 1 & \tau & 0 & 0 & \tau & 0 & \tau \\ 1 & 1 & \tau & 0 & 0 & \tau & 0 \\ 0 & 1 & 1 & \tau & 0 & 0 & \tau \\ 1 & 0 & 1 & 1 & 1 & 0 & 0 \\ 0 & \tau & 0 & \tau & 1 & 1 & 0 \end{bmatrix} .$$

In the $\lambda = 3$ column we have an affirmative answer for $k = 4$; the relevant matrix (with $\omega = (123)$) is

$$\begin{bmatrix} 0 & 1 & 1 & 1 & 1 \\ 1 & 0 & 1 & \omega & \omega^2 \\ 1 & 1 & 0 & \omega^2 & \omega \\ 1 & \omega & \omega^2 & 0 & 1 \\ 1 & \omega^2 & \omega & 1 & 0 \end{bmatrix} .$$

The negative answer for $k = 11$ is justified by a direct argument (Theorem 4 is useful here). The positive answer for $k = 15$ is due to Ito [9]. Here there are several possible (15, 7, 3)-designs, and Ito's matrix uses the 'classical' one formed by the points and planes of $PG(3,2)$. Finally, the entry 6✓ in the $\lambda = 4$ column is justified by a matrix X which can be found by hand without much difficulty. In this case also $\Gamma(X)$ is regular.

From our table, we see the answer to our main question, concerning the smallest graph with girth 6 and valency k, is non-trivial when $k = 7, 11, 13$, and 15. In these cases we seek the leftmost affirmative answer in the relevant row of the table. The case $k = 7$ was originally settled by O'Keefe and Wong [10], who discovered the graph on a computer. Ito [9] showed that their graph is a 3-fold covering of $D(7,3)$, and proved that its automorphism group is a 3-fold covering group of A_7. The other three cases are still in doubt. If we adopt a pessimistic view about $PG(2,10)$ and $PG(2,12)$, we are led to study coverings of $D(11,2)$ and $D(13,4)$ - in both cases there are several designs. Lastly, since there is no plane or biplane with $k = 15$, we must study coverings of the incidence graphs of Haemers' symmetric (71,15,3)-designs.

In a more general vein, one might ask for an infinite family of graphs $D(k, \lambda)$ which have λ - fold coverings of girth 6. Suitable candidates might be the graphs $D(q^2 + q + 1, q + 1)$ associated with the points and planes in $PG(3,q)$ and the graphs $D(2t - 1, t - 1)$ associated with the quadratic residues over $GF(q)$, where q is a prime power, equal to $4t - 1$ in the latter case. Currently, the best general estimates of excess in the girth 6 case are due to Brown [4], but constructions of the kind proposed would yield significant improvements.

BIBLIOGRAPHY

1. H. Beker and W. Haemers, "2-designs having an intersection number k-n", *J. Combin. Theory Ser. A*, 28 (1980), 64-81.
2. N.L. Biggs, "Covering biplanes", *Proc. International Graph Theory Conference*, W. Michigan, 1980, to appear.
3. N.L. Biggs and T. Ito, "Graphs with even girth and small excess", *Math. Proc. Cambridge Philos. Soc.*, 88 (1980), 1-10.
4. W.G. Brown, "On Hamiltonian regular graphs of girth 6", *J. London Math. Soc.* 42 (1967), 514-520.
5. W. Haemers, *Eigenvalue Techniques in Design and Graph Theory*, (Math. Centrum, Amsterdam, 1979).

6. J.I. Hall, A.E.J.M. Jansen, A.W.J. Kolen and J.H. van Lint, "Equidistant codes with distance 12", *Discrete Math.*, 17 (1977), 71-83.

7. M. Hall Jr., *Combinatorial Theory* (Blaisdell, Waltham, Mass., 1967).

8. M. Hall Jr., "Construction of block designs". *A Survey of Combinatorial Theory*, Colorado State Univ., 1971 (ed. J.N. Srivastava, North Holland, Amsterdam, 1973), pp.251-258.

9. T. Ito, "On a graph of O'Keefe and Wong", *J. Graph Theory*, to appear.

10. M. O'Keefe and P.K. Wong, "A smallest graph of girth 6 and valency 7", *J. Graph Theory*, to appear.

11. S.E. Payne, "On the non-existence of a class of configurations which are nearly generalized n-gons", *J. Combin. Theory Ser. A.*, 12 (1972), 268-282.

12. R.R. Singleton, "On minimal regular graphs of maximum even girth", *J. Combin. Theory*, 1 (1966), 306-332.

Mathematics Department
Royal Holloway College
University of London
Egham, Surrey TW20 0EX
U.K.

ARCS AND BLOCKING SETS

A.A. Bruen* and R. Silverman

In Section 1 we concentrate on arcs in finite planes of order n with particular emphasis on the case $n = 10$. In Section 2 we are mainly concerned with arcs and blocking sets in the classical planes.

1. PLANES OF ORDER 10

Let π be a finite projective plane of order n. Recall that a k-arc K is a set of k points in π with no three collinear. K is *complete* if K is not contained in any k'-arc with $k' > k$. We are interested in studying complete k-arcs for small values of k (in particular for $k < n+1$). The following result is well-known (Segre [15]).

THEOREM 1.1: *If* K *is a complete* k-*arc in* π *then* $\binom{k-1}{2} \geq n$.

PROOF: Take any point P on K and draw a tangent t to K at P (if no such tangent exists then $k \geq n+2$). There are n points on t apart from P and, since K is complete, each such point must lie on a secant to K. Moreover, since such a secant cannot contain P, there are only $\binom{k-1}{2}$ such secants. Thus $n \geq \binom{k-1}{2}$. □

A *blocking set* is a subset S of the points of π such that each line in π contains at least one point in S and at least one point not in S. We define a *dual blocking set* in the obvious way. A result similar to Theorem 1.1 is as follows.

THEOREM 1.2 : *If* K *is a complete* k-*arc then* $\binom{k}{2} \geq n + \sqrt{n} + 1$.

PROOF : The secants to K yield a dual blocking set in π. The result now follows by duality from the main result in [4]. □

With reference to the bound in Theorem 1.1 we examine the case $n = 10$, $k = 6$. A computer result of Denniston [9] shows that complete 6-arcs do not exist in planes of order 10. In a related paper MacWilliams, Sloane and Thompson [13] discuss codewords in a plane of order 10. In particular, using a computer, they show that there are no codewords of weight 15. However, as we indicate in [6], [7] these computer results in [9], [13] are equivalent by duality. In fact in [7], using Denniston's result, we prove a result which is stronger than that in [13]. Namely, we show that *in a projective plane of order 10 a blocking set has to have at least 16 points.* This result may be used to show that *a net of order 10 which contains six or more parallel classes can be completed to a plane of order 10 in at most one way.* This improves on the classical result of Bruck [3, Theorem 3.1]).

One possible attack on the existence problem for a plane of order n with $n = 10$ is to try to show the non-existence of complete k-arcs with $k < n + 2 = 12$. (The analagous method can be used to give an elementary proof of the non-existence of a plane of order six.) Our main results in this section (Theorems 1.6, 1.7) indicate that the task of ruling out complete 9-arcs or complete 10-arcs in a plane of order 10 may also require the use of a computer.

To begin, let K be a complete 6-arc in a projective plane π of order 10. (Of course, by Denniston's result, no such arc exists: but, let us proceed.) Let B_i, $0 \leq i \leq 5$, denote the number of points of π that lie on exactly i secants of K. Since K is complete, $B_0 = 0$. Also the 6 points on K yield $B_5 = 6$. It is easy to see that $B_4 = 0$. Solving the usual 3 incidence equations (see [13]), we get $B_1 = 90$, $B_2 = 0$, $B_3 = 15$. We refer to the 15 points of π through which exactly three secants of K pass as the *F-points*.

LEMMA 1.3 : *Each secant of* K *contains exactly* 3 F*-points.*

PROOF : Let x be any secant and let A be any point of $K - x$. Join A to the three other points of $K - x$. These three lines meet x in distinct points U, V, W. Since $B_2 = 0$, each of U, V, W is an F-point. These are all the F-points on x. □

LEMMA 1.4 : *Any tangent to* K *contains no* F*-points, and any non-secant contains just two* F*-points.*

PROOF : Let t be a tangent to K at P. Each of the 10 points of $t - P$ has either 1 or 3 secants to K on it. Since these secants cannot contain P there are just $\binom{5}{2} = 10$ of them. Thus each of the 10 points of $t - P$ has exactly one secant on it, so t contains no F-points. Similarly any non-secant to K contains exactly two F-points. □

We now have that any line of π contains at most three F-points and that the lines of π containing exactly three F-points are the secants to K. We refer to the 15 secants of K as *F-lines.*

Let us now examine the structure of these F-points and F-lines. Take any F-line x. On x are three F-points A, B, C. Take any two of these three points, say A and B. Let $y \neq x$ be an F-line on A and $z \neq x$ be an F-line on B. Note that y, z are secants to K. Let y meet z in X. If X were not on K then, since $B_2 = 0$, there would be a third secant to K on X. This would force X on x which is impossible. We conclude that X is not in K. Now, apart from x, there are two F-lines on A yielding four F-points off x : the same is true for B and C. As pointed out above, y cannot meet z in an F-point. Thus we have accounted for exactly 12 distinct F-points off x. Together with A, B, C this accounts for all 15 F-points. In particular, we have shown

LEMMA 1.5 : *Given an* F*-line* x *and an* F*-point* Y *off* x *there is a unique* F*-point* Z *on* x *such that the line joining* Y *to* Z *is an* F*-line.* □

By a special case of the Buekenhout-Shult theorem (see Shult [16]) this means that the F-points and F-lines are the points and lines of the symplectic geometry in PG(3,2). Another model of this geometry is obtained as follows. Take a complete oval (6-arc) θ in the plane $\sigma = PG(2,4)$. Then the points of the symplectic geometry are defined to be the $21 - 6 = 15$ points of σ off θ. The lines of the symplectic geometry are the 15 lines of σ that meet θ in two points. We then have an incidence-preserving isomorphism f from the F-points and F-lines of π to the above 15 points and 15 lines of the plane σ. Three F-points are collinear if and only if their three images under f lie on a secant line to θ in σ.

THEOREM 1.6 : *The existence of a complete* 6*-arc* K *in a projective plane* π *of order* 10 *implies the existence of a collection of six complete* 10*-arcs in* π*, with any two of them meeting in a* 6*-arc. The graph corresponding to any of the six complete* 10*-arcs is the Petersen graph.*

PROOF : Let x be any line of σ that misses the oval θ. Let M denote the 10 points of $\sigma - \theta$ that lie off x. Suppose that some three points of M were to lie on a line t of σ, such that t was also a secant to θ. Now t meets x in a point of σ not on θ. Thus the secant line t of σ would contain six points, which is impossible. We conclude that no three points of M lie on a secant to θ in σ. Therefore the 10 F-points of π corresponding to M under the isomorphism f yield a 10-arc K in π. Using previous results here one can show by an easy argument that K is actually complete. There are six possible choices for x and this yields the six complete 10-arcs in the theorem. The statement concerning the Petersen graph follows from the definition of that graph. □

Next, take any point X of σ not on θ. There are exactly two lines y, z of σ on X which miss θ. The nine F-points of π corresponding under f to the nine points on y and z in σ yield a 9-arc in π, which can also be shown to be complete. Since, in the construction, there are 15 possibilities for X, we have shown

THEOREM 1.7: *The existence of a complete* 6*-arc* K *in a projective plane* π *of order* 10 *implies the existence of a set of* 15 *complete* 9*-arcs in* π. □

REMARKS: (1) Theorem 1.6 (1.7) shows that if one can rule out complete 10-arcs (9-arcs) in π then the non-existence of complete 6-arcs follows.

(2) Let π' denote the plane dual to π. Let K be *any* k-arc in π (complete or not). The set of tangents to K in π yields a set S of points in π'. If k is odd then *every* point of π is on an odd number of tangents to K. Thus every line of π' contains an odd number of points of S. Similarly, if k is even then every line of π' contains an even number of points of S. We can form the extended code D for π' as in [13]. Since the code D is self-dual it follows that the set of points in S always yields a codeword in D. In this context it is of interest that the 20 tangents to any of the six complete 10-arcs M of Theorem 1.6 yield a *primitive* codeword $S = M_{20}$ of weight 20 in the language of Hall [10] and Anstee, Hall and Thompson [1].

2. CLASSICAL PLANES

Let π denote $PG(2,q)$ the projective plane over the finite field $GF(q)$ of order q. Let S be a blocking set in π containing exactly $q+\lambda$ points, so $|S| = q+\lambda$. Since S is a blocking set we can show that no line of π contains more than λ points of S. If some line of π contains exactly

λ points of S we say that S is of *Rédei type*. An explanation for this terminology is as follows. In [5, p.246] (see also [8]) we showed how one of the main results in Rédei's book [14, p.226], having to do with difference quotients in finite fields, can be phrased entirely in terms of blocking sets of Rédei type. The blocking set S is said to be *irreducible* if no proper subset of S is also a blocking set. This is equivalent (see [11, p.367]) to saying that at every point P of S there is at least one tangent to S. Note that if S is of Rédei type then S must be irreducible. We proceed to examine the case of irreducible blocking sets which are not of Rédei type.

THEOREM 2.1 : *In* $\pi = PG(2,q)$ *let* S *be a blocking set with* $|S| = q + \lambda$. *Assume that some line* x *of* π *contains exactly* $\lambda - 1$ *points of* S. *Then, if* S *is irreducible, it follows that* $|S| \geq q + \frac{1}{2}(q+3)$.

PROOF : Let P be any point of x not in S, so P is in $x - S$. Of the q lines of π on P different from x, one line $y = y(P)$ contains two points of S and the rest contain just one point of S. We now consider two cases.

CASE 1 : For any two points $P_1 \neq P_2$ on $x - S$ the lines $y(P_1)$ and $y(P_2)$ always meet in a point not in S.

CASE 2 : For some two points $P_1 \neq P_2$ on $x - S$ the lines $y(P_1)$, $y(P_2)$ meet in a point Z of S. Now $|x - S| = q + 1 - (\lambda - 1)$. Thus, in Case 1, we get that $|S| \geq (\lambda - 1) + 2(q - \lambda + 2) = 2q - \lambda + 3$. Since $|S| = q + \lambda$ there follows $\lambda \geq (q+3)/2$ and we are done.

We examine Case 2. S being irreducible, there exists a tangent line z to S at Z. Let J denote the set of those lines of π other than z which pass through Z and which are tangent to S at Z. The lines ZP_1 and ZP_2 are not tangents to S. Thus

$$|J| \leq (q+1) - 1 - 2 - (\lambda - 1) = q - \lambda - 1 .$$

Let $\pi_A = AG(2,q)$ be the affine plane got by removing z from $\pi = PG(2,q)$. We construct a set S_A of points in π_A as follows. On each line α of J choose any one point $R = R(\alpha)$ with $R(\alpha) \neq Z$ for all lines α in J. Denote by H the set of all such points $R(\alpha)$. Thus $|H| = |J| \leq q - \lambda - 1$. Finally, put

$$S_A = S - \{Z\} \cup H .$$

Since S is a blocking set in π it follows that S_A intersects all lines of π_A. Also

$$|S_A| \leq (q+\lambda) - 1 + (q - \lambda - 1).$$

Therefore

$$|S_A| \leq 2q - 2 .$$

But, an important result of Jamison [12] and Brouwer-Schrijver (see [2] for their beautiful proof) asserts that $|S_A| \geq 2q - 1$. It follows that Case 2 cannot occur, and the theorem is proved. □

It is possible to generalize Theorem 2.1 .

THEOREM 2.2 : *In* $\pi = PG(2,q)$ *let* S *be an irreducible blocking set with* $|S| = q + \lambda$. *Assume that some line of* π *contains exactly* $\lambda - \theta$ *points of* S. *Then* $\lambda \geq (q + 1 + 2\theta)/(\theta + 1)$. □

The proof again uses the result on intersection sets in $AG(2,q)$ that was utilized in Theorem 2.1 .

REMARKS : (1) The proofs of Theorems 2.1 and 2.2 will remain valid in any projective plane π of order n having the following property.

Let z be any line of π and let S_A be a set of points in the affine plane $\pi_A = \pi - z$ such that all lines of π_A contain at least one point of S_A. Then $|S_A| \geq 2n - 1$.

Thus it is of interest to study such intersection sets S_A in arbitrary finite affine planes π_A. The best known results on this problem are in [8], where the theorem on blocking sets in finite projective planes [4] is exploited. We have recently obtained some slight improvements on certain results in [8]. They will be reported elsewhere.

(2) In Chapter 9 of Hirschfeld [11] some improvements on the bound in Theorem 1.1 are described. In the proof of Theorem 1.2 we have mentioned a connection between complete arcs and dual blocking sets. Using this idea, and the result of Theorem 2.2, one can rule out the existence of certain small arcs in PG(2,q) in cases which do not seem to be covered by any of the theorems in [11]. For example, using Theorem 2.2, one can prove the non-existence of a complete 11-arc in PG(2, 43).

ACKNOWLEDGEMENTS : The authors are very grateful to Professor E. Shult for a valuable conversation at the conference relating to Theorem 1.6.

This research was supported by the Natural Sciences and Engineering Research Council of Canada.

BIBLIOGRAPHY

1. R.P. Anstee, M. Hall Jr. and J.G. Thompson, "Planes of order 10 do not have a collineation of order 5", *J. Combin. Theory Ser. A*, 29 (1980), 39-58.
2. A.E. Brouwer and A. Schrijver, "The blocking number of an affine space", *J. Combin. Theory Ser. A*, 24 (1978), 251-253.
3. R.H. Bruck, "Finite nets. II. Uniqueness and imbedding", *Pacific J. Math.*, 13 (1963), 421-457.
4. A.A. Bruen, "Blocking sets in finite projective planes", *SIAM J. Appl. Math.*, 21 (1971), 380-392.
5. A.A. Bruen, "Collineations and extensions of translation nets", *Math. Z.*, 145 (1975), 243-249.
6. A.A. Bruen and J.C. Fisher, "Blocking sets, k-arcs and nets of order ten", *Adv. in Math.*, 10 (1973), 317-320.
7. A.A. Bruen and J.C. Fisher, "Blocking sets and complete k-arcs", *Pacific J. Math.*, 53 (1974), 73-84.
8. A.A. Bruen and J.A. Thas, "Blocking sets", *Geom. Dedicata*, 6 (1977), 193-203.

9. R.H.F. Denniston, "Non-existence of a certain projective plane", *J. Austral. Math. Soc.*, 10 (1969), 214-218.
10. M. Hall Jr., "Configurations in a plane of order ten", to appear.
11. J.W.P. Hirschfeld, *Projective Geometries over Finite Fields* (Oxford Univ. Press, Oxford, 1979).
12. R.E. Jamison, "Covering finite fields with cosets of subspaces", *J. Combin. Theory Ser. A*, 22 (1977), 253-266.
13. F.J. MacWilliams, N.J.A. Sloane and J.G. Thompson, "On the existence of a projective plane of order 10", *J. Combin. Theory Ser. A*, 14 (1973), 66-79.
14. L. Redéi, *Lancunary Polynomials over Finite Fields* (North-Holland, Amsterdam, 1973).
15. B. Segre, *Lectures on Modern Geometry* (Cremonese, Rome, 1961).
16. E.E. Shult, "Characterizations of certain classes of graphs", *J. Combin. Theory Ser. B*, 13 (1972), 142-167.

Department of Mathematics
University of Western Ontario
London
Ontario N6A 5B9
Canada

Department of Mathematics
Wright State University
Dayton
Ohio 45431
U.S.A.

FLAT EMBEDDINGS OF NEAR 2n-GONS

Peter J. Cameron

1. INTRODUCTION

Near 2n-gons, as defined by Shult and Yanushka [9], form a class of geometries including the familiar generalised 2n-gons. It also includes the dual polar spaces, for which a system of axioms was given by Cameron [3] and refined by Shult [8].

The concept of a flat embedding of a geometry in a projective space is abstracted from the classification of antiflag transitive collineation groups of projective spaces by Cameron and Kantor [4]. Geometries possessing flat embeddings include symplectic geometries, generalised hexagons of type $G_2(q)$, and dual orthogonal geometries of type $O(2n+1, q)$. The embeddings in the last case can be constructed using spinors. Note that the second and third types are near 2n-gons.

This paper is directed towards determining the flat embeddings of near 2n-gons. Theorems 4.1 and 4.2 give this determination under additional hypotheses. As preliminaries, a summary of the theory of near 2n-gons and the axioms for dual polar spaces is given in Section 2, and the embedding theorem of Cameron and Kantor is stated in Section 3. The final section gives an account of the spinor embedding of dual $O(2n+1, q)$.

I am grateful to A.L. Wells Jr. for discussions about the topic of this paper.

2. NEAR 2n-GONS AND DUAL POLAR SPACES

This section outlines the definition and theory of near 2n-gons, due to Shult and Yanushka [9] and Shult [8], and the axiomatisation of dual polar spaces given by Cameron [3] and

Shult [8]. All our geometries will be finite.

A *near 2n-gon* $(n \geq 2)$ is a connected incidence structure of diameter n having the property that, given any point x and line L, there is a unique point of L nearest to x. (We identify each line with the set of points incident with it. The distance between two points is the minimum number of lines in a path joining them.) Note that, in a near 2n-gon, two points lie on at most one line. For if L and M are lines containing x and y, and z is another point of L, then z is at distance 1 from the two points x and y of M, hence $z \in M$; thus $L \subset M$, and similarly $M \subset L$. A similar argument shows that a near 2n-gon contains no triangles. A near 4-gon is the same thing as a (possibly degenerate) generalised quadrangle - see Higman [7]. A near 2n-gon with just two points on every line is the same thing as a connected bipartite graph of diameter n. However, as we shall see, the structure of near 2n-gons with more than two points per line is much more restricted.

In a near 2n-gon, a *quad* is a set Δ of points with the properties

(i) Δ has diameter 2;

(ii) a point collinear with two points of Δ lies in Δ;

(iii) Δ contains an ordinary quadrilateral.

The geometry induced on a quad by the lines it contains is a generalised quadrangle - hence the name. Of course, there may not be any quads; but Shult and Yanushka [9] showed the following:

PROPOSITION 2.1 : *Let* G *be a near 2n-gon in which every line has more than two points. Then any ordinary quadrilateral in* G *is contained in a unique quad.*

(In fact, their result is stronger: in any near 2n-gon, a quadrilateral one of whose sides has more than two points lies in a unique quad.)

Let x be a point and Δ a quad in a near 2n-gon. The pair (x, Δ) is said to be *classical* if there is a unique point y of Δ nearest to x, and any point $z \in \Delta$ satisfying $d(x, z) = d(x, y) + 1$ is collinear with y. An *ovoid* in a generalised quadrangle is a set of points meeting every line in a single point. The point-quad pair (x, Δ) is called *ovoidal* if the set of points of Δ nearest to x is an ovoid of Δ. Now Shult [8] showed:

PROPOSITION 2.2 : *Let* G *be a near 2n-gon in which every line has more than two points. Then any point-quad pair in* G *is either classical or ovoidal.*

A *dual polar space* is a geometry whose points and lines are the maximal and next-to-maximal subspaces of a polar space, with reverse inclusion as incidence. (A polar space is defined by the axioms of Tits [10], and is not required to be "thick"; but note that Tits has shown that the only thick finite polar spaces of rank at least 3 are the classical symplectic, orthogonal and unitary geometries.) Cameron [3] axiomatised dual polar spaces; his axioms were streamlined by Shult [8] as follows.

THEOREM 2.3 : *An incidence structure* G *is a dual polar space of rank* n *if and only if the following conditions hold:*

(i) G *is a near 2n-gon ;*

(ii) *any two points at distance* 2 *in* G *are contained in a quad;*

(iii) *any point-quad pair in* G *is classical.*

3. FLAT EMBEDDINGS

An incidence structure G is said to be *embedded* in a projective space P if the following conditions hold:

(i) the point and line sets of G are subsets of the point and line sets of P, and incidence in G is induced from P ;

(ii) any point of P on a line of G is a point of G .

(Condition (ii) ensures that G-lines are complete projective lines. This is the concept of embedding used, for example, by Buekenhout and Lefèvre [2]. For discussion of a more general concept, in which (ii) is relaxed, see the article by Percsy in this volume.) There is no loss in assuming that the point set of G spans P, and we shall always do so.

An incidence structure G of diameter n is said to be *metrically regular* if there are integers k, c_i, a_i, b_i $(0 \leq i \leq n)$ such that the following conditions hold:

(i) any line of G is incident with k points;

(ii) if x and y are points at distance i $(0 \leq i \leq n)$, then the number of points collinear with x and distant $i-1$, i, $i+1$ from y is c_i, a_i, b_i respectively. (Thus the point graph of G is distance-regular, in the sense of Biggs [1].)

Several classes of incidence structures which have been studied in the literature are metrically regular; for example, classical polar and dual polar spaces, regular generalised polygons, partial and semi-partial geometries, subspace geometries of projective spaces, 1½-designs (see Neumaier's article in this volume).

The following result is due to Cameron and Kantor [4].

THEOREM 3.1: *Let* G *be a metrically regular geometry of diameter* $n \geq 2$, *embedded in* $PG(m-1, q)$. *Let* S *be the point set of* G, *and for* $0 \leq i \leq n$ *and* $x \in S$ *let* $W_i(x)$ *be the set of points of* G *distant* i *or less from* x. *Suppose that the following conditions hold:*

(A) *For all* $x \in S$, $W_1(x)$ *is a subspace of* $PG(m-1, q)$.

(B) *For all* $x \in S$ *and* $0 \leq i \leq n$, *there is a subspace* $U_i(x)$ *of* $PG(m-1, q)$ *such that* $W_i(x) = S \cap U_i(x)$.

(C) *For some* $x \in S$ *and some integer* h, $|W_2(x)| = (q^h - 1)/(q - 1)$.

Then either

(i) $n=2$, m *is even, and* G *is the symplectic geometry consisting of all the points and the totally isotropic lines relative to a symplectic polarity of* $PG(m-1, q)$; *or*

(ii) $n=3$, $m=6$ *or* 7, *and* G *is the generalised hexagon of type* $G_2(q)$, *with a uniquely determined embedding (and* q *is even if* $m=6$ *).*

Of the three conditions in the theorem, (C) is the most unnatural. Accordingly, we call an embedding *flat* if it satisfies hypotheses (A) and (B) above. The rest of this paper is directed towards the question of determining which metrically regular geometries or near 2n-gons possess flat embeddings. We show that the dual orthogonal spaces of type $O(2n+1, q)$ do so, and give two characterisation theorems including them.

4. CHARACTERISATION THEOREMS

THEOREM 4.1 : *Let* G *be a near 2n-gon with a flat embedding in a projective space over* $GF(q)$. *Suppose that, whenever two points of* G *lie at distance* 2, *there are at least two points collinear with both. Then* G *is a dual orthogonal space of type* $O(2n+1, q)$.

PROOF : Let x, y be two points of G, with $d(x, y) = 2$. Then there are points u and v collinear with x and y, by hypothesis. Proposition 2.1 shows that x and y lie in a unique quad Δ. Moreover, the points of the line xy of the projective space are contained in $W_1(u) \cap W_1(v)$ (since this set is a subspace), and hence in Δ.

It follows that Δ contains the (projective) line through any two of its points, and so Δ is a subspace.

Now $|\Delta| = (q^d - 1)/(q-1)$ for some integer d. Let s and t be the parameters of Δ as generalised quadrangle (see Higman [7] : $s+1$ and $t+1$ are respectively the number of points on a line and the number of lines through a point). Then

$|\Delta| = (s+1)(st+1)$, and $s = q$, $t = |W_1(u) \cap W_1(v)| - 1 = q(q^e - 1)/(q-1)$ for some integer e. This is only possible if $e = 1$, $d = 4$. Now, for any $x \in \Delta$, $W_1(x) \cap \Delta$ is a plane of Δ, and $x \longleftrightarrow W_1(x) \cap \Delta$ is a symplectic polarity; so Δ is of type $Sp(4, q)$.

Now let x be any point, i the minimum distance from x to a point of Δ. Then $W_i(x) \cap \Delta = U_i(x) \cap \Delta$ is a subspace. Since an ovoid of Δ has $st+1 = q^2+1$ points, (x, Δ) cannot be ovoidal, and is necessarily classical, by Proposition 2.2.

We have now verified the hypothesis of Theorem 2.3, and conclude that G is a dual polar space. We appeal to the classification of polar spaces by Tits [10]: inspection of the parameters shows that G is dual $Sp(2n, q)$ or dual $O(2n+1, q)$. If q is even, the two types are isomorphic. If q is odd, the quads of dual $Sp(2n, q)$ are dual $Sp(4, q)$, so this type cannot occur (since $Sp(4, q)$ is not self-dual if q is odd). This proves the theorem.

THEOREM 4.2: *Let G be a metrically regular near 6-gon with a flat embedding in a projective space over* $GF(q)$. *Then G is a dual orthogonal space of type* $O(7, q)$ *or a generalised hexagon of type* $G_2(q)$.

PROOF: If $c_2 > 1$ then Theorem 4.1 applies. If $c_2 = 1$ then G is a generalised hexagon with $s = q$, $t = (|W_1(x)| - (q+1))/q = q(q^f - 1)/(q-1)$ for some integer f. The inequality $t \le s^3$ of Haemers and Roos for generalised hexagons [6] shows that $f \le 2$. If $f = 1$, then $|W_2(x)| = (q^5 - 1)/(q-1)$ and Theorem 3.1 applies. If $f = 2$, then $t = q(q+1)$. But st is a square (see Higman [7]); so $q = n^2 - 1$ for some integer n. For prime power q, this is only possible if $q = 3$ or 8, $(s, t) = (3, 12)$ or $(8, 72)$. But both cases are ruled out by the divisibility conditions given in Higman [7].

REMARK : For further progress on this question, it would be very helpful to have a generalisation of Proposition 2.1 to near 2n-gons of arbitrary girth. The structure of the near 8-gon described by Cohen in this volume gives some hope that such a result might be true.

5. THE SPINOR EMBEDDING OF DUAL $O(2n+1, q)$

THEOREM 5.1 : *For any* $n \geq 2$, *dual* $O(2n+1, q)$ *has a flat embedding in* $PG(2^n - 1, q)$.

PROOF : First, we outline the facts we require about spinors. For proofs and further details, see Chevalley [5].

Let V be a 2m-dimensional vector space over $GF(q)$, and Q a quadratic form on V of index m, for example

$$Q(x_1, \ldots, x_{2m}) = x_1 x_{m+1} + \ldots + x_m x_{2m} .$$

The *Clifford algebra* $C(Q)$ of Q is the associative algebra with 1 generated by V subject to the relations $v^2 = Q(v) \cdot 1$, i.e. the quotient of the tensor algebra on V by the ideal generated by $\{ v \otimes v - Q(v) \cdot 1 \mid v \in V \}$. (Compare the exterior algebra of V, obtained by using the identically zero form.) The dimension of $C(Q)$ is 2^{2m}, and $C(Q)$ is isomorphic to the algebra of endomorphisms of a vector space S of dimension 2^m; the space S (or the projective space $PG(2^m - 1, q)$ based on S) is the *space of spinors* associated with Q.

To obtain a more concrete representation of S, we want a process for recovering S from $End(S)$. Note that any minimal left ideal of $End(S)$ is the set of endomorphisms whose range is contained in a given 1-dimensional subspace of S; dually, a minimal right ideal is the set of endomorphisms whose kernel contains a given subspace of codimension 1; and the intersection of minimal left and right ideals has dimension 1. Let I be a fixed minimal right ideal. Then End(S) acts on I by right multiplication, and I may be identified with S. We then have a bijection between points of the projective space on I and minimal left ideals of End(S).

Now, let W be a maximal totally singular subspace of V. Then the subalgebra of C(Q) generated by W is just the exterior algebra of W (since $Q|_W$ is identically zero). Let f_W be the product of the elements in a basis of W : this is determined up to a scalar multiple by W. We use for $I = S$ the ideal $f_W C(Q)$.

If U is any other maximal totally singular subspace then, as above, $\langle C(Q) f_U \cap f_W C(Q) \rangle$ is a point of the projective space on S. Thus we have an injection from the set of points of the dual polar space of type $O^+(2m, q)$ (the totally singular m-subspaces of V) to the set of points of $PG(2^m - 1, q)$ (the 1-subspaces of S). The image of this injection is the set of *pure spinors*. It has the following properties.

1. $S = S^+ \oplus S^-$, where $\dim S^+ = \dim S^- = 2^{m-1}$ and any pure spinor lies in either S^+ or S^-. Moreover, the intersection of two totally singular (t.s.) m-spaces has even codimension in each if and only if the corresponding pure spinors lie in the same *half-spinor space* $S^{\pm}$. (We speak of the two *families* of t.s. m-spaces defined in this way.)

2. A t.s. (m-1)-space is contained in a unique t.s. m-space of each family.

3. A t.s. (m-2)-space is contained in exactly q+1 t.s. m-spaces of each family; the corresponding sets of q+1 pure spinors are lines of spinor spaces; and any line of spinor space which contains more than two pure spinors is of this form.

Thus we have a (metrically regular) *half-spinor geometry* embedded in $PG(2^{m-1} - 1, q)$, consisting of the set of pure spinors in a half-spinor space together with the lines contained in this set. It can be shown, using the coordinatisation of spinors given by Chevalley, that this geometry satisfies condition

B of Theorem 3.1 (but not A or C in general).

Let v be a non-singular vector in V. Then $v^\perp$ carries an orthogonal geometry of type $O(2m-1, q)$. If U is a t.s. $(m-1)$-space in $v^\perp$, then each family of t.s. m-spaces has a unique member containing U; we call these subspaces U^+, U^-. Moreover, if W is a t.s. m-space, then $W \cap v^\perp$ is a t.s. $(m-1)$-space in $v^\perp$. Thus we have a bijection between the points of dual $O(2m-1, q)$ and the t.s. m-spaces in each family, i.e. the pure spinors in each half-spinor space. Now any line of dual $O(2m-1, q)$ consists of all the t.s. $(m-1)$-spaces containing a fixed t.s. $(m-2)$-space; it is mapped by the bijection to a line of the corresponding half-spinor geometry. So the lines of dual $O(2m-1, q)$ are identified with some of the lines of the half-spinor geometry, and we have an embedding of dual $O(2m-1, q)$ in $PG(2^{m-1} - 1, q)$. Much as before, this embedding satisfies B of Theorem 3.1.

In fact it also satisfies A. For let U be a t.s. $(m-1)$-space in $v^\perp$. As in Theorem 3.1, $W(U)$ denotes the set of those t.s. $(m-1)$-spaces meeting U in an $(m-2)$-space. Let U_1 and U_2 be two members of $W_1(U)$, not collinear in dual $O(2m-1, q)$. Then $\dim(U \cap U_1) = \dim(U \cap U_2) = m - 2$, so $\dim(U_1 \cap U_2) = m - 3$, and $\dim(U_1^+ \cap U_2^+) \geq m - 3$. Since U_1^+ and U_2^+ belong to the same family, their intersection has even codimension, whence $\dim(U_1^+ \cap U_2^+) = m - 2$. Now the line joining the corresponding pure spinors consists entirely of pure spinors. Let U_3^+ be the subspace corresponding to one such pure spinor, and $U_3 = U_3^+ \cap v^\perp$. It is enough to show that $U_3 \in W_1(U)$. But, since $\langle U_1^+ \cap U, U_2^+ \cap U \rangle = U$, we have that $T = \langle U_1^+ \cap U_2^+, U \rangle$ is a t.s. m-space and belongs to the other family. Now $\dim(T \cap U_3^+) = m - 1$, so $\dim(U \cap U_3^+) = m - 2$. Since $U \cap U_3^+ \subseteq v^\perp$, we have $U \cap U_3^+ \subseteq U_3$, giving $\dim(U \cap U_3) = m - 2$, or $U_3 \in W_1(U)$ as required.

REMARK : For $m = 3$, each half-spinor geometry of type $O^+(6, q)$ is the whole of $PG(3, q)$; we have the reverse of the *Klein correspondence* between lines of $PG(3, q)$ and points of the Klein quadric in $PG(5, q)$. The dual $O(5, q)$ geometry is isomorphic to the $Sp(4, q)$ geometry, embedded in the usual way in $PG(3, q)$.

For $m = 4$, each half-spinor geometry of type $O^+(8, q)$ is isomorphic to the original $O^+(8, q)$; these three copies of $O^+(8, q)$ are permuted transitively by the *Study triality* automorphism (see [5]). The embedding of G = dual $O(7, q)$ is that described in [4], involving all the points and some of the lines of $O^+(8, q)$. The subgeometry of G induced on $w^\perp$, for a non-singular vector w, is the $G_2(q)$ hexagon, consisting of all the points and some of the lines of the $O(7, q)$ geometry. This is the embedding referred to in Theorem 3.1.

BIBLIOGRAPHY

1. N.L. Biggs, *Algebraic Graph Theory* (Cambridge Univ. Press, Cambridge, 1974).
2. F. Buekenhout and C. Lefèvre, "Generalized quadrangles in projective spaces", *Arch. Math.*, 25 (1974), 540-552.
3. P.J. Cameron, "Dual polar spaces", *Geom. Dedicata*, to appear.
4. P.J. Cameron and W.M. Kantor, "2-transitive and antiflag transitive collineation groups of finite projective spaces", *J. Algebra*, 60 (1979).
5. C. Chevalley, *The Algebraic Theory of Spinors* (Columbia Univ. Press, Morningside Heights, N.Y., 1954).
6. W. Haemers and C. Roos, "An inequality for generalized hexagons", *Geom. Dedicata*, to appear.
7. D.G. Higman, "Invariant relations, coherent configurations and generalized polygons", *Combinatorics* (eds. M. Hall Jr. and J.H. van Lint, Reidel, Dordrecht, 1975), pp. 347-363.
8. E.E. Shult, Lecture at Int. Conf. Finite Geometries and Groups, Han-sur-Lesse, 1979.
9. E.E. Shult and A. Yanushka, "Near n-gons and line systems", *Geom. Dedicata*, 9 (1980), 1-72.

10. J. Tits, *Buildings of Spherical Type and Finite BN-pairs*, Lecture Notes in Math. 386 (Springer-Verlag, Berlin, 1974).

Merton College
Oxford OX1 4JD
U.K.

CODES, CAPS AND LINEAR SPACES

P.V. Ceccherini and G. Tallini

I. CAPS OF PG(r,q) AND LINEAR CODES

1. NOTATION

Let $V = V_{r+1,q}$ be the $(r+1)$-dimensional vector space over the Galois field $GF(q)$ and let $S = S_{r,q} = PG(r,q)$ be the related projective space of dimension r.

If $x \in V \setminus \{0\}$, then we denote by $[x]$ the point of S related to x. Let us denote by the same symbol K the following:

$$K = (x^{(1)}, x^{(2)}, \ldots, x^{(k)}) \quad \text{(ordered k-set of } V) , \qquad (1)$$

$$K = ([x^{(1)}],[x^{(2)}],\ldots,[x^{(k)}]) \quad \text{(ordered k-set of } S) , \qquad (2)$$

$$K = [x^{(1)}\ x^{(2)}\ \ldots\ x^{(k)}] \quad ((r+1) \times k \text{ matrix over } GF(q)), \qquad (3)$$

where $x^{(1)}, x^{(2)}, \ldots, x^{(k)}$ are (column) vectors pairwise independent and spanning V. The latter condition implies

$$r + 1 \le k , \qquad (4)$$

and we have

$$\langle K \rangle = \langle x^{(1)}, \ldots, x^{(k)} \rangle = V , \qquad (5)$$

$$\langle K \rangle = \langle [x^{(1)}], \ldots, [x^{(k)}] \rangle = S , \qquad (6)$$

$$\text{rank } K = r + 1 . \qquad (7)$$

2. CODES AND ORDERED SETS OF POINTS

With K as above, let $C = C(K)$ be the linear code of $V_{k,q}$ defined by

$$C(K) = \left\{ x \in V_{k,q} \mid Kx = 0 \right\} .$$

By (7) we have that

$$\dim C(K) = k - (r+1) .$$

Moreover, in order that each column of the matrix K is a non-zero vector, the code $C(K)$ satisfies the following condition:

(C) The code C does not contain any basis vector, i.e. it does not contain any fundamental subspace.

Conversely, if $r+1 \leq k$ and C is any $(k, k-(r+1))$-code satisfying condition (C), then we may give, in several ways, a cartesian representation for C of the type

$$C = \left\{ x \in V_{k,q} \mid Kx = 0 \right\} , \tag{8}$$

where K is a suitable $(r+1) \times k$ matrix with $\operatorname{rank} K = r+1$ and with columns $x^{(1)}, \ldots, x^{(k)}$ which are non-zero vectors, by condition (C).

As in §1, let us denote by the same symbol, $K = K(C)$, the above matrix and the ordered k-sets (1) and (2). Obviously

$$C(K(C)) = C . \tag{9}$$

3. PERFECT CODES AND THE SIZE OF A CAP

K is a k-cap of kind s in S, denoted $k^s_{r,q}$, if $s+2$ is the minimum number of dependent points of K and K spans S.

LEMMA 3.1 :

(i) *Let* K *be as in* (1)-(7) *and let* $C = C(K)$ *be the code related to* K *as in* § 2. *The following conditions are equivalent:*

(a) K *is an (ordered)* $k^s_{r,q}$;

(b) $r+2$ *is the minimum number of distinct dependent columns of* K ;

(c) *the code* C *has distance* $s+2$, *so that* C *is an* $[(s+1)/2]$*-error correcting code;*

(ii) *Conversely, if* C *is an* $(k,k-(r+1))$*-code satisfying condition* (C) *and if* $Kx = 0$ *represents* C *as in* (8), *with*

K *necessarily as in* (1)-(7), *then the above conditions* (a), (b) *and* (c) *are equivalent.*

PROOF: Because of (9), it is enough to prove (i). (a) $\Leftrightarrow$ (b), by the definition of $k^s_{r,q}$. (c) is equivalent to the minimum weight of a non-zero vector $v = (v_i)$ in $C(K)$ being equal to $s+2$. This is equivalent to the fact that $v \neq 0$ and $Kv = 0$ implies $w(v) \geq s+2$, where equality occurs for some v. This in turn is equivalent to $v \neq 0$ and $\sum_{i=1}^{k} v_i x^i = 0$ implying $w(v) \geq s+2$, where equality occurs for some v, which is equivalent to (b). □

LEMMA 3.2: *If* C *is an* e-*error correcting code in* $V_{k,q}$, *then*

$$\sum_{i=0}^{e} \binom{k}{i} (q-1)^i \leq q^{k-\dim C} ,$$

where equality occurs if and only if C *is a perfect* e-*error correcting code.*

PROOF: If $v \in V_{k,q}$ and $0 \leq i \leq k$, then $\left|\{x \in V_{k,q} \mid d(v,x) = i\}\right| = \binom{k}{i} (q-1)^i$, so that $\left|S(v,e)\right| = \sum_{i=0}^{e} \binom{k}{i} (q-1)^i$.
In order that C is an e-error correcting code, the $q^{\dim C}$ spheres with centre in C and radius e are pairwise disjoint. So $q^{\dim C} \left|S(v,e)\right| \leq q^k$, where equality occurs if and only if C is a perfect e-error correcting code. □

THEOREM 3.3: *The number* k *of points of a* k-*cap* $K = k^s_{r,q}$ *satisfies*

$$\sum_{1=0}^{[(s+1)/2]} \binom{k}{i} (q-1)^i \leq q^{r+1} ,$$

where equality occurs if and only if $C(K)$ *is an* $[(s+1)/2]$-*error correcting perfect code. In particular, for* $q = 2$,

$$\sum_{i=0}^{[(s+1)/2]} \binom{k}{i} \le 2^{r+1}$$

PROOF : This follows from Lemmas 3.1 and 3.2 . □

COROLLARY 3.4 : *Let* $K = k^s_{r,q}$ *be a* k*-cap of kind* s *of* $S_{r,q}$. *If* $C(K)$ *is a perfect code, then* K *is a complete cap, i.e.* K *is not contained in any* $(k+1)^s_{r,q}$. □

COROLLARY 3.5 : *The maximum number* $M^3_{r,q}$ *of points of a cap* $K = k^3_{r,q}$ *satisfies*

$$M^3_{r,q} \le \frac{(q-3) + \sqrt{[(q-3)^2 + 8(q^{r+1} - 1)]}}{2(q-1)} .$$

In particular,

$$M^3_{r,2} \le \frac{-1 + \sqrt{(2^{r+4} - 7)}}{2} , \quad M^3_{r,3} \le \sqrt{[(3^{r+1} - 1)/2]} ,$$

$$M^3_{r,5} \le \frac{1 + \sqrt{(2.5^{r+1} - 1)}}{4} ,$$

where equality occurs if and only if the code $C(K)$ *is perfect.*

PROOF : Put $s = 3$ in Theorem 3.3 . □

Let K_1 be the $5^3_{3,2}$ in $S_{3,2}$ consisting of the four fundamental points and the unit point. Let K_2 be the $11^3_{4,3}$ in $S_{4,3}$ consisting of the five fundamental points, the unit point and the five points

[0, 1, -1, -1, 1], [1, 0, -1, 1, -1], [1, 1, 0, -1, -1],

[1, -1, -1, 0, 1], [1, -1, 1, -1, 0] :

see Tallini [2], §3 .

COROLLARY 3.6 : *Let* $K = k^3_{r,q}$ *be a cap of kind* 3 *of* $S_{r,q}$ *and let* $C(K)$ *be the code related to* K . *The following conditions are equivalent:*

(a) K *is not contained in a cap of kind* 2 ;

(b) K *is one of the caps* K_1 *or* K_2 ;

(c) k *is equal to the upper bound for* $M^3_{r,q}$ *in Corollary* 3.5 ;

(d) $C(K)$ *is perfect.*

PROOF : By [2], §3, we have (a) $\Leftrightarrow$ (c), and by [3] we have (a) $\Leftrightarrow$ (b). By Corollary 3.5, (c) $\Leftrightarrow$ (d). □

COROLLARY 3.7 : *The only* 2-*error correcting perfect codes (with minimum distance* 5 *) not containing basis vectors are the codes* $C(K_1)$, *a repetition code, and* $C(K_2)$, *a ternary Golay code.*

PROOF : If C is any $(k, k-(r+1))$-code not containing any basis vector, then $C = C(K)$ where $K = k^s_{r,q}$ is a cap of kind s, by Lemma 3.1 .

In order that C has distance 5, again by Lemma 3.1, it follows that $s = 3$. Since C is perfect, the equivalence of (b) and (d) of Corollary 3.6 gives the result. □

II. LINE SPACES AND BINARY LINEAR CODES

4. THE INCIDENCE MATRIX OF A GEOMETRIC SPACE

Let (S, G) be a finite *geometric space* (i.e. a finite set S of points with a set G of non-empty subsets (blocks) of S), with $|S| = r$, $|G| = k$. Let us order S and G : $S = (P_1, \ldots, P_r)$, $G = (G_1, \ldots, G_k)$ and let M denote the $r \times k$ incidence matrix over $GF(2)$ of (S, G) defined by

$$M = (m_{ij}), \quad m_{ij} = 1 \text{ if } P_i \in G_j, \quad m_{ij} = 0 \text{ if } P_i \notin G_j . \qquad (10)$$

THEOREM 4.1 : *If the geometric space* (S, G) *is such that every* h-*subset of* S, *with* $h \geq 2$, *is contained in at most one block of* G, *then every* $s+1$ *columns of the incidence matrix* M *of* (S, G) *are linearly independent where*

$s = [(\rho - 1)/(h - 1)]$ *and* $\rho = \min_{G \in \mathcal{G}} |G|$.

PROOF : It is enough to prove that, when $2 \leq m \leq s$, that is $m \geq 2$ and

$$\rho - (h - 1)(m - 1) \geq h , \tag{11}$$

the sum of m columns of M cannot be a column of M . Let $x, y, z_1, \ldots, z_{m-2}$ be m distinct columns of M and let us denote by the same symbols the corresponding blocks of $\mathcal{G}$. If $x + y + z_1 + \ldots + z_{m-2} = e$ is a column of M (block of $\mathcal{G}$), we obtain, using $|x \cap y| \leq h - 1$ and $|x \cap z_i| \leq h - 1$ $(i = 1, \ldots, m-2)$, that x contains at least $\rho - (h - 1)(m - 1) \geq h$ points, $P_{i_1}, P_{i_2}, \ldots, P_{i_h}$ say, belonging neither to y nor to z_i $(i = 1,2,\ldots, m-2)$. It follows that $c \cap x \supseteq \{P_{i_1}, P_{i_2}, \ldots, P_{i_h}\}$, and therefore $c = x$. Interchanging x and y , we get $c = y$, so that $x = y$, which is impossible. □

COROLLARY 4.2 : *Let* M *be the* $(q^2 + q + 1) \times (q^5 + q + 1)$ *incidence matrix of the geometric space* $(S_{2,q}, \mathcal{C} \cup \mathcal{R})$, *where* $\mathcal{C}$ *is the set of all the non-singular conics and* $\mathcal{R}$ *is the set of all the lines of* $S_{2,q}$. *Then, every* $[q/4] + 1$ *columns of* M *are linearly independent.*

PROOF : Put $\rho = q + 1$ and $h = 5$ in Theorem 4.1 . □

5. THE CAP RELATED TO A LINE SPACE

A finite partial line space $(S, \mathcal{R})$ is a finite set S of points provided with a set $\mathcal{R}$ of subsets (lines), such that any line contains at least two points and every two points of S are contained in at most one line, and $S = \bigcup_{R \in \mathcal{R}} R$.

LEMMA 5.1 : *Let* M *be the* $r \times k$ *incidence matrix of a finite partial line space* $(S, \mathcal{R})$ *such that* $|S| = r$, $|\mathcal{R}| = k$ *and* $|R| = n$ *for every* R *in* $\mathcal{R}$. *Then every* n *columns of* M *are* *linearly independent.*

PROOF : Put $\rho = n$ and $h = 2$ in Theorem 4.1 . □

Let $(S, \mathcal{R})$ and M be as in Lemma 5.1 . If $\mathcal{L} \subseteq \mathcal{R}$ is any set of lines, define the *multiplicity* $m_P = m_{P,\mathcal{L}}$ of a point P in S with respect to $\mathcal{L}$ as the number of lines R in $\mathcal{L}$ containing P. Then $\mathcal{L}$ is of *even type* if, for all P in S, the multiplicity $m_{P,\mathcal{L}}$ is even. If $R \in \mathcal{R}$, let us denote the corresponding column of M (vector of Z_2^r) and the corresponding point of $PG(r-1, 2)$ by the same symbol R. It is easy to prove the following statements:

$$\sum_{R\in\mathcal{L}} R = 0 \Leftrightarrow \mathcal{L} \text{ is of even type;} \qquad (12)$$

$\mathcal{L}$ is a dependent set of columns $\Leftrightarrow$ $\mathcal{L}$ contains a non-empty subset $\mathcal{L}'$ of even type; (13)

$|\mathcal{L}| \leq n \Rightarrow \mathcal{L}$ does not contain any non-empty subset $\mathcal{L}'$ of even type (i.e. for every non empty $\mathcal{L}' \subseteq \mathcal{L}$, the set $\bigcup_{R\in\mathcal{L}'} R$ contains at least a point P with $m_{P,\mathcal{L}'}$ odd). (14)

One may use (13) and Lemma 5.1 to prove (14). Therefore, if we define

$$s+2 = \min\{\, |\mathcal{L}| \mid \emptyset \neq \mathcal{L} \subseteq \mathcal{R},\ \mathcal{L} \text{ of even type}\,\},$$

then $s+2 > n$, and the columns of M give rise, as points of $S_{r-1,2}$, to a k-cap of kind s of $S_{r-1,2}$ spanning a space of dimension $r' - 1$, where $r' = \text{rank}\, M \leq r$. Actually

$r' = \text{rank}\, M = \max\{\, |H| \mid H \subseteq S,\ H$ contains no non-empty subset H' of even type) }, (15)

where a set $H' \subseteq S$ of points is said to be of *even type* if every R in $\mathcal{R}$ meets H' in an even number of points.

We note incidentally that, from (15), it follows that: $\max\{\, |H| \mid H \subseteq S,\ H$ contains no non-empty subset H' of even type } = $\max\{\, |\mathcal{L}| \mid \mathcal{L} \subseteq \mathcal{R},\ \mathcal{L}$ contains no non-empty subset $\mathcal{L}'$ of even type }. This property holds more generally for every geometric space $(S, \mathcal{G})$, when subsets $H \subseteq S$ and $L \subseteq \mathcal{G}$ of even type are similarly defined.

The above statement is summarized in part (a) of the following.

THEOREM 5.2 :

(a) *The columns of the* $r \times k$ *incidence matrix of a partial line space* $(S, \mathcal{R})$, *with* $|S| = r$, $|\mathcal{R}| = k$, $|R| = n$ *for every* R *in* $\mathcal{R}$, *give rise to a* k-*cap* $k^{s}_{r'-1,2}$ *in* $S_{r-1,2}$, *where* $s \geq n-1$ *and* $s+2$ *is precisely the minimal cardinality of a non-empty set of even type of lines and* r' *is the maximal cardinality of a set of points which contains no non-empty subset of even type.*

(b) *Furthermore, if* $(S, \mathcal{R})$ *is a line space,* i.e. *actually a Steiner system* $S(2,n,r)$, *then, denoting by* m *the number* $(r-1)/(n-1)$ *of lines through a point, we have:*

(i) *For any set* E *of even type of points of* $(S, \mathcal{R})$,
$m \equiv n \equiv 0$ *(mod 2)* $\Longrightarrow E = \emptyset$ *or* $E = S$;
$m \equiv 0$, $n \equiv 1$ *(mod 2)* $\Longrightarrow E = \emptyset$.

(ii) *The dimension of the projective space spanned by the cap related to* $(S, \mathcal{R})$ *is* $r'-1$ *where* $r' = r-1$ *if* $m \equiv n \equiv 0$ *(mod 2);* $r' = r$ *if* $m \equiv 0$ *and* $n \equiv 1$ *(mod 2);* $r' \leq r-1$ *if* $m \equiv 1$ *and* $n \equiv 0$ *(mod 2).*

PROOF :

(a) It is contained in the argument stated after Lemma 5.1.

(b) (i) Let $m \equiv 0 \pmod 2$. If $\emptyset \neq E \neq S$, then there exist points $A, B \in S$ such that $A \notin E$ and $B \in E$. Because E is of even type and $m \equiv 0 \pmod 2$, we find that $|E|$ is both an even and an odd number, when we count $|E|$ by the lines through A and also by the lines through B. In conclusion, $E = \emptyset$ or $E = S$; the second case is obviously excluded if n is odd.

(ii) This follows immediately from (15) and from (i), assuming $H = S$ if $m \equiv 0$ and $n \equiv 1 \pmod 2$, and $H = S \setminus \{x\}$, the complementary set of a point $x \in S$, if $m \equiv n \equiv 0 \pmod 2$;

if $m \equiv 1$ and $n \equiv 0 \pmod 2$, then S is of even type so that $r' \leq r-1$. □

BIBLIOGRAPHY

1. B. Segre, "Le geometrie di Galois", *Ann. Mat. Pura. Appl.*, 48 (1959), 1-97.
2. G. Tallini, "On caps of kind s in a Galois r-dimensional space", *Acta Arith.*, 7 (1961), 19-28.
3. G. Migliori, "Calotte di specie s in uno spazio r-dimensionale di Galois", *Atti. Accad. Naz. Lincei Rend.*, 60 (1976), 789-792.

Istituto Matematico
Università di Roma
00185 Roma
Italy.

GEOMETRIES ORIGINATING FROM CERTAIN DISTANCE-REGULAR GRAPHS

Arjeh M. Cohen

Distance-regular graphs having intersection number $c_2 = 1$ are point graphs of linear incidence systems. This simple observation plays a crucial role in both the existence proof of a regular near octagon (associated with the Hall-Janko group) whose point graph is the unique automorphic graph with intersection array $\{10, 8, 8, 2; 1, 1, 4, 5\}$ and the non-existence proof of a distance-regular graph with intersection array $\{12, 8, 6, \ldots; 1, 1, 2, \ldots\}$. These results imply a partial answer to a problem put forward by Biggs in 1976.

1. BASIC NOTIONS

Let $\Gamma = (V,E)$ be a connected graph of diameter d and let $\Gamma_i(\alpha)$ for $\alpha \in V$ denote the set of vertices at distance i from α. We recall from [2] that Γ is *distance-regular* if for any i $(0 \le i \le d)$ the numbers $b_i = |\Gamma_{i+1}(\alpha) \cap \Gamma_1(\beta)|$ and $c_i = |\Gamma_{i-1}(\alpha) \cap \Gamma_1(\beta)|$ do not depend on the choice of α, β such that $\beta \in \Gamma_i(\alpha)$. Of course, $b_d = 0$ and $c_1 = 1$. Write $k = |\Gamma_1(\alpha)|$. The array $\{k, b_1, b_2, \ldots, b_{d-1}; c_1, c_2, \ldots, c_d\}$ is called the *intersection array* of Γ.

The graph Γ is called *distance-transitive* if its automorphism group $\mathrm{Aut}(\Gamma)$ is transitive on each of the classes $\{\{\alpha, \beta\} \subset V \mid \beta \in \Gamma_i(\alpha)\}$ $(0 \le i \le d)$, and Γ is called *automorphic* (cf. [3]) whenever it is distance-transitive, not a complete graph or a line graph, and has an automorphism group which is primitive on V.

Given a linear incidence system (V,L) of points V and lines L, its *point graph* has the points V for vertices and the pairs of adjacent points for edges. For any two points of (V,L)

the distance is meant to be the distance in the point graph. The notion of a regular near 2d-gon is introduced in [9]. A linear incidence system (V,L) is called a *regular near* 2d*-gon of order* $(s, t_d, t_2, \ldots, t_{d-1})$ and of *diameter* d if

(i) a line is incident with s+1 points ; $s \geq 2$.

(ii) For any point $p \in V$ and line $\ell \in L$ there is a unique point q on ℓ nearest p .

(iii) The point graph of (V,L) is connected with diameter d .

(iv) For any i $(1 \leq i \leq d)$ and any two points p, q of distance i, there are precisely $1 + t_i$ lines through q bearing a point of distance i-1 to p .

Note that $1 + t_d$ is the number of lines through a point. A regular near 2d-gon has a distance-regular point graph. The following lemma (in the spirit of [4] and [7]) provides a partial converse to this phenomenon.

2. A LEMMA

LEMMA : *Let* L *be the set of maximal cliques in a distance-regular graph* Γ *with intersection array* $\{k, b_1, b_2, \ldots, b_{d-1} ; 1, 1, c_3, \ldots, c_d\}$. *Then* (V,L) *is a linear incidence system with constant line size* $k - b_1 + 1$. *Moreover, if there are natural numbers* $s, t, t_3, \ldots, t_{d-1}$ *such that* Γ *has intersection array* $\{s(t+1), st, st, s(t-t_3), \ldots, s(t-t_{d-1}); 1, 1, t_3+1, \ldots, t_{d-1}+1, t+1\}$, *then* (V,L) *is a regular near* 2d*-gon of order* $(s, t, 0, t_3, \ldots, t_{d-1})$.

PROOF : Given two mutually adjacent points $\alpha, \beta \in V$, there are $k - b_1 - 1$ points adjacent to both α, β. If γ, δ are two such points, they must be adjacent, for otherwise $\alpha, \beta \in \Gamma_1(\delta) \cap \Gamma_1(\gamma)$ would contradict $c_2 = 1$. This implies that all maximal cliques have size $k - b_1 + 1$ and that any two adjacent points are contained in precisely one maximal clique. As to the second statement of the lemma, fix $\alpha \in V$. By induction, we may assume that each line bearing points in both $\Gamma_{i-1}(\alpha)$ and $\Gamma_i(\alpha)$ has exactly one point in $\Gamma_{i-1}(\alpha)$. Thus there are precisely $1 + t_i$ lines through a point $\beta \in \Gamma_i(\alpha)$ bearing a point of $\Gamma_{i-1}(\alpha)$.

On these lines there are $(1+t_i)(s-1)$ points in $\Gamma_i(\alpha) \cap \Gamma_1(\beta)$. According to the intersection numbers these are all points of $\Gamma_i(\alpha) \cap \Gamma_1(\beta)$. Hence the remaining $t - t_i$ lines through β have all points but β in $\Gamma_{i+1}(\alpha)$. □

3. THE NEAR OCTAGON ASSOCIATED WITH THE HALL-JANKO GROUP

The Hall-Janko simple group HJ of order 604800 has a conjugacy class V_0 of 315 involutions whose centralizers contain Sylow 2- subgroups. Setting E_0 for the pairs of involutions of V_0 whose product is again an element of V_0, a distance-regular graph (V_0, E_0) with intersection array $\{10, 8, 8, 2; 1, 1, 4, 5\}$ results. By the previous lemma, the maximal cliques are the lines of a regular near octagon of order $(2,4,0,3)$.

It is possible to present this near octagon without assuming any previous knowledge on HJ. In order to do so we exploit the 'root system' given in [5]. Let $H = \mathbb{R}1 + \mathbb{R}i + \mathbb{R}j + \mathbb{R}k$ be the skew field of real quaternions whose multiplication is determined by $ij = k = -ji$ and $i^2 = j^2 = k^2 = -1$. The multiplicative subgroup $Q = \{\pm 1, \pm i, \pm j, \pm k\}$ is the well-known quaternion group. Set $\tau = (1+\sqrt{5})/2$ and $\zeta = (-1-i-j-k)/2$.

The projective plane P over H is represented by homogeneous coordinates $[x]$ for $x \in H^3 \setminus \{0\}$. Here H^3 stands for the right vector space over H of dimension 3. Let A be the collineation group of P of order 1152 generated by the permutations of the homogeneous coordinates, scalar multiplication by ζ (acting on the left) and the transformations

$$\begin{bmatrix} x_1 \\ x_2 \\ x_3 \end{bmatrix} \longrightarrow \begin{bmatrix} px_1 \\ qx_2 \\ rx_3 \end{bmatrix} \quad \text{for } p, q, r \in Q \text{ with } pqr \in \{\pm 1\}.$$

Write

$$V_1 = A\begin{bmatrix} 1 \\ 0 \\ 0 \end{bmatrix} \cup A\begin{bmatrix} 1 \\ 1 \\ 0 \end{bmatrix} \cup A\begin{bmatrix} 1 \\ \zeta^2+\tau \\ 1 \end{bmatrix} \cup A\begin{bmatrix} 1 \\ \zeta(1-\tau) \\ \zeta^2\tau \end{bmatrix}.$$

Then V_1 consists of four A-orbits having lengths 3, 24, 192, 96 respectively, so $|V_1| = 315$. Furthermore, the angles between elements [x], [y] of V_1 (i.e. $\frac{(x|y)(y|x)}{(x|x)(y|y)}$, where $(x|y)$ stands for the standard unitary inner product on H^3 linear in y) attain a limited number of values. The first A-orbit of V_1 consists of an *orthogonal triple,* i.e. a triple in which the angle of each pair is 0. Denote by L_1 the set of all orthogonal triples of V_1.

PROPOSITION 3.1 : *Let* (V_1, L_1) *be as above. Then* (V_1, L_1) *is a regular near octagon of order* (2,4,0,3).

PROOF : In view of the lemma, it suffices to establish that the point graph Γ^1 of (V_1, L_1) is distance-regular with intersection array $\{10, 8, 8, 2; 1, 1, 4, 5\}$. Consider the collineation s_X of P for $X = [x] \in V_1$ defined by

$$s_X(y) = y - 2x(x|y)/(x|x) \quad (y \in H^3).$$

Each of the s_X $(X \in V_1)$ stabilizes V_1 and leaves the angles invariant, whence $s_X \in \mathrm{Aut}(\Gamma^1)$. Let H be the subgroup of $\mathrm{Aut}(\Gamma^1)$ generated by the s_X $(X \in V_1)$ and A. It is straightforward that H acts transitively of rank 6 on V_1, the orbitals being indexed by the angles. In fact, the distances and angles of two points $X, Y \in V_1$ are related to the order of $s_X s_Y$ as indicated in the table

distance	0	1	2	3	4
angle	1	0	$1/\sqrt{2}$	1/2	$\tau/2, (\tau-1)/2$
order of $s_X s_Y$	1	2	4	3	5

The proof of the proposition is thus reduced to the check whether the intersection number c_4 for a pair of points from the orbital indexed by angle $\tau/2$ coincides with the number c_4 for a pair from the $(\tau-1)/2$-orbital. This is left for the reader to verify.□

COROLLARY 3.2 : *The point graph* Γ^1 *of the near octagon in* (3.1) *is an automorphic graph.*

PROOF : The group H of the proof of (3.1) is the simple group of Hall-Janko (cf. [5] or [11]). It has index 2 in its automorphism group (cf. [8]). But Γ^1 can be described fully in terms of the conjugacy class of involutions $\{ s_X | X \in P_1 \}$ which is left invariant by Aut(H), so Aut(H) is a subgroup of $\mathrm{Aut}(\Gamma^1)$. In the permutation character of H on V_1 (cf. [8]) the two irreducible constituents of degree 14 are permuted by an outer automorphism of H. Thus Aut(H) has rank 5 on V_1. Finally, inspection of the subdegrees (which are 1, 10, 80, 160, 64) yields that the action of Aut(H) on V_1 is primitive. □

THEOREM 3.3 : *The automorphic graph* Γ^1 *of* (3.1) *is the unique automorphic graph with intersection array* $\{10, 8, 8, 2; 1, 1, 4, 5\}$ *and has automorphism group* Aut(HJ) .

PROOF : Let $\Gamma = (V,E)$ be an automorphic graph with intersection array as given above and set $G = \mathrm{Aut}(\Gamma)$. By the lemma $G = \mathrm{Aut}(V,L)$ for L the set of maximal cliques in Γ. Fix $\alpha \in V$. As G_α permutes the five lines through α, the order of its restriction to $\Gamma_1(\alpha)$ divides $5! \times 2^5$. From a theorem by Jordan (cf. [12]; Theorem 18.4) and from distance transitivity of Γ, it follows that $|G| = 2^a . 3^b . 5^c . 7$ with $a \geq 6$, $b \geq 2$, $c \geq 2$. Let N be a minimal normal subgroup of G. Since G is primitive, N is transitive on V, so 7 divides $|N|$. But 7^2 does not, so N is nonabelian simple. The centralizer in G of N, being normal in G, must be trivial, whence up to isomorphism $N \leq G \leq \mathrm{Aut}(N)$ holds. Fix $\delta \in \Gamma_4(\alpha)$. There is a 1-1 correspondence between the lines through α and the lines through δ such that a line ℓ through α corresponds to the unique line through δ that bears no point of distance 2 to any point of ℓ (this is a consequence of $c_3 = 4$). Denote by K the subgroup of $G_{\alpha,\delta}$ fixing the lines through α. We claim that K is trivial. For, K fixes the lines through δ (because of the 1-1 correspondence with the fixed lines through α), but each line through δ bears a unique point nearest α. These points are therefore fixed by K and so are the third points on the lines through δ. It follows that all neighbours within $\Gamma_4(\alpha)$ of any fixed point of K in $\Gamma_4(\alpha)$ are K-fixed points too. As there are at least 46 points in the connected component of δ within the induced subgraph on $\Gamma_4(\alpha)$ of Γ and as this connected component is a block of imprimitivity for G_α in $\Gamma_4(\alpha)$, it follows that all points of $\Gamma_4(\alpha)$ are fixed by K. But G_α is faithful on $\Gamma_4(\alpha)$, whence $K = 1$, as claimed.

The result is that G has order $[G:G_\alpha][G_\alpha:G_{\alpha,\delta}]\,|G_{\alpha,\delta}|$ dividing $315 \times 64 \times 5! = 2^9 . 3^3 . 5^2 . 7$. On the other hand, $|N|$ is a multiple of $3^2 . 5^2 . 7$, as N_α must be transitive on the five lines through α. From [6], it is readily seen that the only possible order for N less than 10^6 is $2^7 . 3^3 . 5^2 . 7$, that N must be isomorphic to HJ, and that the corresponding permutation representation of degree 315 is unique. $Aut(HJ)$ is then a subgroup of G (cf. 3.2) of index at most 2. Consequently $G = Aut(HJ)$. The remaining values for $|N|$ are $2^a . 3^3 . 5^2 . 7$ with $a \in \{8,9\}$. According to the work of Beisiegel [1] and Stingl [10], there is no simple group of such an order. □

4. A NONEXISTENCE RESULT

PROPOSITION: *There does not exist a distance-regular graph with intersection array* $\{12, 8, 6, \ldots; 1, 1, 2, \ldots\}$.

PROOF: Let (V,L) be the linear incidence system obtained from such a distance-regular graph in the way described by the lemma. Fix $\alpha \in V$ and $\gamma \in \Gamma_2(\alpha)$. As there are 3 points on the unique line through γ containing a point of $\Gamma_1(\alpha)$ and as $k - b_2 - c_2 = 5$ there are two lines through γ containing points of $\Gamma_3(\alpha)$. From $c_3 = 2$, it results that each of these two lines contains precisely one point distinct from γ inside $\Gamma_3(\alpha)$. Thus through any given point $\delta \in \Gamma_3(\alpha)$, there is exactly one line through δ bearing a point in $\Gamma_2(\alpha)$ and this line has precisely two points, say z_1, z_2 in $\Gamma_2(\alpha)$. By symmetry of the argument, there is precisely one line through α bearing two distinct points w_1, w_2 in $\Gamma_1(\alpha)$ of distance 2 to δ. Now w_i must be adjacent to z_i ($i = 1,2$; up to a permutation of indices). This implies that w_1, w_2, z_1, z_2 span a minimal 4-circuit, conflicting $c_2 = 1$. □

5. CONCLUDING REMARKS

(i) The results of Gordon and Levinston put together with those above account for a complete solution of the existence problems connected with Biggs' list in [3, pp. 125, 126].

(ii) The author is grateful to P. Rowlinson for bringing to his attention the problems dealt with in Theorem 3.3 and §4 and for explaining to him some of the known techniques.

(iii) In Theorem 3.3, the adjective 'automorphic' may be replaced by 'distance-regular'. A proof of the stronger statement thus obtained will appear elsewhere.

BIBLIOGRAPHY

1. B. Beisiegel, "Über einfache endliche Gruppen mit Sylow 2-Gruppen der Ordnung höchstens 2^{10}", *Comm. Algebra,* 5 (1977), 113-170.
2. N.L. Biggs, *Algebraic Graph Theory,* Cambridge Math. Tracts 67 (Cambridge Univ. Press, Cambridge 1974).
3. N.L. Biggs, "Automorphic graphs and the Krein condition", *Geom. Dedicata,* 5 (1976), 117-127.
4. R.C. Bose and T.A. Dowling, "A generalization of Moore graphs of diameter two", *J. Combin. Theory,* 11 (1971), 213-226.
5. A.M. Cohen, "Finite quaternionic reflection groups", *J. Algebra,* 64 (1980), 293-324.
6. J. Fischer and J. McKay, "The nonabelian simple groups G, $|G| < 10^6$ - maximal subgroups", *Math. Comp.,* 32 (1978), 1293-1302.
7. L.M. Gordon and R. Levinston, "The construction of some automorphic graphs", preprint.
8. M. Hall Jr. and D.B. Wales, "The simple group of order 604800", *J. Algebra,* 9 (1968), 417-450.
9. E.E. Shult and A. Yanushka, "Near n-gons and line systems", *Geom. Dedicata,* 9 (1980), 1-72.
10. V. Stingl, "Endliche einfache component-type Gruppen deren Ordnung nicht durch 2^{11} teilbar ist", Thesis, Universität Mainz, 1976.
11. J. Tits, "Quaternions over $Q(\sqrt{5})$, Leech's lattice and the sporadic group of Hall-Janko", *J. Algebra,* 63 (1980), 56-75.
12. H. Wielandt, *Finite Permutation Groups* (Academic Press, New York, 1964).

Mathematisch Centrum
Kruislaan 413
1098 SJ Amsterdam
The Netherlands

TRANSITIVE AUTOMORPHISM GROUPS OF FINITE QUASIFIELDS

Stephen D. Cohen, Michael J. Ganley* and Vikram Jha

1. INTRODUCTION

Over the last few years several authors have been concerned with problems of the following type:

"Let $\pi(Q)$ be a finite translation plane, coordinatized by a quasifield Q, and let $\pi_0(Q_0)$ be a subplane, coordinatized by a sub-quasifield Q_0. Suppose that π admits a group of collineations G which fixes π_0 and acts transitively on the points of $\ell_\infty \setminus (\ell_\infty \cap \pi_0)$. What then can be said about π, π_0, Q, Q_0 or G?"

(See, for instance, Jha [4], Johnson [5], Kirkpatrick [6], Lorimer [7], Walker [9]).

If G fixes π_0 element-wise, then we say that G acts *tangentially transitive* (t.t.) w.r.t. π_0. The main results of the above authors can be summarised as follows:

THEOREM A: *If* G *is t.t. w.r.t.* π_0, *then*

(i) Q *is a field and* $|Q| = |Q_0|^2$ *or else* $|Q| = 16$ *and* $|Q_0| = 2$.

(ii) *There is a unique translation plane of order* 16 *admitting a group t.t. w.r.t. a subplane of order* 2.

(iii) π *is derived from a semifield plane of dimension* ≤ 2 *over its middle nucleus.*

In [4], Jha also studied another variation of the basic problem. This time we take $Q_0 = K$, the kern of Q and assume that G fixes pointwise the quadrangle of reference in π_0. Such a group is called *transitive*. Alternatively, we could say that a group of automorphisms of Q is *transitive* if it acts transitively on $Q \setminus K$.

THEOREM B : *If* G *is a transitive group of automorphisms of* Q, *then*

(i) $|Q| = |K|^2$ *or else* $|Q| = 16$ *and* $|K| = 2$,

and (ii) each element of K *commutes with each element of* Q.

This result is the starting point for our present investigation. *From now on, we shall assume that* Q *is a (right) quasifield with kern* K. G *will be an automorphism group of* Q *which acts transitively on* $Q \setminus K$. *In view of Theorem B we shall assume that* $|K| = q > 2$ *and* $|Q| = q^2$, *and also that* K *is central in* Q.

2. PRELIMINARIES

DEFINITION: Let $\alpha, \beta \in K$ $(\alpha \neq 0)$ and let $\sigma \in \mathrm{Aut}\, K$. Denote by $\theta_{\alpha,\beta,\sigma}$ the mapping of $K \to K$ defined by $\theta_{\alpha,\beta,\sigma} : k \to \alpha k^{\sigma} + \beta$. The set of all such mappings, under composition, forms a group *(the 1-dimensional affine group of* K*)* which we shall denote by $AG(K)$.

As $|Q| = |K|^2$, then we may regard Q as a (right) vector space of dimension 2 over K, and if $t \in Q \setminus K$ then $\{1,t\}$ is a basis for Q.

DEFINITION : Let Γ denote the full group of non-singular semi-linear maps of the vector space Q, which fix the element $1 \in K$. Suppose $\gamma \in \Gamma$ such that $\gamma : t \to t\alpha + \beta$ $(\alpha, \beta \in K, \alpha \neq 0)$ and the restriction $\gamma|_K = \sigma \in \mathrm{Aut}\, K$. Let $\overline{\gamma} \in AG(K)$ be defined by $\overline{\gamma} : k \to \alpha^{-1} k^{\sigma} + \alpha^{-1}\beta$ for all $k \in K$. (i.e. $\overline{\gamma} = \theta_{\alpha^{-1},\alpha^{-1}\beta,\sigma}$). Finally define a mapping $\phi : \Gamma \to AG(K)$ via $\phi : \gamma \to \overline{\gamma}$.

LEMMA 2.1 : *Let* M *be a minimal* 2*-transitive subgroup of* $AG(K)$ *which is not sharply* 2*-transitive. Let* $M_{0,1}$ *denote the stabilizer (in* M*) of* $0, 1 \in K$ *and let* $F(M_{0,1})$ *denote the set of fixed elements of* $M_{0,1}$*. Then* $F(M_{0,1})$ *is a subfield of* K *and* $[K : F(M_{0,1})] = 2^m$ *for some* $m \geq 1$.

PROOF : See Corollary 15.4 and Proposition 15.5 of [1].

LEMMA 2.2 : *With the above notation,* $F(G_t)$ *is a subquasifield of* Q*. Also, if* $K_1 = K \cap F(G_t)$*, then* $|F(G_t)| = |K_1|^2$*, so that* $F(G_t)$ *is a quadratic extension of* K_1.

LEMMA 2.3 : (a) ϕ *is an isomorphism between* Γ *and* $AG(K)$.
(b) *If* Σ *is a subgroup of* Γ*, then*

(i) $\overline{\Sigma}$ *is* 2*-transitive if and only if* Σ *is transitive on* $Q \setminus K$*, and*

(ii) $\overline{\Sigma}$ *is sharply* 2*-transitive if and only if* Σ *is regular on* $Q \setminus K$.

DEFINITION : If G is transitive, then we say that G is *minimal* if no proper subgroup of G is also transitive.

PROPOSITION 2.4 : *If* G *is minimal, then* G *acts regularly on* $Q \setminus K$ *and so, in particular,* $|G| = q(q-1)$.

PROOF : Suppose that G is minimal and transitive, but not regular. As $G \leq \Gamma$, we have (by lemma 2.3) that $\overline{G}$ is a minimal 2-transitive subgroup of $AG(K)$ which is not sharply 2-transitive. So, by lemma 2.1, $[K : F(\overline{G}_{0,1})] = 2^m$ $(m \geq 1)$. Now, clearly $\overline{G}_t = \overline{G}_{0,1}$ and $F(\overline{G}_{0,1}) = K_1 = F(G_t) \cap K$ (see lemma 2.2). As $t \in F(G_t)$, then t lies in a quadratic extension of K_1 and so $t^2 = ta + b$ for some $a, b \in K_1$. Let $f(x) \equiv x^2 - ax - b$, then $f(x)$ is reducible over K as $[K : K_1] = 2^m$ $(m \geq 1)$. But $t \in Q \setminus K$ is a root of $f(x)$, which contradicts a result of Jha [4, Proposition 4.2(b)]. Thus

G must act regularly, as claimed.

COROLLARY 2.5 : *If G is minimal, then G is isomorphic to a sharply 2-transitive group acting on q letters. In particular, G has q elements of order 2 if q is odd and q-1 elements of order 2 if q is even.*

PROOF : See above, and also Hall [3, p.382-3].

From now on, we shall assume that G is minimal.

3. MULTIPLICATION IN Q

LEMMA 3.1 : $(t\xi)\lambda = \xi(t\lambda) = t(\xi\lambda)$ *for all* $\xi, \lambda \in K$.

LEMMA 3.2 : *Let* $g \in G$ *such that* $g : t \to t\alpha + \beta (\alpha, \beta \in K, \alpha \neq 0)$ *and suppose that* $g|_K = \sigma$. *Then* $g : t\xi + \lambda \to t(\alpha\xi^\sigma) + (\beta\xi^\sigma + \lambda^\sigma)$.

NOTATION : In Lemma 3.2, the automorphism σ depends on α and β. For the moment, we shall denote this automorphism by $\sigma_{\alpha,\beta}$. Similarly we shall denote g by $g_{\alpha,\beta}$. Note that $\sigma_{\alpha,\beta}$ and $g_{\alpha,\beta}$ are both well-defined.

PROPOSITION 3.3 : *Multiplication in Q satisfies the following rule:*

$$(t\gamma + \delta)(t\alpha + \beta) = \begin{cases} t(\gamma\beta) + (\delta\beta) & \text{if } \alpha = 0 \\ t(\delta\alpha - \gamma\beta + \gamma a^\sigma) + (\delta\beta + \gamma\alpha^{-1}(b^\sigma + a^\sigma\beta - \beta^2)) & \\ \text{if } \alpha \neq 0, \text{ where } \sigma = \sigma_{\alpha,\beta} \text{ and } a, b \in K. & \end{cases}$$

PROOF : Begin with $t^2 = ta + b$ for some $a, b \in K$. Assume that $\alpha \neq 0$, and apply the automorphism $g_{\alpha,\beta}$ to the above equation. This yields

$$t(t\alpha + \beta) = t(a^\sigma - \beta) + \alpha^{-1}(b^\sigma + a^\sigma\beta - \beta^2).$$

Then consider $(t\gamma + \delta)(t\alpha + \beta) = \gamma(t(t\alpha + \beta)) + t(\alpha\delta) + \beta\delta$. Combining these two equations gives the required result.

PROPOSITION 3.4 : *(i) With the multiplication given above,* Q *is a quasifield if and only if the equations*

$$\left.\begin{aligned} x &= \alpha^2(\alpha\lambda - \beta\xi)^{-1}\ (b^\sigma + a^\sigma(\xi\alpha^{-1}) - (\xi\alpha^{-1})^2) \\ \text{and} \quad y &= (\alpha\lambda - \beta\xi)^{-1}\ (\alpha\beta b^\sigma + \alpha\lambda a^\sigma - \xi\lambda) \end{aligned}\right\} \text{ where } \sigma = \sigma_{x,y}$$

have a unique solution $(x,y) \in K^2$, *with* $x \neq 0$, *for all choices of* $\alpha, \beta, \xi, \lambda \in K$, *with* $\alpha \neq 0$ *and* $\alpha\lambda - \beta\xi \neq 0$.

(ii) If Q *is a quasifield, then* $f_0(x) \equiv x^2 - ax - b$ *is irreducible over* K.

PROOF : It is readily checked that all the axioms for a quasifield are satisfied, except for the axiom that says that the equation $cx = d$ must have a unique solution $x \in Q$, for all $c, d \in Q$ with $c \neq 0$. Then (i) is obtained by considering the equation $(t\alpha + \beta)(tx + y) = (t\xi + \lambda)$ for $t\alpha + \beta \neq 0$. Part (ii) follows from Jha [4, Proposition 4.2(b)].

REMARK : Clearly, if $a^\sigma = a$, $b^\sigma = b$ for all $\sigma = \sigma_{\alpha,\beta}$, then from Proposition 3.4, we have that Q is a quasifield. Referring back to Proposition 3.3 shows that Q is a Hall quasifield. This is the case, for instance, when $\sigma_{\alpha,\beta} = I$ (for all $\alpha, \beta \in K$, $\alpha \neq 0$), so that G is tangentially transitive.

4. THE AUTOMORPHISMS $\{\sigma_{\alpha,\beta}\}$

LEMMA 4.1 : *If* $g_{\alpha,\beta}$ *has order* 2, *then* $\sigma_{\alpha,\beta} = I$.

LEMMA 4.2 : *The elements of order* 2 *in* G *are* $\{g_{1,\beta} : \beta \in K, \beta \neq 0\}$ *if* q *is even and* $\{g_{-1,\beta} : \beta \in K\}$ *if* q *is odd.*

LEMMA 4.3 : $\sigma_{1,\beta} = I$ *for all* $\beta \in K$.

LEMMA 4.4 : $\sigma_{\alpha,\beta} = \sigma_{\alpha,0}$ *for all* $\alpha, \beta \in K$, *with* $\alpha \neq 0$.

NOTATION : Lemma 4.4 shows that $\sigma_{\alpha,\beta}$ depends only on the first component α. So, from now on, we shall denote $\sigma_{\alpha,\beta}$ by $\sigma(\alpha)$.

LEMMA 4.5 : $\sigma(\alpha)\,\sigma(\beta) = \sigma(\beta\alpha^{\sigma(\beta)})$ *for all* $\alpha,\ \beta \in K\setminus\{0\}$.

DEFINITION : Define a new binary operation "o" on the elements of K by

$$x \circ y = \begin{cases} 0 & \text{if } xy = 0 \\ yx^{\sigma(y)} & \text{if } xy \neq 0 . \end{cases}$$

LEMMA 4.6 : *The system* $N(K) = (K, +, \circ)$ *is a nearfield (i.e. a quasifield having associative multiplication).*

A considerable amount of information is available concerning nearfields - indeed the finite ones have been completely classified by Zassenhaus [10].

PROPOSITION 4.7 : *Let* $H = \{h \in K^* : \sigma(h) = I\}$. *Then*

(a) H *is a subgroup of* K^*.

(b) $\sigma(x) = \sigma(y)$ *if and only if* x,y *belong to the same coset of* H.

(c) *If* $[K^* : H] = s$, *then* $|K| = q = u^s$ *and* $N(K)$ *has its centre isomorphic to* $GF(u)$.

(d) *If* w *is a generator of* K^* *and if* $w_i = w^{(u^i-1)/(u-1)}$ $(i = 0,1,\ldots,s-1)$, *then* $\{w_0,w_1,\ldots,w_{s-1}\}$ *is a set of coset representatives of* H *and* $\sigma(w_i) : x \to x^{u^i}$.

The nearfields we have constructed are the so-called *regular nearfields*. Such nearfields, of order u^s with centre $GF(u)$, exist if and only if each prime divisor of s divides $u-1$, with the additional restriction that $u \equiv 1 \pmod 4$ if $s \equiv 0 \pmod 4$. This means, for instance, that the only nearfields of certain orders are in fact fields, or equivalently, that $\sigma(x) = I$ for all $x \in K^*$. These cases yield Hall quasifields, as already noted. For example

COROLLARY 4.8 : Q *is a Hall quasifield if either of the following conditions is satisfied:*

(a) $q = p^{p^n}$, p *prime,* $n \geq 0$.

or *(b)* $q = p_1^{p_2}$, *where* p_1, p_2 *are primes and* $p_1 \not\equiv 1 \pmod{p_2}$.

Using the results of this section, we can re-phrase Proposition 3.4 .

PROPOSITION 4.9 : *The multiplication given in Proposition 3.3 makes* Q *a quasifield if and only if the equation*

$$x = \alpha(\beta^2 - a^{\sigma(x)}\beta - b^{\sigma(x)})$$

has a unique solution $x \in K^*$ *for each choice of* $\alpha, \beta \in K$, $\alpha \neq 0$.

5. MAIN RESULTS

NOTATION : Now suppose that $|N(K)| = q = u^s$ and that $N(K)$ has centre $GF(u)$. Let $f_i(x) \equiv x^2 - a^{u^i}x - b^{u^i}$ $(i = 0,1,\ldots, s-1)$, where a,b are as earlier defined. Let w be a generator of K^* and let

$$H_0 = \{w^{is} : i = 0,1,\ldots, (u^s - 1)/s - 1\}, \quad \text{(i.e. the subgroup of } s^{th} \text{ powers in } K^*).$$

Finally, let H_i denote the coset $w^i H_0$ $(i = 0,1,\ldots, s-1)$.

LEMMA 5.1 : *Precisely one element from the set* $\{\alpha w^{s-i} f_i(\beta) : i = 0,1,\ldots, s-1\}$ *is an element of* H_0, *for each choice of* $\alpha, \beta \in K$, $\alpha \neq 0$.

LEMMA 5.2 : *If* $F(\beta) = \prod_{i=0}^{s-1} f_i(\beta)$, *then* $F(\beta) \in H_0$ *for each* $\beta \in K$.

PROPOSITION 5.3 : $a^u = a$ *and* $b^u = b$.

PROOF : Suppose that the result is false. Let s_1 $(s_1 > 1)$ be the least divisor of s such that $a^{u^{s_1}} = a$ and $b^{u^{s_1}} = b$.
Let $G(x) = f_0(x)\, f_1(x) \ldots f_{s_1 - 1}(x)$, then $G(x)$ is a square-free polynomial of degree $2s_1$ in $K[x]$, and $F(x) = (G(x))^{s/s_1}$ where $F(x)$ is defined in Lemma 5.2. Hence $G(\beta)$ is a non-zero s_1-th power for all $\beta \in K$. So, the number of solution pairs (x,y) in K^2 of

$$y^{s_1} = G(x) \qquad \text{(i)}$$

is precisely $s_1 u^s$.

To each solution (x,y) of (i) there corresponds a point $(x,y,1)$ on the projective curve

$$y^{s_1} z^{s_1} = z^{2s_1} G(x/z) \qquad \text{(ii)}$$

over K . The point $P(0,1,0)$ also lies on (ii). Now since (i) is absolutely irreducible in $K[x,y]$, by Weil's Theorem (e.g. [8, p.175]), the number of points on (ii) is less than $u^s + 1 + 2\, g u^{s/2}$, where g is the genus of (ii) over K .

Since P is a singularity of (ii), of multiplicity s_1 , it follows that

$$g \leqslant \tfrac{1}{2}\{(2s_1 - 1)(2s_1 - 2) - s_1(s_1 - 1)\} = \tfrac{1}{2}(3s_1 - 2)(s_1 - 1)$$

(see, for instance [2, p.201]).

Consequently

$$s_1 u^s + 1 < u^s + 1 + (3s_1 - 2)(s_1 - 1)\, u^{s/2}$$

which implies that

$$u^{s/2} < (3s_1 - 2) < 3s - 2 .$$

Thus, we are left with two possibilities. Either $s_1 = 1$ and the result is proved, or else $u^s = 3^2$. This final possibility can easily be discarded.

THEOREM C: *If* G *is a transitive automorphism group of a finite quasifield* Q, *with kern* K, *then either*

(i) $|Q| = 16$, $|K| = 2$ *and the associated translation plane is the Lorimer-Rahilly plane, or*

(ii) Q *is a Hall quasifield.*

THEOREM D: *(i) If a quasifield* Q *(of order* $\neq 16$*) admits a minimally transitive group of automorphisms, which induces* s *distinct automorphisms on its kern* K, *then* s *is odd.*

(ii) If s *is odd and there exists a nearfield of order* u^s, *having centre* GF(u), *then there exists a Hall quasifield of order* u^{2s} *having the property described in (i).*

REMARK: The case $s = 2$ can be dealt with by purely elementary means. Unfortunately, we have been unable to discover a simple method for $s > 2$.

6. RELATED PROBLEMS

As mentioned in the introduction, there are many problems similar to the one discussed in this paper. Some of these problems have been looked at recently, mainly by Jha.

If one insists that π_0 is the *kern* subplane then progress seems to be possible. For instance Jha has shown that if π has order $\neq 16$, if π_0 is the kern subplane and G fixes $\pi_0 \cap \ell_\infty$ pointwise, then π is a Hall plane. More recently, we have studied the case when π_0 is the kern subplane and G acts as an *autotopism* group of π. This case is, as yet, unfinished. On the basis of our results so far, the following conjecture would seem reasonable.

CONJECTURE : If π has order $\neq 16$, if π_0 is the kern subplane and if G fixes π_0 and acts transitively on $\ell_\infty \setminus (\ell_\infty \cap \pi_0)$, then π is a Hall plane.

As soon as one moves away from π_0 being the kern subplane, then progress appears to be rather slow. In the cases looked at so far, it has been shown that π_0 is a Baer subplane of π (provided π has order $\neq 16$) and that π is derivable with respect to $\pi_0 \cap \ell_\infty$. Only when G is tangentially transitive has the analysis been completed.

BIBLIOGRAPHY

1. D.A. Foulser, "The flag-transitive collineation groups of the finite desarguesian affine planes", *Canad. J. Math.*, 16 (1964), 443-472.
2. W. Fulton, *Algebraic Curves* (Benjamin, New York, 1969).
3. M. Hall, Jr., *The Theory of Groups* (Macmillan, New York, 1959).
4. V. Jha, "On tangentially transitive translation planes and related systems", *Geom. Dedicata*, 4 (1975), 457-483.
5. N.L. Johnson, "A characterization of generalized Hall planes" *Bull. Austral. Math. Soc.*, 6 (1972), 61-67.
6. P.B. Kirkpatrick, "Generalization of Hall planes of odd order", *Bull. Austral. Math. Soc.*, 4 (1971), 205-209.
7. P. Lorimer, "A projective plane of order 16", *J. Combin. Theory Ser. A*, 16 (1974), 334-347.
8. A.D. Thomas, *Zeta Functions: an Introduction to Algebraic Geometry* (Pitman, 1977).
9. M. Walker, "A note on tangentially transitive affine planes", *Bull. London Math. Soc.*, 8 (1976), 273-277.
10. H. Zassenhaus, "Über endliche Fastkörper", *Abh. Math. Sem. Univ. Hamburg*, 11 (1935), 187-220.

Department of Mathematics
University of Glasgow
Glasgow G12 8QW
U.K.

ON k-SETS OF TYPE (m,n) IN PROJECTIVE PLANES OF SQUARE ORDER

Massimo de Finis

Let π be a projective plane of order q and let K be a subset of its point-set, $|K| = k$; the set K will be called a *two character* k-*set of type* (m,n) (or briefly a k-*set of type* (m,n)) if a line in π either meets K in m or in n points and m-secants and n-secants actually exist; hence $0 \leq m < n \leq q+1$.

In 1966, Tallini Scafati [6] gave necessary arithmetical conditions for a k-set of type (m,n) to exist in a projective plane of order q, $q = p^h$, p a prime, h a non-negative integer. Namely, she proved that k must be a root of the equation

$$k^2 - k(n+m+q(n+m-1)) + mn(q^2+q+1) = 0, \qquad (1)$$

so that its discriminant must be a non-negative square; moreover, $n-m$ must divide q. As a special case, she characterized k-sets of type (1,n), $n \neq q+1$, proving that q must be a square, $n = \sqrt{q}+1$ and such an arc either is a Baer subplane or a Hermitian arc (a Hermitian curve if a certain reciprocity condition holds). Thus sets of type (m,q), $m \neq 0$, are also characterized, being the complements of the preceding ones. A set of type (1,q+1) is a line; a set of type (0,q) is an affine plane.

In 1973, Antonelli [1], investigating the case $2 \leq m < n \leq q-1$, found two infinite classes (depending on q) of triples k, m, n satisfying the necessary arithmetical existence conditions; moreover, using a computer, she found other triples, not belonging to those classes, for q up to 2^{10}.

In 1977, G. Tallini posed the problem of whether the necessary conditions were also sufficient for the existence of such sets. These first results were the subject of a talk given in 1978 at Ferrara (during a meeting of G.N.S.A.G.A., a research group supported by Italian National Research Council) and appeared in [2].

Let t_m and t_n be the numbers of m-secants and n-secants of K, respectively. Then

$$\begin{aligned} t_m + t_n &= q^2 + q + 1 , \\ m\,t_m + n\,t_n &= k(q+1) , \\ m(m-1)\,t_m + n(n-1)\,t_n &= k(k-1) . \end{aligned} \tag{2}$$

Also, if u_m and u_n are the numbers of m-secants and n-secants, respectively, through a point not belonging to K and v_m and v_n the numbers of m-secants and n-secants through a point in K, then

$$u_m + u_n = q + 1 , \qquad m u_m + n u_n = k ; \tag{3}$$

$$v_m + v_n = q + 1 , \qquad (m-1)v_m + (n-1)v_n = k - 1 . \tag{4}$$

Therefore, (1), (2), (3), (4) and the requirement $(n - m) | q$ are all necessary conditions for a k-set of type (m,n) to exist.

The complement K' of a k-set of type (m,n) is a k'-set of type (m',n'), where

$$k' = q^2 + q + 1 - k , \quad m' = q + 1 - n , \quad n' = q + 1 - m .$$

In the dual plane π^* of π, the m-secants and the n-secants of K form the sets K_1 and K_2 dual to K and K', having the following parameters:

$$k_1 = t_m = \frac{n(q^2+q+1) - k(q+1)}{n-m} , \quad n_1 = u_m = \frac{n(q+1) - k}{n-m} ,$$

$$m_1 = v_m = \frac{n(q+1) - k - q}{n-m} ;$$

$$k_2 = t_n = \frac{k(q+1) - m(q^2+q+1)}{n-m} , \quad n_2 = v_n = \frac{k + q - (q+1)m}{n-m} ,$$

$$m_2 = u_n = \frac{k - m(q+1)}{n-m} .$$

PROPOSITION 1 : In a Galois plane $PG(2,q^2)$, $q = p^h$, p a prime, $h \geq 1$ an integer, there exist k-sets of type (m,n), where

$$k = s(q^2 + q + 1) , \quad m = s , \quad n = q + s ,$$

for $s = 2,3,\ldots, q^2 - q - 2$; hence, also their complements and dual sets exist.

PROOF : It is easy to verify that the given parameters satisfy all necessary conditions; so we prove that such sets actually exist. Let π and π' be two Baer subplanes of $PG(2,q^2)$ having no common point (we call them skew). Such planes exist, since it is well-known that $PG(2,q^2)$ can be partitioned into $q^2 - q + 1$ Baer subplanes. Let K be the set-theoretic union of the points in π and π'; we prove that K is a $2(q^2+q+1)$-set of type $(2, q+2)$. By definition, K is of class $[2, q+2, 2q+2]$; therefore, we need only to prove that $t_{2q+2} = 0$, t_{2q+2} being the number of $(2q+2)$-secants. We get this result by solving the simultaneous equations:

$$t_2 + t_{q+2} + t_{2q+2} = q^4 + q^2 + 1$$

$$2t_2 + (q+2)t_{q+2} + (2q+2)t_{2q+2} = 2(q^2+1)(q^2+q+1)$$

$$2t_2 + (q+2)(q+1)t_{q+2} + (2q+2)(2q+1)t_{2q+2} = $$
$$= 2(q^2+q+1)(2q^2+2q+1) .$$

Indeed, $t_2 = q^4 - q^2 - 2q - 1$, $t_{q+2} = 2q^2 + 2q + 2$, $t_{2q+2} = 0$. Now, let $\pi_1, \pi_2, \ldots, \pi_s$ be any s pairwise skew Baer subplanes of $PG(2,q^2)$. By the previous argument, no line in $PG(2,q^2)$ has $q+1$ points in more than one such subplane. Consequently, the set $K = \pi_1 \cup \pi_2 \cup \ldots \cup \pi_s$ is an $(s(q^2+q+1))$-set of type $(s, q+s)$ and the statement is proved.

We remark that the complements and the dual sets of the sets constructed in Proposition 1 have again parameters in the same family.

Furthermore, since a k-set of type (m,n) is a $PBD(v, \{m,n\}, 1)$, $v = k$, Proposition 1 gives an infinite class of PBD's which are embeddable in a projective plane.

PROPOSITION 2 : Let π be a projective plane of order q. The following are necessary conditions for $s \geq 2$ pairwise skew maximal $\{qn - q + n; n\}$-arcs to exist:

(i) q is a square;

(ii) $n = \sqrt{q}$.

Furthermore, if s pairwise skew $\{q\sqrt{q} - q + \sqrt{q}; \sqrt{q}\}$-arcs exist, then the set-theoretic union of their points is an $s(q\sqrt{q} - q + \sqrt{q})$-set of type $((s-1)\sqrt{q}, s\sqrt{q})$, $s = 2,3,\ldots,\sqrt{q}-1$.

PROOF: Let us suppose that $s = 2$ and let K be the $2(qn - q + n)$-set consisting of the points belonging to two skew maximal arcs. K is of class $[0, n, 2n]$. Thus

$$t_0 + t_n + t_{2n} = q^2 + q + 1$$

$$nt_n + 2nt_{2n} = 2(q+1)(qn - q + n)$$

$$n(n-1)t_n + 2n(2n-1)t_{2n} = 2(qn - q + n)(2qn - 2q + 2n - 1).$$

Solving these simultaneous equations, we obtain

$$t_0 = q^2/n^2 - q, \quad t_n = 2q(qn - q + n)/n^2, \quad t_{2n} = (qn - q + n)^2/n^2.$$

Therefore, when $n = \sqrt{q}$, we have that $t_0 = 0$.

Hence, if two skew maximal $(q\sqrt{q} - q + \sqrt{q}; \sqrt{q})$-arcs exist, then the set-theoretic union of their points is a $2(q\sqrt{q} - q + \sqrt{q})$-set of type $(\sqrt{q}, 2\sqrt{q})$. The necessary arithmetical conditions are obviously satisfied. Consequently, if s pairwise skew maximal arcs exist, then the set-theoretic union of their points is an $s(q\sqrt{q} - q + \sqrt{q})$-set of type $((s-1)\sqrt{q}, s\sqrt{q})$, as there is no line in π external to more than one arc. Moreover,

$$s\sqrt{q} \le q - 1 \implies s \le \sqrt{q} - 1/\sqrt{q} \implies s \le \sqrt{q} - 1,$$

and the statement is proved.

A similar result is given by

PROPOSITION 3: Let π be a projective plane of order q, q a square. If π contains s pairwise skew Hermitian arcs, $s = 2,3,\ldots,\sqrt{q}-1$, then π contains a k-set of type (m,n) where

$$k = s(q\sqrt{q} + 1), \quad m = s(\sqrt{q} + 1) - \sqrt{q}, \quad n = s(\sqrt{q} + 1).$$

PROOF : First we remark that the above parameters satisfy all necessary existence conditions. Suppose that π contains two pairwise skew Hermitian arcs and let K be the set-theoretic union of their points. Then K is a $2(q\sqrt{q}+1)$-set of class $[2, \sqrt{q}+2, 2(\sqrt{q}+1)]$, for which

$$t_2 + t_{\sqrt{q}+2} + t_{2(\sqrt{q}+1)} = q^2 + q + 1$$

$$2t_2 + (\sqrt{q}+2)t_{\sqrt{q}+2} + 2(\sqrt{q}+1)\, t_{2(\sqrt{q}+1)} = 2(q\sqrt{q}+1)(q+1)$$

$$2t_2 + (\sqrt{q}+2)(\sqrt{q}+1)t_{\sqrt{q}+2} + 2(\sqrt{q}+1)(2\sqrt{q}+1)t_{2(\sqrt{q}+1)} = 2(q\sqrt{q}+1)(2q\sqrt{q}+1) .$$

Hence

$$t_2 = 0 ,\quad t_{\sqrt{q}+2} = 2(q\sqrt{q}+1),\quad t_{2(\sqrt{q}+1)} = (q - \sqrt{q})^2 - 1 .$$

Therefore, there is no line in π tangent to both Hermitian arcs; hence K has two characters and is of type $(\sqrt{q}+2, 2(\sqrt{q}+1))$. Thus, given s pairwise skew Hermitian arcs in π, a line in π can be tangent to at most one of them and the statement is proved.

As to the existence of the k-sets considered in Proposition 3, it is easy to prove that in PG(2,q), q a square, there are no pairwise skew Hermitian curves, but nothing seems to be known for Hermitian arcs which are not Hermitian curves.

BIBLIOGRAPHY

1. S. Antonelli, "Sugli archi di tipo (m,n) in un piano grafico π", Relazione 23 (Istituto Matematico, Università di Napoli, 1973).
2. M. de Finis, "Sui k-insiemi a due caratteri in un piano di Galois", Semininario di Geometrie Combinatorie 23 (Università di Roma, 1979).
3. J.W.P. Hirschfeld, "Cyclic Projectivities in PG(n,q)", *Teorie Combinatorie*, Tomo I, Rome, 1973. Atti dei Convegni Lincei 17 (Accad. Naz. Lincei, Rome, 1976), pp.201-211.
4. B. Segre, "Forme e geometrie hermitiane, con particolare riguardo al caso finito", *Ann. Mat. Pura Appl.*, 70 (1965), 1-201.
5. J. Singer, "A theorem in finite projective geometry and some applications to number theory", *Trans. Amer. Math. Soc.*, 43 (1938), 377-385.

6. M. Tallini Scafati, "{k,n} - archi in un piano grafico finito, con particolare riguardo a quelli a due caratteri", *Atti. Accad. Naz. Lincei Rend.*, 40 (1966), 812-818; 1020-1025.
7. M. Tallini Scafati, "The k-sets of type (m,n) in a Galois space $S_{r,q}$ $(r \geq 2)$", *Teorie Combinatorie*, Tomo II, Rome, 1973. Atti dei Convegni Lincei 17 (Accad. Naz. Lincei, Rome, 1976), pp.459-463.
8. G. Tallini, "Problemi e risultati sulle geometrie di Galois", Relazione 30 (Istituto Matematico, Università di Napoli, 1973).

Istituto Matematico
Università di Roma
00185 Roma
Italy

ON k-SETS OF TYPE (m,n) IN A STEINER SYSTEM $S(2, \ell, v)$

Marialuisa J. de Resmini

A Steiner system $S(2, \ell, v)$ is a pair $(S, \mathcal{B})$, where S is a v-set, whose elements are called *points*, and $\mathcal{B}$ is a family of ℓ-subsets of S called *blocks*, such that any two distinct points belong to exactly one block.

If $b = |\mathcal{B}|$ is the number of blocks and r is the number of blocks through a point, then

$$b = v(v-1)/[\ell(\ell-1)] \quad \text{and} \quad r = (v-1)/(\ell-1),$$

from which the well-known necessary conditions for the existence of a $S(2, \ell, v)$ follow; namely,

$$v(v-1) \equiv 0 \pmod{\ell(\ell-1)} \quad \text{and} \quad v-1 \equiv 0 \pmod{\ell-1}$$

When $\ell = 3,4,5$, such conditions are also sufficient.

Tallini [8] defined k-sets of type $(m_1, \ldots, m_s)$ in a $S(2, \ell, v)$ and proved some necessary conditions for their existence. He also suggested the problem of investigating such sets.

A k-subset K of S will be called of *class* $[m_1, \ldots, m_s]$ if a block of $S(2, \ell, v)$ meets K either in $m_1, m_2, \ldots,$ or in m_s points. Let t_{m_j} be the number of m_j - secants of K; then

$$\sum_{j=0}^{\ell} t_{m_j} = b, \quad \sum_{j=1}^{\ell} m_j t_{m_j} = kr, \quad \sum_{j=2}^{\ell} m_j(m_j-1)t_{m_j} = k(k-1). \tag{1}$$

A k-set K will be said of *type* $(m_1, \ldots, m_s)$ if it is of class $[m_1, \ldots, m_s]$ and $t_{m_j} \neq 0$ for $j = 1, \ldots, \ell$. (The m_j's are also called the *characters* of K.)

Here k-sets of type (m,n), $0 \le m < n \le \ell - 1$, briefly $(k\,; m,n)$-sets, will be considered. Equations (1) now become

$$t_m + t_n = b,$$
$$m t_m + n t_n = rk, \quad (2)$$
$$m(m-1)t_m + n(n-1)t_n = k(k-1).$$

These equations are consistent if k is a solution of

$$k^2 - k(r(m+n-1)+1) + mnb = 0, \quad (3)$$

so that its discriminant must be a non-negative square.

Let u_n and u_m be the numbers of n-secants and m-secants respectively through a point not belonging to K and v_n and v_m the numbers of n-secants and m-secants through a point in K. Then

$$u_m + u_n = r, \quad mu_m + nu_n = k; \quad (4)$$

$$v_m + v_n = r, \quad (m-1)v_m + (n-1)v_n = k-1. \quad (5)$$

From (4) and (5), a necessary existence condition follows; namely,

$$r \equiv 1 \pmod{n-m}, \quad (6)$$

and when $1 \leq m < n \leq \ell - 1$, $n - m$ must be a proper divisor of $r - 1$.

When $(k; 0,n)$-sets are considered, (3) and (6) become respectively:

$$k = 1 + r(n-1), \quad (7)$$

$$r \equiv 1 \pmod{n}. \quad (8)$$

Now a general result will be proved which completely characterizes $(k; 0, \ell-1)$-sets in $S(2, \ell, v)$'s and is most interesting when $\ell = 3,4,5$ since the Steiner systems do always exist.

PROPOSITION 1 : *A necessary condition for a* $\left(\frac{v(\ell-2)+1}{\ell-1}; 0, \ell-1\right)$-*set* K *to exist in an* $S(2, \ell, v)$ *is*

$$v \equiv \ell \quad \text{or} \quad \ell(\ell-1)+1 \pmod{\ell(\ell-1)^2}.$$

Furthermore, if K *exists, its complement* $S \setminus K$ *is a subsystem* $S(2, \ell, u)$, $u = (v-1)/(\ell-1)$, *and* K, *with its* $(\ell-1)$*-secants as blocks, is a resolvable* $S(2, \ell-1, \frac{v(\ell-2)+1}{\ell-1})$. *Conversely, if a resolvable* $S(2, \ell-1, w)$ *and a* $S(2, \ell, \frac{w-1}{\ell-2})$ *exist, then there exists an* $S(2, \ell, v)$, $v = (w(\ell-1)-1)/(\ell-2)$, *containing* $S(2, \ell-1, w)$ *as a* $(w; 0, \ell-1)$*-set and* $S(2, \ell, \frac{w-1}{\ell-2})$ *as a subsystem.*

PROOF: For a Steiner system $S(2, \ell, v)$ to exist, v must satisfy

$$v \equiv 1 \text{ or } \ell \pmod{\ell(\ell-1)}.$$

(Sometimes there are other values; e.g. when $\ell = 6$.)
From (8), the necessary condition $(\ell-1) \mid \frac{v-\ell}{\ell-1}$ follows; thus, if $v \equiv 1 \pmod{\ell(\ell-1)}$, then

$$(\ell-1) \mid \frac{v-\ell}{\ell-1} \Longrightarrow v \equiv \ell(\ell-1)+1 \pmod{\ell(\ell-1)^2};$$

if $v \equiv \ell \pmod{\ell(\ell-1)}$, then

$$(\ell-1) \mid \frac{v-\ell}{\ell-1} \Longrightarrow v \equiv \ell \pmod{\ell(\ell-1)^2}.$$

When these conditions are satisfied, $\frac{v-1}{\ell-1} \equiv 1$ or $\ell \pmod{\ell(\ell-1)}$.

We will prove that the complement $\tilde{K} = S \setminus K$ is a subsystem $S(2, \ell, \frac{v-1}{\ell-1})$; indeed, if two points belong to $\tilde{K}$ the block joining them is completely contained in $\tilde{K}$, otherwise K would have n-secants, $1 \le n \le \ell-2$, a contradiction.

Next we prove that K, with its $(\ell-1)$-secants as blocks, is a resolvable $S(2, \ell-1, \frac{v(\ell-2)+1}{\ell-1})$. The necessary condition for a resolvable Steiner system to exist is satisfied. The number of $(\ell-1)$-secants of K is equal to the number of blocks of such a system and the number of 0-secants of K is equal to the number of blocks in an $S(2, \ell, \frac{v-1}{\ell-1})$. Through a point not belonging to K there are u_0 0-secants and $u_{\ell-1}$ $(\ell-1)$-secants; $u_{\ell-1}$ is equal to the number of blocks in each parallel class, the number of these classes being $\frac{v-1}{\ell-1}$, and u_0 is the number of blocks through a point in an $S(2, \ell, \frac{v-1}{\ell-1})$.

Finally, if a resolvable $S(2, \ell - 1, \frac{v(\ell - 2) + 1}{\ell - 1})$ on K and an $S(2, \ell, \frac{v - 1}{\ell - 1})$ on S' are given, then an $S(2, \ell, v)$ on $S = K \cup S'$, containing the former as $(\frac{v(\ell - 2) + 1}{\ell - 1}; 0, \ell - 1)$-set, the latter as subsystem, can be constructed as follows. The blocks of the resolvable system are partitioned into parallel classes, then to all blocks belonging to a class the same point of $S(2, \ell, \frac{v - 1}{\ell - 1})$ is added. The blocks of the system on S are the blocks of $S(2, \ell, \frac{v - 1}{\ell - 1})$ together with the extended blocks of the resolvable system. An easy computation shows that the number of blocks is correct. Moreover, given any two points $x, y \in S$, there is exactly one block through them: (i) if $x, y \in S'$, the block is the one belonging to $S(2, \ell, \frac{v - 1}{\ell - 1})$; (ii) if $x, y \in K$, the block is the one belonging to the resolvable system, with the new point added; (iii) if $x \in S'$, $y \in K$, there is a unique block through y belonging to the parallel class to which x is added. The proof is completed.

When $\ell = 3, 4, 5$, Proposition 1 becomes the following respective corollaries.

COROLLARY 1 : (STS's). *For all* $v \equiv 3$ *or* $7 \pmod{12}$, *a* STS *containing a* $(\frac{v+1}{2}; 0, 2)$*-set and an* $S(2, 3, \frac{v-1}{2})$ *subsystem can be constructed, and this construction goes back to Kirkman* (1847) [6]. *Therefore, a headless* STS *contains no* $(k; 0, 2)$*-set; for instance, this is true for any non-degenerate plane.*

Furthermore, it is easy to prove that, if an STS $S(v)$ *contains a* $(\frac{v+1}{2}; 0, 2)$*-set, then the* $S(2v + 1)$ *obtained by Horner's construction* [2], [5], *contains a* $(2\frac{(v+1)}{2}; 0, 2)$*-set.*

COROLLARY 2 : $(S(2, 4, v)$*'s). Let* S *be a* v*-set* $v \equiv 4$ *or* 13 $\pmod{36}$. *If on a* $\frac{2v + 1}{3}$*-subset of* S *a resolvable* STS *is constructed and on its complement a* $S(2, 4, \frac{v - 1}{3})$ *is constructed, then, when the same point of the latter is added to all blocks in a parallel class of the former, an* $S(2, 4, v)$ *is obtained containing a set of type* $(0,3)$ *and a maximal subset. This construction extends to* $S(2, 4, v)$*'s Kirkman's construction for*

STS's.

COROLLARY 3 : (S(2, 5, v)'s). *When* $v \equiv 5$ *or* $21 \pmod{80}$, *a* $S(2, 5, v)$ *containing both a* $(\frac{3v+1}{4}; 0, 4)$*-set and a maximal subsystem* $S(2, 5, \frac{v-1}{4})$ *exists. As a resolvable* $S(2, 4, \frac{3v+1}{4})$ *and an* $S(2, 5, \frac{v-1}{4})$ *exist, the construction is the one already proved, and extends the one by Kirkman.*

When $\ell \geq 6$, the construction works provided the required systems exist.

Now an infinite class of Steiner systems containing $(k; 0, n)$-sets will be constructed.

PROPOSITION 2 : *For any positive integers* h, m, $1 \leq m \leq h-1$, $h \geq 2$, *there exists a Steiner system* $S(2, 2^m, 2^{h+m} - 2^h + 2^m)$ *containing* k*-sets of type* $(0, 2^s)$, *for all* $s = 1,2,\ldots,m-1$, *where* $k = 2^{h+s} - 2^h + 2^s$.

PROOF : Denniston [1] constructed maximal $(2^{h+m} - 2^h + 2^m; 2^m)$-arcs in $PG(2, 2^h)$, $1 \leq m \leq h$, taking a quadratic form $f(x,y)$, irreducible over $GF(2^h)$, and defining the arc as the set of points (X,Y) in $AG(2, 2^h)$ for which $f(X,Y) \in A$, where A is a subgroup of $GF(2^h)^+$, $|A| = 2^m$. With its 2^m-secants as blocks, this arc is a Steiner system $\mathcal{S}$. Now, if H is a subgroup of A, with $|H| = 2^s$, $1 \leq s \leq m-1$, then the Steiner system $S(2, 2^s, 2^{h+s} - 2^h + 2^s)$ on $\{(X,Y) \in AG(2, 2^h) : f(X,Y) \in H\}$, the blocks being the 2^s-secants, is a set of type $(0, 2^s)$ in $\mathcal{S}$, and the statement is proved. The necessary conditions (3) and (6) are obviously satisfied.

When $m = 2$, Proposition 2 gives a class of $S(2, 4, 12.2^{h-2} + 4)$'s, $h \geq 3$, each containing a $(2^h + 2; 0, 2)$-set (when $h = 2$, then $S(2, 4, 16) = AG(2,4)$ and a conic (for which the line at infinity is external) plus its nucleus is a $(6; 0, 2)$-set). When $m = 3$, Proposition 2 gives a class of $S(2, 8, 2^{h-3}.56 + 8)$'s each containing a $(3.2^h + 4; 0, 4)$-set, $h \geq 4$ (if $h = 3$, a maximal $(28; 4)$-arc in $AG(2,8) = S(2, 8, 64)$ is a solution).

From the conditions given at the beginning, some necessary existence conditions can be derived; the main results, together with some non-existence cases, will be stated below.

S(2, 4, v)'s.

$v \equiv 4 \pmod{12}$ is a necessary condition for a $(\frac{v+2}{3}; 0,2)$-set to exist, and, by Proposition 2, some of such sets exist.

Type (1,3) sets exist if: (i) $v \equiv 4 \pmod{24}$, and (ii) v is a square; then $k = (v \pm \sqrt{v})/2$. (The smallest values are $v = 100, 196, 484, 676$.)

S(2, 5, v)'s.

$v \equiv 5$ or $21 \pmod{40}$ is a necessary condition for a $(\frac{v+3}{4}; 0, 2)$-set to exist and $v \equiv 5$ or $41 \pmod{60}$ for a $(\frac{v+1}{2}; 0, 3)$-set.

No sets of type (1,3), (1,4) and (2,3) (and their complements) exist in an S(2, 5, v), $v > 5$.

Other results are summed up in Table 1.

No k-sets of types (1,3), (1,4), (1,5), and (2,4) (and their complements) exist in S(2, 6, v)'s. No k-sets of types (1,n), $n = 3,4,5,6$, (2,4), (2,5) (and their complements) exist in S(2, 7, v)'s. No k-sets of types (1,3), (1,7), (2,4), (2,6), (3,5) (and their complements) exist in S(2, 8, v)'s. On the other hand, by Proposition 2, there exists S(2, 8, v)'s, $v \equiv 8 \pmod{56}$, containing sets of types (0,2) and (0,4). As to sets of type (1,5) in S(2, 8, v)'s, the necessary existence condition is $v \equiv 8 \pmod{56}$ and the smallest values for which the discriminant of (2) is a square are $v = 64$ and $v = 4992$. Therefore, there could exist (36; 1, 5)-sets and (10; 1, 5)-sets (and so their complements) in an S(2, 8, 64); moreover, if a S(2, 8, 4992) exists, then it could contain (2760; 1, 5)- and (806; 1, 5)-sets (and their complements). For type (1,6) sets in S(2, 8, v)'s the necessary existence condition is $v \equiv 8$ or $113 \pmod{280}$; the discriminant of (2) is never a square when $v \equiv 8 \pmod{56}$ and for small v's when $v \equiv 113 \pmod{56}$, and the same is true for type (2,5), the necessary existence condition being $v \equiv 8$ or $113 \pmod{168}$.

A Steiner system $S(2, \ell, v)$ will be called *s-fold-resolvable* if its blocks can be partitioned in s different ways into $\frac{v-1}{\ell-1}$ parallel classes, each class containing v/ℓ blocks no two of which have a common point. Of course, 1-fold-resolvable Steiner systems are the usual resolvable systems and the necessary condition for them to exist is $v \equiv \ell \pmod{\ell(\ell-1)}$; this condition is also sufficient when $\ell = 3$ and 4. Any partition of the blocks into parallel classes will be called a *parallelism* of the system; thus, an s-fold-resolvable Steiner system has s distinct parallelisms. Two or more different parallelisms may share a class; if this happens, a less restrictive definition of s-fold-resolvability can be given.

TABLE 1

$S(2, \ell, v)$	(k; m, n)-set	necessary existence condition
$S(2, 6, v)$	$(\frac{v+4}{5}; 0, 2)$-set	$v \equiv 6$ or $16 \pmod{30}$
	$(\frac{2v+3}{5}; 0, 3)$-set	$v \equiv 6$ or $21 \pmod{30}$
	$(\frac{3v+2}{5}; 0, 4)$-set	$v \equiv 6$ or $46 \pmod{60}$
$S(2, 7, v)$	$(\frac{v+5}{2}; 0, 2)$-set	$v \equiv 7$ or $43 \pmod{84}$
	$(\frac{v+2}{3}; 0, 3)$-set	$v \equiv 7$ or $43 \pmod{126}$
	$(\frac{v+1}{2}; 0, 4)$-set	$v \equiv 7$ or $127 \pmod{168}$
	$(\frac{2v+1}{3}; 0, 5)$-set	$v \equiv 7$ or $127 \pmod{210}$
$S(2, 8, v)$	$(\frac{v+6}{7}; 0, 2)$-set	$v \equiv 8 \pmod{56}$
	$(\frac{3v+4}{7}; 0, 4)$-set	$v \equiv 8 \pmod{56}$
	$(\frac{2v+5}{7}; 0, 3)$-set	$v \equiv 8$ or $113 \pmod{168}$
	$(\frac{4v+3}{7}; 0, 5)$-set	$v \equiv 8$ or $113 \pmod{280}$
	$(\frac{5v+2}{7}; 0, 6)$-set	$v \equiv 8 \pmod{168}$

If an s-fold-resolvable $S(2, \ell, v)$ is given and the s parallelisms share no class, then adding, for each parallelism, the same point to all blocks in a parallel class (for every class), an $S(2, \ell + s, w)$ containing the given system as a $(v ; 0, \ell)$-set is constructed. Obviously, the necessary arithmetical conditions must be satisfied and new blocks on the new points must be added. For instance, it is easy to construct an $S(2, 2, 6)$ having 6 distinct parallelisms, any two of them sharing a parallel class, and to complete it to an $S(2, 5, 21)$.

If certain s-fold-resolvable Steiner systems exist, then some of the previous necessary existence conditions become sufficient.

Now the best general results on sets of types $(0,n)$, $2 \le n \le \ell - 2$, and $(1,n)$, $3 \le n \le \ell - 1$, will be stated without proofs.

PROPOSITION 3 : *A necessary condition for a* $(k; 0, n)$*-set* K*, with* $k = [1 + r(n-1)]$*, to exist in an* $S(2,\ell,v)$ *is*

$$v \equiv \ell \quad \textit{or} \quad (\ell - 1)(n + 1) + 1 \pmod{\ell n(\ell - 1)},$$

with the further condition $\ell \mid n(n+1)$ *in the second case. Moreover,* K*, with its* n*-secants as blocks, is an* $(\ell - n)$*-fold-resolvable* $S(2, n, k)$*. Conversely, if such an* $S(2, n, k)$ *exists, it is possible to complete it to an* $S(2, \ell, v)$, $v = \dfrac{(k-1)(\ell-1)}{n-1} + 1$, *adding, for each partition into parallel classes, the same point to all blocks in the same parallel class. Furthermore, when* $n \mid \ell$, $(k; 0, n)$*-sets can exist for any* $v \equiv \ell \ (\ell(\ell - 1))$ *and they do not exist if* $v \equiv 1 \pmod{\ell(\ell - 1)}$.

PROPOSITION 4 : *If* $(n - 1) \mid \ell$ *and* $v \equiv \ell \pmod{\ell(\ell - 1)}$*, then a* $S(2, \ell, v)$ *can contain* k*-sets of type* $(1,n)$*; if* $v \equiv 1 \pmod{\ell(\ell - 1)}$*, then no* k*-set of type* $(1,n)$ *exists. If* $(n-1) \nmid \ell$*, then a necessary condition for* $(k; 1,n)$*-sets to exist is* $v \equiv \ell$ *or* $(\ell - 1)n + 1 \pmod{\ell(\ell - 1)(n - 1)}$*, in the second case with the further condition that* $\ell \mid n(n-1)$.

Finally, we turn to a special class of Steiner systems, namely projective planes.

PROPOSITION 5 : *Let* π *be a projective plane of order* q , q *a square, that is a Steiner system* $S(2, q+1, q^2+q+1)$. *If* π *contains a subplane* α *of order* $\sqrt{q}-1$, *then* π *contains a* $(q - \sqrt{q} + 1)^2$*-set of type* $((\sqrt{q} - 1)^2, q - \sqrt{q} + 1)$ *and hence its complement.*

PROOF : Let K be the set consisting of the points in $\pi \setminus \alpha$ belonging to the lines of α. Then $|K| = (q - \sqrt{q} + 1)(q + 1 - \sqrt{q})$. So K is a set of type $((\sqrt{q}-1)^2, q-\sqrt{q}+1)$ in π. Indeed, since α is a three character set of type $(0, 1, \sqrt{q})$ in π, for a line t in π one of the following is true:

(i) t does not meet α, in which case t meets K at $q-\sqrt{q}+1$ points;

(ii) t meets α at one point, in which case t meets K at $q-\sqrt{q}+1-\sqrt{q} = (\sqrt{q}-1)^2$ points;

(iii) t meets α at $\sqrt{q}$ points (i.e. t contains a line in α), in which case t meets K at $q + 1 - \sqrt{q}$ points.

The complement of the set K in Proposition 5 is a $(2\sqrt{q}(q - \sqrt{q} + 1))$-set of type $(\sqrt{q}, 2\sqrt{q})$.

As an example of Proposition 5, consider any non-Desarguesian plane of order 9 ; it contains a Fano subplane. Therefore, it contains a $(49; 4, 7)$-set and its complement, a $(42; 3, 6)$-set.

BIBLIOGRAPHY

1. R.H.F. Denniston, "Some maximal arcs in finite projective planes", *J. Combin. Theory,* 6 (1969), 317-319.
2. M.J. de Resmini, "Sistemi di Steiner e questioni collegate", Seminario di Geometrie Combinatorie 12 (Università di Roma, 1978).
3. M.J. de Resmini, "Sui k-insiemi a due caratteri nei sistemi di Steiner S(2, ℓ, v)", Seminario di Geometrie Combinatorie 25 (Università di Roma, 1980).
4. H. Hanani, "Balanced Incomplete Block Designs and Related Designs", *Discrete Math.,* 11 (1975), 255-369.
5. J. Horner, "On triads of once-paired elements", *Quart. J. Math.,* 9 (1868), 15-18.
6. T.P. Kirkman, "On a problem in combinations", *Cambridge and Dublin Math. J.,* 2 (1847), 191-204.
7. B. Segre, "Forme e geometrie hermitiane, con particolare riguardo al caso finito", *Ann. Mat. Pura Appl.,* 70 (1965), 1-202.
8. G. Tallini, "Problemi di immersione nei sistemi di Steiner", Seminario di Geometrie Combinatorie 21 (Università di Roma, 1979).
9. M. Tallini Scafati, "(k,n)-archi in un piano grafico finito, con particolare riguardo a quelli con due caratteri", *Atti Accad. Naz. Lincei Rend.,* 40 (1966), 812-818; 1020-1025.

Istituto Matematico
Università di Roma
00185 Roma
Italy

SOME TRANSLATION PLANES OF ORDER 81

D.A. Foulser

THE PROBLEM: to find all translation planes A of order 81 which admit two collineations σ and τ of order 3 such that the fixed-point sets $F(\sigma)$ and $F(\tau)$ are Baer subplanes (i.e. σ and τ are *Baer* 3-*collineations*) which properly overlap, $0 \neq F(\sigma) \cap F(\tau) \neq F(\sigma)$.

If A is a translation plane of characteristic $p > 3$, then it is known [2] that no such overlapping Baer p-collineations exist. However, the nearfield plane of order 9 does admit such collineations. The general problem then is the investigation of overlapping Baer 3-collineations in translation planes of order 3^{2e}, for $e > 1$.

THEOREM: *There are (up to isomorphism)* n *planes of order* 81 *admitting such overlapping* 3-*collineations* σ *and* τ, *where* $3 \leq n \leq 6$. *(Note: all examples but not necessarily all isomorphisms, are known.)*

PROOF: Normalize extensively by hand; then compute all normalized cases by machine. In stage 1 of the computing, 96×81^2 cases were checked, resulting in approximately 5,000 successes. During stage 2, for each success in stage 1, approximately 100 cases were checked. Each of the approximately two dozen successes in stage 2 describes a plane of the given type, but there are many obvious isomorphisms, resulting in 3 to 6 isomorphism classes.

PRELIMINARIES: Let A be a translation plane of order 81, represented as a spread S on the vector space $V = V(8,3)$ of dimension 8 over GF(3). That is, S is a set of 82 mutually disjoint 4-dimensional subspaces. If W_0 and $W_\infty \in S$, then $V = W_0 \oplus W_\infty$; and any other element $W \in S$ has an equation $y = Tx$ for $x \in W_0$, $y \in W_\infty$, and T a 4×4 matrix over GF(3). That is, $W = \{(x, Tx) : x \in W_0\}$. The corresponding set $\mathcal{T}$ of 81 matrices T (including 0 for W_0) is the *slope set* of S, and satisfies the condition: if $S \neq T \in \mathcal{T}$, then $\det(S - T) \neq 0$.

From previous work [2], it is known that the overlapping Baer 3-collineations σ and τ have the following form as matrices acting on V:

$$\sigma = \operatorname{diag}(E,E) \quad \text{and} \quad \tau = \operatorname{diag}(E,E'),$$

where $E = \begin{pmatrix} I & I \\ 0 & I \end{pmatrix}$, $E' = \begin{pmatrix} I & 0 \\ -I & I \end{pmatrix}$, and I and 0 are the 2×2 identity and zero matrices, respectively. Then $\sigma\tau^{-1} = \operatorname{diag}(I,Q)$, where I is 4×4, and $Q = E(E')^{-1} = \begin{pmatrix} -I & I \\ I & I \end{pmatrix}$. Thus, $\sigma\tau^{-1}$ is a homology of A with axis W_0 and coaxis W_∞. Similarly, $\langle \sigma, \tau \rangle$ contains all homologies $\operatorname{diag}(I,Q)$, for $Q \in \mathcal{Q} = \pm \left\{ \begin{pmatrix} I & 0 \\ 0 & I \end{pmatrix}, \begin{pmatrix} -I & I \\ I & I \end{pmatrix}, \begin{pmatrix} I & I \\ I & -I \end{pmatrix}, \begin{pmatrix} 0 & I \\ -I & 0 \end{pmatrix} \right\}$, the quaternion group of order 8. Clearly, $F(\sigma) = \{(x_1, 0, y_1, 0)\}$ and $F(\tau) = \{(x_1, 0, 0, y_2)\}$, for all x_1, y_1, and y_2.

Directing our attention to the slope set, it is clear that $\langle \sigma, \tau \rangle$ induces the group G on $\mathcal{T}$, $G = \{T \to QE^{\mu}TE^{-\mu} : Q \in \mathcal{Q}$, E as above, $\mu = 0,1,2\}$. That is, σ induces conjugation by E, and the homologies $\operatorname{diag}(I,Q)$ induce left multiplication by $Q \in \mathcal{Q}$. Also, $G \simeq \langle \sigma, \tau \rangle \simeq SL(2,3)$, $\langle \sigma, \tau \rangle$ acts faithfully on W_∞, and $\langle \sigma, \tau \rangle$ induces $\mathcal{Q}$ on W_0.

The problem may be restated as: *find all slope sets* $\mathcal{T}$ *which are invariant under* G.

THE EXAMPLES: Let $c = \begin{pmatrix} 1 & -1 \\ 1 & 1 \end{pmatrix}$, a 2×2 matrix over GF(3). Then c is a primitive root of a field $K \simeq GF(9)$. The slope set $\mathcal{T}$ of the first example A_1 can be described as follows:

T_1 and T_2 are specific 2×2 matrices over K (i.e., 4×4 matrices over GF(3)); let $M_1 = G(T_1)$ and $M_2 = G(T_2)$ be the G-orbits of T_1 and T_2. It turns out that $|M_1| = |M_2| = 24$. Further, let $N_1 = \mathcal{Q}\{I, c^2 E\}$ and $N_2 = N_1 c$. That is, $N_1 = \{Q, c^2 QE : Q \in \mathcal{Q}\}$, and $N_2 = \{Tc = cT : T \in N_1\}$. These matrices are all 2×2 over K, and c is a scalar. Then $\mathcal{T} = M_1 \cup M_2 \cup N_1 \cup N_2 \cup \{0\}$ describes the plane A_1, and $\mathcal{T}$ admits G. Also, N_1 and N_2 each consist of two G-orbits of length 8.

LEMMA 1: $N_1 = \mathcal{Q}\{I, c^2 E\}$ *and* $N_1 c^2 = \mathcal{Q}\{c^2 I, E\}$ *are replaceable slope sets of degree* 16. *(That is, every line* $y = Tx$, *for* $T \in N_1$, *is completely covered by the lines of* $N_1 c^2$*; and conversely.)*

PROOF: For example, $c^2 I \in N_1 c^2$ is covered by N_1, as follows. If $Q \in \mathcal{Q}\setminus\{\pm I\}$, then $Q^2 = -I$, so Q has minimal polynomial $x^2 + 1$. Since $c^2 = -1$ in K, c^2 is an eigenvalue of Q, so $|Q - c^2 I| = 0$. Thus the line $L : y = c^2 x$ intersects each of the lines $y = Qx$, $Q \neq \pm I$, in a 1-dimensional K-subspace, accounting for 6 of the 10 1-dimensional K-subspaces of L. Further, L intersects each of the lines $y = c^2 QE$ for which QE has order 3 (and hence eigenvalue 1), accounting for the remaining K-subspaces. So L is covered by N_1.

COROLLARY: *By independently replacing* N_1 *and* N_2 *by* $N_1 c^2$ *and* $N_2 c^2$, *respectively, we obtain* 4 *planes of the required type, all of which have kernel* $K = GF(9)$.

REMARK: N_1 is a replaceable partial spread of $16 = 2(q-1)$ lines in PG(3,q), for $q = 9$. Therefore, N meets the minimum bound $2(q-1)$ of Bruen [1] for replaceable partial spreads which are larger than a regulus. Bruen shows that for q not too large, such minimal replaceable partial spreads are constructed from two reguli. However, N_1 cannot be constructed in this way, showing that some restriction on q is essential for Bruen's theorem.

LEMMA 2 : *Let* $N = N_1 \cup N_2$. *Then* N *is replaced by* $N\begin{pmatrix} f & 0 \\ 0 & f \end{pmatrix}$, *for each* $f \in \langle c, i \rangle$, *where* $i = \begin{pmatrix} -1 & 1 \\ 1 & 1 \end{pmatrix}$ *induces the automorphism* $c^i = c^3$ *of* K. *These replacements in* A_1 *result in* 8 *planes,* $A_1, \ldots, A_8$, *of the required type (i.e. admitting* G*).* *The replacements* Nc^ν $(\nu = 0,1,2,3)$ *result in the* 4 *planes of the Corollary. The planes* $Nc^{\nu i}$ *have kernel* $GF(3)$, *since* $i \notin K$. *Hence, there are at least* 2 *and (by some obvious isomorphisms) at most* 5 *non-isomorphic planes among these examples.*

PROOF : Computation.

Note: The planes $A_1, \ldots, A_4$, and in particular the replaceable partial spreads Nc^ν, can all be represented as sets of lines in $PG(3,9)$. But the partial spreads $Nc^{\nu i}$ do not appear in $PG(3,9)$; $PG(7,3)$ must be used instead.

COROLLARY : *If* T *is any slope of* N *(or of* $M_1 \cup M_2$*), then* $\{Tc^\nu : 0 \le \nu \le 7\} \cup \{0\}$ *is a regulus* R *covered by* N *(or by* $M_1 \cup M_2$*). But* $R \not\subset Nc^\nu$ *for any* ν.

LEMMA 3 : *Let* $M = M_1 \cup M_2$ *as above. Then* $MQ = M$, *for all* $Q \in \mathcal{Q}$. *Since* $N_1 \mathcal{Q} = \mathcal{Q} N_1$ *(because* E *normalizes* $\mathcal{Q}$*), then the slope sets of* $A_1, \ldots, A_8$ *admit the collineations* $\mathcal{Q}_R : T \to TQ$, $Q \in \mathcal{Q}$. *These collineations are homologies of the planes,* $(x,y) \to (Q^{-1}x, y)$, *with axis* W_∞ *and coaxis* W_0. *The group* $H = \langle G, \mathcal{Q}_R \rangle$ *has* $\mathcal{Q}_L \times \mathcal{Q}_R$ *as a normal subgroup.* H *contains* 16 3*-subgroups, giving rise to the corresponding* 16 *Baer fixed-point subplanes. These Baer subplanes are Desarguesian, by the way.*

LEMMA 4 : *The planes* $A_1, \ldots, A_8$ *contain many Baer nearfield* 9*-subplanes. In fact, each partial spread* $\mathcal{Q}T \cup \{0\} \cup \{\infty\}$ *and*

$T\mathcal{Q} \cup \{0\} \cup \{\infty\}$ *carries* 4 *such subplanes.*

LEMMA 5: *For* $Q \in \mathcal{Q} \setminus \{\pm I\}$, *the mapping* $\eta_Q : T \to QT^i c$ *is a Baer involution of* A_1. *These* 6 *involutions plus the homology* $T \to -T$ *form* 3 *Klein* 4-*groups. The group* $\langle \dot{G}, \mathcal{Q}_R, \eta_Q \rangle$ *has* 5 *orbits on* ℓ_∞ *in* A_1, *namely:* $M = M_1 \cup M_2$; 2 *orbits in* $N = N_1 \cup N_2$; $\{0\}$, *and* $\{\infty\}$.

LEMMA 6: $M = M_1 \cup M_2$ *can be partitioned into* 3 *replaceable slope sets of degree* 16, $M = R_1 \cup R_2 \cup R_3$. *Conjugation by* E *maps* $R_1 \to R_2 \to R_3$. *For example,* $R_1 \cong \mathcal{Q}\{I, c^2 \begin{pmatrix} I & 0 \\ 0 & -I \end{pmatrix}\}$ *is replaced by* $R_1 c^2$. R_1 *does not seem to be isomorphic to* N_1. *Replacing* R_1 *by* $R_1 c^2$ *in* A_1 *results in a plane which still admits* $\mathcal{Q}_L \times \mathcal{Q}_R$, *but not the Baer* 3-*collineations.*

LEMMA 7: *There is another plane admitting* G, *where* $T = M_1 \cup M_2' \cup N_1' \cup N_2' \cup \{0\}$. *This plane seems different from* $A_1, \ldots, A_8$. *Here* M_1 *is a subset of* T *for* A_1, *but* M_2', N_1', *and* N_2' *are not.* $N_1' \approx N_2'$ *have degree* 16 *and* N_1' *is replaceable, but* $N_1' \neq N_1$. *This plane does not admit* $\mathcal{Q}_R$.

BIBLIOGRAPHY

1. A.A. Bruen, "Some new replaceable translation nets", *Canad. J. Math.*, 29 (1977), 225-237.
2. D.A. Foulser, "Baer p-elements in translation planes", *J. Algebra*, 31 (1974), 354-366.

Department of Mathematics
University of Illinois at Chicago Circle
Chicago
Illinois 60680
U.S.A.

A NEW PARTIAL GEOMETRY CONSTRUCTED FROM THE HOFFMAN-SINGLETON GRAPH

Willem Haemers

We give the construction of a partial geometry with parameters $s = 4$, $t = 17$, $\alpha = 2$. We also obtain two new strongly regular graphs.

A (finite) *partial geometry* with parameters s, t and α is a $1-(v, s+1, t+1)$ design (for which we speak of *lines* rather than blocks), satisfying the following two conditions.

(i) Any two distinct lines have at most one point in common;

(ii) for any non-incident point-line pair (x, L) the number of lines containing x and intersecting L equals α.

A partial geometry is called *proper* if $1 < \alpha < \min\{s, t\}$ (this means that the geometry is not equivalent to a combinatorial object for which another name is more common). Partial geometries were introduced by Bose [2]. At that time no example of a proper one was known. In the meantime some construction methods for proper partial geometries have been found, see [15], [13], [10], [3], [5], [14], [7]. Only one of the known ones has $\alpha = 2$, viz. the sporadic one of van Lint and Schrijver [10]. Here we construct a second proper partial geometry with $\alpha = 2$, which is (up till now) sporadic too.

The *point graph* of a partial geometry is the graph whose vertices are the points, two vertices being adjacent whenever the two corresponding points lie on one line. We need to quote some results. The first one is well-known (see [2]) and easily verified.

RESULT 1 : *The point graph of a partial geometry with parameters* s, t, *and* α *is strongly regular with parameters* $((v, k, \lambda, \mu) = ((s+1)(st+\alpha)/\alpha,\ s(t+1),\ s-1+t(\alpha-1),\ \alpha(t+1)))$. □ (*)

A strongly regular graph is called *pseudo-geometric* if there exist integers s, t and α such that its parameters satisfy (*). The next two results are well-known (see [12] or [2]). For a proof of Result 4 we refer to [6]. For convenience, we call an eigenvalue *restricted* if it has an eigenvector different from the all-one vector.

RESULT 2 : *A graph is strongly regular iff its adjacency matrix has just two distinct restricted eigenvalues.* □

RESULT 3 : *For a strongly regular graph with parameters* v, k, λ, μ *and restricted eigenvalues* r_+, r_- $(r_+ > r_-)$ *we have*

$$\mu = k - r_+ r_- = \lambda - r_+ - r_- = (k - r_+)(k - r_-)/v\ .$$ □

RESULT 4 : *Let* G *be a regular graph of degree* k *with* v *vertices. Let* G_1 *be an induced regular subgraph of* G *with* v_1 *vertices and degree* k_1. *Let* r_+ *be the largest, and* r_- *be the smallest restricted eigenvalue of the adjacency matrix of* G. *Then*

(i) $r_+ \geq (vk_1 - v_1k)/(v - v_1) \geq r_-$,

(ii) *if equality holds on one of the sides then any vertex outside* G_1 *is adjacent to exactly* $v_1(k - k_1)/(v - v_1)$ *vertices of* G_1. □

The following lemma is known, but hard to find in the literature.

LEMMA 1 : *Let* G *be a pseudo-geometric graph corresponding to the parameters* s, t *and* α. *Let* $\sum$ *be a set of* (s+1)*-cliques (cliques with* s+1 *vertices) of* G. *If any two adjacent vertices of* G *are in exactly one clique of* $\sum$, *then the design with the* *vertices of* G *as points and the cliques of* Σ *as lines is a partial geometry with parameters* s, t *and* α, *having* G *as its point graph.*

PROOF: Only axiom (ii) is not trivially fulfilled. To prove this axiom we apply Result 4. Take G_1 to be a clique of Σ, then $k_1 = v_1 - 1 = s$. By use of Result 3 we have $r_+ = s - \alpha$. Now substitution in the left hand inequality shows that equality holds. Hence by (ii) any vertex outside G_1 is adjacent to $v_1(k - k_1)/(v - v_1) = \alpha$ vertices of G_1. This proves the lemma. □

We use a construction of the Hoffman-Singleton graph which is based on the following well-known result (see [4]).

RESULT 5 : *There exists a* 1 - 1 *correspondence between the* 35 *lines of* PG(3, 2) *and the* 35 *triples of a* 7-*set, such that lines intersect iff the corresponding triples have exactly one element in common.* □

Now the Hoffman-Singleton graph H is constructed as follows: The vertices are the 15 points together with the 35 lines of PG(3, 2). Points are mutually non-adjacent. A point is adjacent to a line whenever the point lies on that line. Two lines are adjacent whenever the corresponding two triples are disjoint.

LEMMA 2 : H *is strongly regular with parameters* (50, 7, 0, 1).

PROOF: The only step in the verification that may not be straightforward is to see that for a non-adjacent point-line pair there is exactly one line adjacent to both. This however follows easily after one has verified that a triple of the 7-set together with the four triples disjoint from it correspond to a spread in PG(3, 2). □

Hoffman and Singleton [8] first constructed and proved uniqueness of a strongly regular graph with the above parameters. Next we describe a known construction of another strongly regular graph G (see [9]). The vertices of G are the 175 edges of H. Two vertices of G are adjacent whenever the corresponding edges of H have distance two (i.e. the two edges are disjoint and there exists an edge connecting the two).

LEMMA 3: G *is strongly regular with parameters* (175, 72, 20, 36), *i.e.* G *is pseudo-geometric corresponding to the parameters* $s = 4$, $t = 17$, $\alpha = 2$.

PROOF: Let A, B and C be the adjacency matrices of G, H and the line graph of H respectively (observe that these three graphs are regular). Let N be the incidence matrix of H. Then it is easily seen that

$$NN^t = B + 7I, \quad N^tN = C + 2I,$$

$$A = C^2 - 5C - 12I.$$

B has restricted eigenvalues -3 and 2 (apply Result 3). Because NN^t and N^tN have the same nonzero eigenvalues, it follows that the restricted eigenvalues of C are -2, 2 and 7. Hence A has restricted eigenvalues 2 and -18. So by Result 2 the graph G is strongly regular. The parameters now readily follow by use of Result 3. □

The Hoffman-Singleton graph H contains many Petersen graphs as induced subgraphs. We define such a Petersen subgraph of H to be *special* if it consists of seven lines of PG(3, 2) corresponding to seven triples of the 7-set of the following form:

$$\begin{pmatrix} 1 & 1 & 1 & 0 & 0 & 0 & 0 \\ 1 & 0 & 0 & 1 & 1 & 0 & 0 \\ 0 & 1 & 0 & 1 & 1 & 0 & 0 \\ 0 & 0 & 1 & 1 & 1 & 0 & 0 \\ 1 & 0 & 0 & 0 & 0 & 1 & 1 \\ 0 & 1 & 0 & 0 & 0 & 1 & 1 \\ 0 & 0 & 1 & 0 & 0 & 1 & 1 \end{pmatrix},$$

together with the three points of PG(3, 2) that lie on the first of these seven lines. One readily verifies that these ten vertices of H indeed induce a Petersen graph.

LEMMA 4 : *Each pentagon of* H *is contained in exactly one special Petersen graph.*

PROOF : First we must realize that our description of H gives rise to two types of pentagons. The first type consists of three lines of PG(3, 2), whose triples have the form:

$$\begin{pmatrix} 1 & 1 & 1 & 0 & 0 & 0 & 0 \\ 1 & 0 & 0 & 1 & 1 & 0 & 0 \\ 0 & 1 & 0 & 0 & 0 & 1 & 1 \end{pmatrix},$$

together with two points of PG(3, 2), viz. the point in the intersection of the first and the second line, and the point in the intersection of the first and the third line. It is not hard to see that there is a unique way of completing this pentagon to a special Petersen graph. The second type of pentagon consists of four lines, whose triples have the form

$$\begin{pmatrix} 1 & 0 & 0 & 1 & 1 & 0 & 0 \\ 0 & 1 & 0 & 0 & 0 & 1 & 1 \\ 0 & 0 & 1 & 1 & 1 & 0 & 0 \\ 1 & 0 & 0 & 0 & 0 & 1 & 1 \end{pmatrix},$$

together with the unique vertex (which has to be a point of PG(3, 2)) adjacent to the first and the fourth line (here Lemma 2 is used). It is easy to see that there is a unique special

Petersen graph P containing the four lines. However, inside P there is also a unique vertex adjacent to the first and fourth line, so P contains the whole pentagon. □

A 1-*factor* of a Petersen graph is a set of five mutually disjoint edges.

LEMMA 5 : *The Petersen graph satisfies:*

(i) *any two edges of a* 1-*factor have distance two;*

(ii) *any two edges at distance two lie in a unique* 1-*factor.*

PROOF : By straightforward verification. □

Now we define the geometry S as follows. The points are the vertices of G, i.e. the edges of H. The lines are the 1-factors of the special Petersen graphs in H.

THEOREM : S *is a partial geometry with parameters* $s = 4$, $t = 17$, $\alpha = 2$.

PROOF : By (i) of Lemma 5 the lines of S are 5-cliques of G. By Lemma 3 G is pseudo-geometric corresponding to the above parameters. Hence, by Lemma 1, it suffices to prove that any edge of G lies on a unique line of S. Take two adjacent vertices of G, i.e. two edges x and y of H at distance two from each other. Then, by Lemma 2, x and y lie in a unique pentagon of H. By Lemma 4 this pentagon lies in a unique special Petersen graph. By (ii) of Lemma 5, inside this special Petersen graph, x and y lie in a unique 1-factor. This completes the proof. □

The *dual* geometry of S (the roles of points and lines are interchanged) is a partial geometry with parameters $s = 17$, $t = 4$, $\alpha = 2$. By Result 1 the *line graph* of S, which is the point graph of its dual, is strongly regular with parameters (630, 85, 20, 10). This strongly regular graph seems to be new. But the present setting leads to another new strongly regular graph. To show this we need another lemma.

LEMMA 6 : *Let* G_1 *be an induced regular subgraph of* G *on* v_1 *vertices of degree* k_1 . *Let* A_1 *and*

$$A = \begin{bmatrix} A_1 & M \\ M^t & A_2 \end{bmatrix}$$

be the adjacency matrices of G_1 *and* G . *Define*

$$A' = \begin{bmatrix} o & j^t & o \\ j & A_1 & J-M \\ o & J-M^t & A_2 \end{bmatrix}$$

where j *is the all-one vector, and* J *the all-one matrix. If* (v_1, k_1) *equals* $(70, 18)$ *or* $(90, 38)$ *then* G', *the graph with adjacency matrix* A', *is strongly regular with parameters* $(176, 70, 18, 34)$ *or* $(176, 90, 38, 54)$ *respectively.*

PROOF : Result 4 implies that not only A and A_1 , but also M and A_2 have constant row and column sums. Now the result follows, either by straightforward verification, or more elegantly, from the theory of regular two-graphs, see [11] . □

It is not difficult to see that the 70 edges of H that do not contain a point of $PG(3, 2)$ induce a regular subgraph of G of degree 18 . So, by Lemma 6, we have a strongly regular graph with parameters $(176, 70, 18, 34)$. This graph is known, see [9]. In order to prove existence of the other subgraph of Lemma 6 we need the description of the Hoffman-Singleton graph given in [1], where the vertex set is partitioned into two subsets of size 25 , both inducing 5 disjoint pentagons. Two pentagons, one from each subset, form a Petersen graph and the 1- factor in there connecting the two pentagons forms a 5- clique in G . This gives 25 disjoint 5- cliques in G . By Result 4, any vertex outside such a 5- clique is adjacent to exactly two vertices of that 5- clique. Thus any set of 18 disjoint 5- cliques of G gives a regular induced subgraph with 90 vertices of degree 38. Now, since $18 \le 25$, we have constructed, by Lemma 6, a strongly regular graph with parameters $(176, 90, 38, 54)$. This graph seems to be new. The amusing thing about this graph is that it is pseudo-geometric corresponding to the parameters $s = 5$, $t = 17$, $\alpha = 3$. We tried to construct this partial geometry in a way similar to that used for S, but did not succeed. The reason for this failure is that the set of 18 disjoint 5-cliques that we used does not look natural. If one finds a more natural set of 18 disjoint 5-cliques in G, then there is a good chance of success.

We conclude with a remark about the automorphism group of S. The automorphism group of H that fixes the partition into 15 points and 35 lines is the alternating group A_7. Clearly this group is also an automorphism group of S. It seems interesting that the action of A_7 on S is transitive on the 630 lines, but not transitive on the 175 points of S. We do not know whether A_7 is the full automorphism group of S.

ACKNOWLEDGEMENT: I thank J.H. van Lint for several helpful conversations on the subject.

BIBLIOGRAPHY

1. E.R. Berlekamp, J.H. van Lint and J.J. Seidel, "A strongly regular graph derived from the perfect ternary Golay code", *A Survey of Combinatorial Theory* (ed. J.N. Srivastava, North-Holland, Amsterdam, 1973), pp. 25-30.
2. R.C. Bose, "Strongly regular graphs, partial geometries, and partially balanced designs", *Pacific J. Math.*, 13 (1963), 389-419.
3. A.M. Cohen, "A new partial geometry with parameters $(s, t, \alpha) = (7, 8, 4)$", Report ZN 92/79, (Math. Centrum, Amsterdam, 1979).
4. G.M. Conwell, "The 3-space PG(3, 2) and its group", *Ann. Math.*, 11 (1910), 60-76.
5. F. de Clerck, R.H. Dye and J.A. Thas, "An infinite class of partial geometries associated with the hyperbolic quadric in PG(4n - 1, 2)", preprint.
6. W. Haemers, *Eigenvalue Techniques in Design and Graph Theory*, Tract 121 (Math. Centrum, Amsterdam, 1980).
7. W. Haemers and J.H. van Lint, "A partial geometry pg(9, 8, 4)", preprint.

8. A.J. Hoffman and R.R. Singleton, "On Moore graphs of diameter 2 and 3", *I.B.M. J. Res. Develop.*, 4 (1960), 497-504.
9. X.L. Hubaut, "Strongly regular graphs", *Discrete Math.*, 13 (1975), 357-381.
10. J.H. van Lint and A. Schrijver, "Construction of strongly regular graphs, two-weight codes and partial geometries by finite fields", *Combinatoria*, 1 (1980), to appear.
11. J.J. Seidel, "A survey of two-graphs", *Teorie Combinatorie*, Tomo I, Rome, 1973, Atti dei Convegni Lincei 17, (Accad. Naz. Lincei, Rome, 1976), pp. 481 - 511 .
12. J.J. Seidel, "Strongly regular graphs", *Surveys in Combinatorics*, London Math. Soc. Lecture Note Series 38 (ed. B. Bollobás, Cambridge Univ. Press, Cambridge, 1979), pp. 157 - 180 .
13. J.A. Thas, "Construction of maximal arcs and partial geometries", *Geom. Dedicata*, 3 (1974), 61-64.
14. J.A. Thas, "Some results on quadrics and a new class of partial geometries", preprint.
15. W.D. Wallis, "Configurations arising from maximal arcs", *J. Combin. Theory Ser. A*, 15 (1973), 115-119.

Department of Mathematics
University of Technology
P.O. Box 513
5600 MB Eindhoven
The Netherlands.

LOCALLY COTRIANGULAR GRAPHS

Jonathan I. Hall and Ernest E. Shult*

A graph G is said to be locally $\mathcal{X}$, where $\mathcal{X}$ is any collection of graphs if, for each vertex x in G, the subgraph $\Delta(x)$ of all vertices of G adjacent to x is isomorphic to some member of $\mathcal{X}$. In the case that $\mathcal{X}$ consists of a single graph X, we say G is locally X instead of locally $\{X\}$.

A graph G is said to possess the *cotriangle property* if for every pair (x, y) of non-adjacent vertices in G there exists a third vertex z forming a subgraph $T = \{x, y, z\}$ isomorphic to $\overline{K}_3$ (that is, a "cotriangle") having the property that any vertex u of G not lying in the cotriangle T is adjacent to exactly one or all of the vertices of T.

We write $G = A + B$, if A and B are disjoint subgraphs of G the union of whose vertices comprise the vertices of G, and if every vertex in A is adjacent to every vertex in B. Such a graph G possesses the cotriangle property if and only if both A and B do. We say G is *indecomposable* if it admits no such decomposition $G = A + B$ into non-empty subgraphs A and B.

Finally, for any graph G and vertex x, we write $\Delta(x)$ for the subgraph of vertices adjacent to x, and $x^{\perp}$ for $\{x\} \cup \Delta(x)$. Vertices x and y are said to be *equivalent* (written $x \approx y$) if and only if $x^{\perp} = y^{\perp}$. Clearly $\approx$ is an equivalence relation on the vertices of G and if C_i and C_j are any two of its equivalence classes, then either every vertex in C_i is adjacent to every vertex of C_j or else none are. The C_i are themselves cliques in G and one may form a graph G^* whose vertices are the C_i and whose edges are the pairs (C_i, C_j) for which the former alternative above holds; that is, $C_i \cup C_j = C_i + C_j$.

There is a natural homomorphism

$$G \xrightarrow{\ *\ } G^*$$

as graphs. It is clear that a graph G satisfies the cotriangle property if and only if G^* does. (If $G = G^*$, each of the relevant cotriangles is uniquely determined by any pair of its three non-adjacent vertices.)

It follows from the remarks above that in classifying graphs with the cotriangle property, one need only consider graphs which are *reduced;* that is, indecomposable graphs G satisfying $G = G^*$. The following theorem has been known for some time [3].

COTRIANGLE THEOREM: If G is a reduced graph possessing the cotriangle property, then either G consists of a single vertex, or else G is one of the graphs $\overline{T}_n$ $(n \geq 3)$, $Sp(2n, 2)$ $(n \geq 4)$, or N_n^{ε} $(n \geq 3)$.

Here $\overline{T}_n$ is the complement of the triangular graph; its vertices are all unordered pairs drawn from n letters, two being adjacent if they have no letter in common. $Sp(2n, 2)$ is the graph of the perpendicular relation induced on the non-zero vectors of a vector space V of dimension $2n$ over $GF(2)$ induced by a non-degenerate symplectic form $B : V \times V \longrightarrow GF(2)$. There are quadratic forms $Q : V \longrightarrow GF(2)$ which have B as their associated symplectic form, and for each of these there is a partition of $Sp(2n, 2)$ into two subgraphs N_{2n}^{ε} and S_{2n}^{ε} of vectors achieving value 1 or 0 under Q; respectively. (These are the subgraphs of *non-singular* and *singular* vectors.) For all choices of Q under a fixed B, only two isomorphism types of partitions appear and the type is determined by the symbol $\varepsilon = +$ or $-$ which reflects whether the quadric in $PG(2n-1, 2)$ represented by the subgraph S_{2n} is a polar space of maximal Witt index or not.

Let $\mathcal{D}$ be the class of all indecomposable cotriangular graphs. These are the graphs G such that G^* is one of the graphs appearing in the conclusion of the cotriangle theorem, excluding K_1 and $\overline{K}_3$.

The main result of this work is the complete classification of all graphs which are locally $\mathcal{D}$. (One may note that if $\mathcal{D}'$ denotes the "degenerate" class of graphs G which are K_1 or satisfy $G^* \simeq \overline{K}_3$ (i.e. $G \simeq \overline{K}_{r,s,t}$ for various r, s and t): it is virtually hopeless to classify graphs which are locally $\mathcal{D}'$, for these correspond to the set of all (0, 1)-matrices with row sum 1 or 3 and row inner products 0 or 1. Although the complete list of indecomposable cotriangular graphs is $\mathcal{D} \cup \mathcal{D}'$; it is possible to separate the hopeless degenerate cases from the rest in our local characterization problem because of the following lemma: If G is connected and locally $\mathcal{D} \cup \mathcal{D}'$, then either G is locally $\mathcal{D}$ or G is locally $\mathcal{D}'$.

The classification of locally $\mathcal{D}$ graphs is a consequence of the following theorems.

THEOREM 1: *Let* G *be connected and locally* $\mathcal{D}$. *Then either*

(1) *there is a reduced graph* C *in* $\mathcal{D}$ *with* G *locally* C;

or (2) *there exists an integer* n *such that for each vertex* t *in* G, $\Delta(t)^* \simeq Sp(2n, 2)$, N^+_{2n}, N^-_{2n}, *the choice depending on* t. □

In case (1) of Theorem 1, $C \simeq \overline{T}_n$, $n \geq 5$, N^{ε}_{2n}, $n \geq 3$, or $Sp(2n, 2)$, $n \geq 2$. The first two cases are handled by the next two theorems.

THEOREM 2: *Assume* G *is connected and locally* $\overline{T}_n$, $n \geq 5$. *Then either*

(1) $n = 5$, $G \simeq \overline{T}_7$, $3 \cdot \overline{T}_7$ or $\Gamma L(2, 5^2)$;

(2) $n = 6$, $G \simeq \overline{T}_8$, N^-_6 or $Sp(6, 2)$ *minus a subgraph* $x^{\perp}$ *for any vertex* x;

or (3) $n \geq 7$, $G \simeq \overline{T}_{n+2}$. □

The case $n = 5$ was proved in [2], where the graphs denoted in (1) are fully described.

THEOREM 3: *If* G *is connected and locally* N_n^{ε} *for* $n \geq 3$, *then* $n = 3$ *and only the following two cases occur:*

(1) $\varepsilon = +$, $G \simeq \overline{T}_{10}$;

(2) $\varepsilon = -$, $G \simeq N_6^+(3)$, *the* $\perp$-*graph on the* 117 *points of* PG(5, 3) *lying off a quadric of maximal Witt index, and having square norm.* □

There remains the case $C \simeq Sp(2n, 2)$ in case (1) of Theorem 1. Although this case is covered by the theorem of Buekenhout and Hubaut on locally polar spaces, [1], it is also a consequence of the following more general theorem which includes, for example, the characterization of $Sp(2n, 2)$ as a graph which is locally $Sp(2n-2, 2)$, [2].

THEOREM 4: *Let* G *be connected and suppose that for each vertex* t *in* G, $\Delta(t)^*$ *is the collinearity graph of a non-degenerate polar space of rank at least* 2 *and order* 2. *Then* G *is obtained by removing from the collinearity graph of a polar space* Γ *the subgraph corresponding to a polar subspace* P *of* Γ, *that is* $G = \Gamma - P$. □

Besides handling the last case of Theorem 1 (1), this theorem does double duty in narrowing down the cases to be considered in Theorem 1 (2). It allows us to assume, for example, that an instance in which $\Delta(t)^* \simeq N_{2n}^{\varepsilon}$ actually occurs in part (2) of Theorem 1. In addition, it should be noted that Theorem 4 even surpasses the scope of the graphs being considered. For example, if we subtract from a non-degenerate quadric the points lying in a non-degenerate subspace we obtain the graph $G = S_{2n}^{\varepsilon} - S_{2k}^{'\varepsilon}$ which is not locally cotriangular ($S_7^+ - S_5^+$ contains 100 vertices and is $*$-locally S_3^+, with local valencies 35 and 55 for example.) With the aid of Theorem 4, one obtains:

THEOREM 5: *Let* G *be connected and locally* $\mathcal{D}$ *but not locally* $\mathcal{R} = \{ Sp(2n, 2)\ n \geq 3,\ N_{2n}^{\varepsilon},\ \varepsilon = +, -, n \geq 3 \}$. *Then there is a fixed integer* $n \geq 2$ *such that either*

(1) G *is* $Sp(2n+2, 2)$ *with a polar subspace deleted;*

(2) $G \simeq G_{2n+2}^{\varepsilon}$ *for some* $\varepsilon = +$ *or* $-$;

or (3) $G \simeq H_{2n+2}^{\varepsilon}(T)$ *for some* $\varepsilon = +$ *or* $-$ *and some coclique* T *of* S_{2n+2}^{ε}. □

Here, G_{2n+2}^{ε} is formed in the following way. Set $\Gamma = Sp(2n+2, 2)$ and fix a vertex t in Γ. Then $\Delta(t)^{*} = Sp(2n, 2) = S_{2n}^{\varepsilon} + N_{2n}^{\varepsilon}$ for some orthogonal partition into singular and non-singular elements. Then we may form preimages $Sp(2n, 2)\ [2] = \Delta(t) = \Delta_S(t) + \Delta_N(t)$ with $\Delta_S(t)^{*} = S_{2n}^{\varepsilon}$. Then $P = t \cup \Delta_S(t)$ is a polar subspace of Γ. Now G_{2n+2}^{ε} is defined to be the graph $\{t\} \cup (\Gamma - P)$. To form H_{2n+2}^{ε}, let $\Gamma = Sp(2n+2, 2) = S_{2n+2}^{\varepsilon} + N_{2n+2}^{\varepsilon}$ be a partition of Γ into its subgraphs of singular and non-singular vectors with respect to a quadratic form Q of type ε. Let T be a coclique chosen within S_{2n+2}^{ε}. Then H_{2n+2}^{ε} is the subgraph of Γ induced on the vertex set $T \cup N_{2n+2}^{\varepsilon}$.

These theorems together imply the

MAIN THEOREM: *If* G *is a graph which is locally* $\mathcal{D}$, *then* G *has its connected components isomorphic to one of*

(1) $\overline{T}_n$, $n \geq 7$;

(2) $Sp(2n, 2)$ *minus a polar subspace (possibly empty);*

(3) $H_{2n}^{\varepsilon}(T)$, *for some* T ;

(4) G_{2n}^{ε} ;

(5) $3 \cdot \overline{T}_7$, $\Gamma L(2, 5^2)$ *or* $N_6^{+}(3)$. □

(Note that case (2) includes the graphs $\hat{Sp}(2n, 2)$, N_{2n}^{ε} and $Sp(2n, 2)$ itself.)

BIBLIOGRAPHY

1. F. Buekenhout and X. Hubaut, "Locally polar spaces and related rank 3 groups", *J. Algebra,* 45 (1977), 391-434.
2. J.I. Hall, "Locally Petersen graphs", to appear.
3. E.E. Shult, "Groups, polar spaces and related structures", *Combinatorics,* Tract 57 (eds. M. Hall and J.H. van Lint, Math. Centre, Amsterdam), pp. 130-161.

Department of Mathematics
Michigan State University
East Lansing
Michigan 48824
U.S.A.

Department of Mathematics
Kansas State University
Manhattan
Kansas 66506
U.S.A.

CODING THEORY OF DESIGNS

Marshall Hall Jr.

1. INTRODUCTION

The amount of information which coding theory gives about a design can vary enormously from one design to another. As it is now known that a plane of order 10 can have only the identical collineation, its binary code is now our main source of information on it.

Section 2 gives the definition of the code of a design in terms of its incidence matrix.

For a (v, b, r, k, λ) design D it is advantageous to consider codes over finite fields F_q, where $q = p^t$, p a prime dividing $r - \lambda$, and in Section 3 theorems dealing with this are given. The Thompson-Hayden theorems of this section deal with a vector space V over a field F and the action of a finite orthogonal group G whose order has an inverse in F. We are concerned with the action of G on a subspace C of V which is a G-module. This is particularly appropriate when C is a code and G a group of automorphisms of C.

In Section 4 the role of coding theory is dicussed in connection with the plane of order 10 and the construction of a $(41, 16, 6)$ design.

2. THE CODE OF A DESIGN

A balanced incomplete block design D, or more briefly design, is a system of v points $a_1, \ldots, a_v$ and b blocks $B_1, \ldots, B_b$ together with an incidence relation $a_i \in B_j$ (read a_i belongs to B_j or B_j contains a_i) between certain points a_i and blocks B_j. Each block contains exactly k points and each point is on exactly r blocks. Every pair of distinct points

a_i, a_s occur together in exactly λ blocks. The general theory of designs will be presupposed here and can be found in Chapter 10 of the writer's book [3] or in other standard sources.

The five parameters satisfy two simple relations

$$bk = vr, \quad r(k-1) = \lambda(v-1). \tag{2.1}$$

The incidence matrix A of the design D is given by

$$A = [a_{ij}], \quad i = 1,\dots,v, \quad j = 1,\dots,b; \tag{2.2}$$

$$a_{ij} = 1 \text{ if } a_i \in B_j, \quad a_{ij} = 0 \text{ if } a_i \notin B_j.$$

The matrix A satisfies the relations

$$\begin{aligned} AA^T &= B = (r-\lambda)I + \lambda J_v, \\ J_v A &= k J_{v,b}, \quad A J_b = r J_{v,b}. \end{aligned} \tag{2.3}$$

Here A^T is the transpose of A, I is the identity matrix of order v, J_v and J_b are the square matrices of orders v and b respectively with all entries 1, while $J_{v,b}$ is the v by b matrix with all entries 1.

Let $F_q = GF(q)$ be the finite field with q elements where $q = p^s$ is a prime power. Let $V = F_q^b$ be a vector space of dimension b over F_q. If A is the v by b incidence matrix of a (v, b, r, k, λ) design D then we define the *code of* D *over* F_q as the linear subspace C of V spanned by the rows of A considered as vectors over F_q.

3. PROPERTIES OF LINEAR CODES AND GENERAL APPLICATIONS TO DESIGNS

Let V be the vector space F_q^n of dimension n over $GF(q)$ so that a typical vector x has the shape

$$x = (x_1,\dots,x_n), \quad x_i \in GF(q), \quad i = 1,\dots,n. \tag{3.1}$$

An $[n, s]$ *code* C over F_q is a linear subspace of F_q^n of dimension s. The vectors of C are called *codewords* or more briefly *words*.

The *(Hamming) weight* of a vector $x = (x_1,\dots,x_n) \in F_q^n$ denoted by $wt(x)$ is the number of non-zero x_i and the

(Hamming) distance between vectors x, y is $d_H(x,y) = wt(x-y)$. If every non-zero codeword in C has weight $\geq d$, the code is said to have minimum weight d.

The *dual code* $C^\perp$ is the subspace orthogonal to C.

$$C^\perp = \left\{ u \mid (u,v) = \sum_{i=1}^{n} u_i v_i = 0 \quad \text{for all} \quad v \in C \right\} . \tag{3.2}$$

Here $C^\perp$ is an $[n, n-s]$ code.

If $C \subseteq C^\perp$, C is called *self-orthogonal* while if $C = C^\perp$ it is called *self-dual*.

Let A_i be the number of codewords of C of weight i. Then the set $(A_0, A_1, \ldots, A_n)$ is called the *weight distribution* of C. It is convenient to associate a polynomial with these numbers, the *weight enumerator* $W_C(x,y)$ where

$$W_C(x,y) = A_0 x^n + A_1 x^{n-1} y + \ldots + A_i x^{n-i} y^i + \ldots + A_n y^n . \tag{3.3}$$

An amazing result, due to MacWilliams is that the weight enumerator of $C^\perp$ is completely determined by the weight enumerator of C.

THEOREM 3.1 : (MacWilliams) *Let* C *be an* $[n,s]$ *code over* $GF(q)$. *Then*

$$W_{C^\perp}(x,y) = W_C(x + (q-1)y, x-y) / q^s . \tag{3.4}$$

The special case with $GF(q) = GF(2)$ for the binary codes $[n,s]$ gives

$$W_{C^\perp}(x,y) = W_C(x+y, x-y) / 2^s . \tag{3.5}$$

From (2.3) for the incidence matrix A of a design D

$$AA^T = B = (r - \lambda) I + \lambda J_v , \tag{3.6}$$

we can easily find the determinant of B

$$\det B = (r-\lambda)^{v-1} (r + v\lambda - \lambda) = (r - \lambda)^{v-1} r k . \tag{3.7}$$

We consider the code C of D over $GF(q)$ where $q = p^t$ and the prime p divides $r - \lambda$. Let r_i, r_j, r_s be any three rows of A. Then

$$(r_i, r_i) \equiv (r_i, r_j) \equiv \lambda \pmod{p}, \qquad (3.8)$$
$$(r_i - r_j, r_s) \equiv 0 \pmod{p} \quad \text{all } i, j, s.$$

From this we readily find

LEMMA 3.1 : *In the code* C *of a* (v, b, r, k, λ) *design* D *over* F_q, *with* $q = p^t$, *if* p *divides* $r - \lambda$ *then*

(i) *if* $p \mid r$, $p \mid \lambda$, *then* $C \subseteq C^{\perp}$

and (ii) *if* $p \nmid r$, $p \nmid \lambda$, *then* $C \cap C^{\perp}$ *is of codimension one in* C.

A design D is called symmetric if $v = b$ and so also from (2.1) $r = k$, and we speak of a symmetric (v, k, λ) design D. If $q = k^t$ and p divides $k - \lambda$ to exactly the first power, then Lemma 3.1 gives valuable information on the code C, which we state as a theorem.

THEOREM 3.2 : *Let* D *be a symmetric* (v, k, λ) *design and let* C *be its code over* F_q *where* $q = p^t$ *and the prime* p *divides* $k - \lambda$ *to exactly the first power. Then if* $p \nmid k$, $p \nmid \lambda$

$$\dim C = (v+1)/2, \quad \dim C^{\perp} = (v-1)/2 \quad \text{and} \quad C \supset C^{\perp}.$$

If $p \mid k$, $p \mid \lambda$ *then*

$$\dim C = (v-1)/2, \quad \dim C^{\perp} = (v+1)/2 \quad \text{and} \quad C \subset C^{\perp}.$$

Here $\dim C$ *is the dimension of* C *over* F_q.

Given a (v, b, r, k, λ) design D, an automorphism α of D is a one to one mapping of points onto points and of blocks onto blocks such that if $a_i \in B_j$ then $(a_i)\alpha \in (B_j)\alpha$. Clearly α takes the code C of D over F_q into itself.

The words of C fixed by every automorphism of a group G of automorphisms will themselves form a code C_0. Here the words of the entire vector space V fixed by G form a code V_0. Clearly we may consider C_0 as a subcode of V_0 in a natural way, and this has been called a "collapsed code". A result by Thompson on collapsed codes was useful in showing that a plane of order 10 cannot have a collineation of order 5, [1]. This result has been greatly generalized by Hayden and so it seems appropriate to call these the Thompson-Hayden theorems.

A group G of linear transformations of V is said to be orthogonal if it leaves inner products unchanged

$$(xg, yg) = (x,y) \quad \text{for all vectors } x, y \text{ and all } g \in G. \tag{3.9}$$

(THOMPSON-HAYDEN THEOREMS)

THEOREM 3.3 : *Let* V *be an* n*-dimensional vector space over a field* F. *Assume that* G *is a finite orthogonal group acting on* V *and that its order* $|G|$ *has an inverse in* F. *Suppose* $C \subset V$ *is a* G*-module. Define* θ *on* V *by*

$$v\theta = \sum_{g \in G} vg/|G| \tag{3.10}$$

and set $H = C\theta$. *Then*

$$H^{\perp} = C^{\perp}\theta \oplus \operatorname{Ker}\theta . \tag{3.11}$$

Here $\operatorname{Ker}\theta$ *is the subspace of* $V\theta$ *mapped onto zero by* θ.

For the next theorem a stronger assumption is needed, namely that G is a permutation group on the coordinates of V. If C is the code of a design D and G is an automorphism group of D, then this assumption holds, since the coordinates may be associated with the blocks of the design.

THEOREM 3.4 : *Let* V *be an* n*-dimensional vector space over a field* F. *Assume that* G *is a permutation group on the coordinates of* V *and that* $|G|$ *has an inverse in* F. *Suppose that* $C \subset V$ *is a* G *module. With* θ *as in Theorem 3.3,*

$$v\theta = vg/|G| \quad \textit{and} \quad H = C\theta .$$

Let $W = V\theta$. *Then with an appropriate orthonormal base for* W *(extending* F *if necessary) we have (where* H_W *is the dual in terms of this basis)*

$$(C\theta)^{\perp}_{W} = H^{\perp}_{W} = (C^{\perp})\theta .$$

Here the orbits of G on V partition the coordinates of V and if C_i is the ith orbit of length m_i , then putting C as the vector with coordinate $1/\sqrt{m_i}$ for every point of the ith orbit, the C_i will be an orthonormal basis for W . Since m_i is a divisor of $|G|$, so $\sqrt{m_i}$ has an inverse in F , extended if necessary.

4. EXAMPLE OF CODING THEORY OF DESIGNS

The paper by MacWilliams, Sloane and Thompson [5] was an important milestone in the application of coding theory to designs. The projective plane of order 10 is the symmetric design $\pi(111, 11, 1)$ and its order $11 - 1 = 10$ is the smallest order of a plane whose existence is still unknown. It is convenient to think of the columns of the incidence matrix A as corresponding to the 111 points of π and the 111 rows as corresponding to the lines of π . Theorem 3.2 applies as 2 divides 10 to exactly the first power and so the code C of π over F_2 has dimension 56 over F_2 and $C^{\perp}$ has dimension 55 and $C \supset C^{\perp}$. It is not difficult to show that if a word x of C is the sum of an odd number of lines, then $wt(x) \equiv 3 \pmod 4$ and every line contains an odd number of points of x , and $x \notin C^{\perp}$. If x is the sum of an even number of lines then $wt(x) \equiv 0\ (4)$ and every line contains an even number of points of x (including zero points), and $x \in C$. Since every point is on 11 lines the sum of all rows of A is a vector of all 11's and over F_2 the vector $e = (1, \dots, 1)$. Since $e \in C$, then for any $x \in C$, $e + x \in C$ but $e + x = \bar{x}$ is the complement of x, and $wt(x) + wt(\bar{x}) = 111$. In particular for the weight distribution $(A_0, A_1, \dots, A_{111})$ we have $A_{111-i} = A_i$. It is also easily shown that there are no non-zero words of weight less than 11 and that for weight 11 there are precisely the 111 lines. Using this information and the MacWilliams identity of Theorem 3.1, they were able to express the entire weight distribution in terms of A_{12}, A_{15} and A_{16}. Every word x of C corresponds to the set of points K_x for which the coordinate of x is 1. The lines containing 2 or more points of K_x together with these points form a configuration. For a word of weight 15 there is up to isomorphism only one configuration. A machine search showed that this configuration could not be completed to a full plane. Hence $A_{15} = 0$.

Whiteside [7] has eliminated the possibility of a collineation of order 11 in a plane of order 10. She has also [8] eliminated the orders 9, 25 and 15. This leaves as the only possible orders of collineation groups 1, 3 or 5. In [1], it is shown that a plane of order 10 cannot have a collineation of order 5. Hughes [4] has shown that a collineation of order 5 must have exactly one fixed point and exactly one fixed line. Thus on points and also on lines there are 22 orbits of length 5 and one of length 1. We can adjoin to the 111 points of π a new coordinate and use it as a parity check assigning to the coordinate the value 0 or 1 for each word to make the new weight even. We find that the new code C^* is a self dual code of dimension 56 in V^* of dimension 112. Here the new point is fixed by the collineation and so V^* has 22 orbits of length 5 and 2 of length 1. Hence the collapsed code is a self dual code on a space of dimension 24, following the result of Thompson referred to in the previous section. All such codes have been listed by Pless and Sloane [6] and it was possible to decide exactly which one it had to be. A consequence was that one of the words of the original code C of weight 20 must be made up of four of the orbits of length 5.

Without loss we may consider the points of this word to be 1,..., 20 and the action of the collineation on these 20 to be

$$\alpha = (1,2,3,4,5,)(6,7,8,9,10)(11,12,13,14,15)(16,17,18,19,20) \tag{4.1}$$

Given a word x of C we shall call a line a *heavy line* if it contains more than two points of x. If $wt(x) > 12$ then there are always some heavy lines, and given the heavy lines corresponding to x the intersections of the remaining lines with x are fully determined.

The heavy lines of one such word of weight 20 are

1, 2, 6, 8,11,16	11,12,18,20	
2, 3, 7, 9,12,17	12,13,19,16	
3, 4, 8,10,13,18	13,14,20,17	(4.2)
4, 5, 9, 6,14,19	14,15,16,18	
5, 1,10, 7,15,20	15,11,17,19	

All the remaining possible cases are words x for which there are 20 heavy lines each containing 4 points of x. Furthermore, each point of x is on exactly 4 of the heavy lines. One such case is

1, 2, 6, 8	1, 3,11,12	16,17,11,13	16,18, 8, 9
2, 3, 7, 9	2, 4,12,13	17,18,12,14	17,19, 9,10
3, 4, 8,10	3, 5,13,14	18,19,13,15	18,20,10, 6
4, 5, 9, 6	4, 1,14,15	19,20,14,11	19,16, 6, 7
5, 1,10, 7	5, 2,15,11	20,16,15,12	20,17, 7, 8

(4.3)

There were 91 cases, up to isomorphism, similar to (4.3). A computer search showed that neither (4.2) nor any of the 91 cases could be completed to a full plane. This showed that there could not be a collineation of order 5.

In a personal communication, Zvonimir Janko informs me that he has shown that a plane of order 10 cannot have a collineation of order 3. This means that a plane of order 10 has only the identity collineation. This leaves coding theory as the main tool in proving or disproving the existence of a plane of order 10.

It is gratifying that in the opposite direction coding theory has let to the construction [2] of a symmetric (41,16,6) design. In this case Theorem 3.2 applies over F_2. Since every element of C has weight a multiple of 4, it will not contain $e_{41} = (1,\dots,1)$. But clearly $e_{41} \in C^{\perp}$ so that $C^{\perp} = \langle C, e_{41} \rangle$. The weight distribution is given by

$$\begin{aligned}
A_0 &= 1 \\
A_4 &= A_4 \\
A_8 &= 164 + 13A_4 + 11A_{36} \\
A_{12} &= 15088 + 125A_4 + 33A_{36} \\
A_{16} &= 196718 - 599A_4 - A_{36} \\
A_{20} &= 512992 + 851A_4 + 115A_{36} \\
A_{24} &= 289460 - 425A_4 - 175A_{36} \\
A_{28} &= 33456 - 17A_4 + 109A_{36} \\
A_{32} &= 697 + 51A_4 - 27A_{36} \\
A_{36} &= A_{36}
\end{aligned} \tag{4.4}$$

It can be shown that a (41,16,6) design D cannot have a collineation α of prime order p if $p > 5$, and if $p = 5$, α fixes exactly one block B_0, and one point X. Since α cannot fix a word of weight 4 or 8 we have $A_4 \equiv 0(5)$, $A_8 \equiv 0(5)$. For the entire weight distribution we have modulo 5

$$\begin{aligned}
A_0 &\equiv 1 \\
A_4 &\equiv 0 \\
A_8 &\equiv 0 \\
A_{12} &\equiv 0 \\
A_{16} &\equiv 2 \\
A_{20} &\equiv 2 \\
A_{24} &\equiv 0 \\
A_{28} &\equiv 0 \\
A_{32} &\equiv 0 \\
A_{36} &\equiv 1
\end{aligned} \tag{4.5}$$

We find in terms of Theorems 3.3 and 3.4 that the orbit space W is of dimension 9 consisting of X and eight 5 orbits $C_1,\dots,C_8$. There is a unique word W_{36} fixed by α which will be $X, C_1, \ldots, C_7$ with appropriate numbering. In $H = C\theta$ we will have, with the C's appropriately numbered:

$$\begin{array}{lcllllllll}
W_{36} & = & X & C_1 & C_2 & C_3 & C_4 & C_5 & C_6 & C_7 \\
B_0 & = & X & C_1 & C_2 & C_3 & & & & \\
F_2 & = & X & C_1 & & & C_4 & C_5 & & \\
F_3 & = & X & C_1 & & & & & C_6 & C_7 \\
F_4 & = & X & & C_2 & & C_4 & & C_6 & \\
F_5 & = & X & & C_2 & & & C_5 & & C_7 \\
F_6 & = & X & & & C_3 & C_4 & & & C_7 \\
F_7 & = & X & & & C_3 & & C_5 & C_6 &
\end{array} \qquad (4.6)$$

H is of dimension 4 over F_2 and apart from the 8 elements in (4.6) contains the sum of W_{36} with these. Here $H^{\perp} = \langle H, e_{41} \rangle$.

For the dual design, the same considerations apply to blocks as well as to points. Let the blocks be $B_0, B_1, \ldots, B_{35}, B_{36}, \ldots, B_{40}$ where $B_0, B_1, \ldots, B_{35}$ is the unique fixed word of weight 36. We can construct our *orbit matrix* $K = [k_{ij}]$ where k_{ij} is the number of points of the ith representative block which lie in the jth orbit. One of these is

	X	C_1	C_2	C_3	C_4	C_5	C_6	C_7	C_8
B_0	1	5	5	5	0	0	0	0	0
B_1	1	1	2	2	1	1	4	2	2
B_6	1	2	1	2	4	1	1	2	2
B_{11}	1	2	2	1	1	4	1	2	2
B_{16}	0	0	3	3	2	2	1	3	2
B_{21}	0	3	0	3	1	2	2	3	2
B_{26}	0	3	3	0	2	1	2	3	2
B_{31}	0	2	2	2	3	3	3	1	0
B_{36}	0	2	2	2	2	2	2	0	4

(4.7)

The appearance of this matrix suggests a possible further collineation β of order 3 commuting with α. On the points these collineations are

$$\alpha = (X)(\ 1,\ 2,\ 3,\ 4,\ 5)(\ 6,\ 7,\ 8,\ 9,10)(11,12,13,14,15)(16,17,18,19,20)$$
$$(21,22,23,24,25)(26,27,28,29,30)(31,32,33,34,35)(36,37,38,39,40). \qquad (4.8)$$

$$\beta = (X)(1,6,11)(2,7,12)(3,8,13)(4,9,14)(5,10,15)(16,21,26)(17,22,27)$$
$$(18,23,28)(19,24,29)(20,25,30)(31)(32)(33)(34)(35)(36)(37)(38)(39)(40).$$

It is found that a (41,16,6) design D exists with these collineations fitting the orbit matrix (4.7).

Representative blocks are

B_0 = X, 1, 2, 3, 4, 5, 6, 7, 8, 9,10,11,12,13,14,15
B_1 = X, 4, 6, 7,13,15,16,21,27,28,29,30,31,32,36,38
B_6 = X, 3, 5, 9,11,12,17,18,19,20,21,26,31,32,36,38
B_{11} = X, 1, 2, 8,10,14,16,22,23,24,25,26,31,32,36,38
B_{16} = 6, 7, 9,13,14,15,18,19,22,25,26,31,32,34,39,40
B_{21} = 3, 4, 5,11,12,14,16,23,24,27,30,31,32,34,39,40
B_{26} = 1, 2, 4, 8, 9,10,17,20,21,28,29,31,32,34,39,40
B_{31} = 1, 3, 6, 8,11,13,18,19,20,23,24,25,28,29,30,32
B_{36} = 1, 2, 6, 7,11,12,17,20,22,25,27,30,36,38,39,40 .

BIBLIOGRAPHY

1. R.P. Anstee, M. Hall Jr., and J.G. Thompson, "Planes of order 10 do not have a collineation of order 5", *J. Combin. Theory Ser. A*, 29 (1980), 39-58.
2. W.G. Bridges, M. Hall Jr., and J.L. Hayden, "Codes and designs", *J. Combin. Theory Ser. A*, to appear.
3. M. Hall Jr., *Combinatorial Theory* (John Wiley, New York, 1967).
4. D.R. Hughes, "Collineations and generalized incidence matrices", *Trans. Amer. Math. Soc.*, 86 (1957), 284-296.
5. F.J. MacWilliams, N.J.A. Sloane and J.G. Thompson, "On the existence of a projective plane of order 10", *J. Combin. Theory Ser. A.*, 14 (1973), 66-78.
6. V. Pless and N.J.A. Sloane, "On the classification and enumeration of self-dual codes", *J. Combin. Theory Ser. A.*, 18 (1975), 313-335.

7. S.H. Whitesides, "Projective planes of order 10 have no collineations of order 11", *Seventh S.E. Conf. on Combinatorics, Graph Theory and Computing*, Louisiana State Univ. 1976, (Utilitas Math., Winnipeg, 1976), pp.65-76.

8. S.H. Whitesides, "Collineations of projective planes of order 10, Parts I, II", *J. Combin. Theory Ser. A.*, 26 (1979), 249-268; 269-277.

A.P. Sloan Laboratory of Mathematics and Physics
California Institute of Technology
Pasadena
California 91125
U.S.A.

ON SHEARS IN FIXED-POINT-FREE AFFINE GROUPS

Christoph Hering

Let A be an affine plane and G a finite group of automorphisms of A. In this paper we consider the case that G contains a non-trivial shear and fixes no point and no direction. Let S be the subgroup generated by all elations and T the group consisting of all translations in G. If $T \neq 1$, then the structure of S is determined in [3]. It follows that T is a p-group for some prime p, and if q is the p-part of $|S/T|$, then S/T is isomorphic to SL(2,q) or $Sz(\sqrt{q})$ unless $q = 3$ and $S/T \cong SL(2,5)$, or $q = 2$. Also, it follows that A contains Desarguesian subplanes or Lüneburg planes of order q. We assume here that $T = 1$, which apparently is a quite rare occurrence. Let $(\overline{P}, \overline{L})$ be the substructure generated by all centres and axes of non-trivial shears in G, and let the representation of G on $(\overline{P}, \overline{L})$ be denoted by $^{-}$. We obtain a structure theorem analogous to [4, Theorem 5.5]. It follows that $\overline{G}$ contains exactly one minimal normal subgroup M, and that M is a non-abelian simple group. Also, $C_{\overline{G}}M = 1$. If M is isomorphic to PSL(2,q) or Sz(q), where $2|q$, then all shears in G are involutions. If $M \cong PSL(2,q)$, where $2|q$, then actually $(\overline{P}, \overline{L}) \cong PG(2,q)$, and the lines in $\overline{L}$, which are not axes of non-trivial shears form a dual oval in $(\overline{P}, \overline{L})$, whose knot is the line at infinity.

NOTATION: Let (P, L) be a projective plane, $P \in P$, $\ell \in L$ and G a group of automorphisms of (P, L). Then [P] is the set of lines incident with P, $L(G)$ the lattice consisting of all substructures of (P, L) which are left invariant by G, G(P,P) the set of elations with centre P and $G(\ell,\ell)$ the set of elations with axis ℓ in G. Also, $F(G)$ is the substructure determined by all fixed points and fixed lines of G.

LEMMA 1: *Let* (P, L) *be a projective plane and* G *a group of automorphisms of* (P, L) *which leaves invariant a line* ℓ *but no point.*

(i) *Each element of* $L(G)$ *is of one of the following types:*

(A) $(\emptyset, \emptyset)$;

(B_i) $(\overline{P}, \{\ell\})$, *where* $\overline{P} \subseteq \ell$ *and* $|\overline{P}| = i$, $i \neq 1$;

(D_2) *a triangle intersecting* ℓ *trivially*;

(E_a) *a subplane containing* ℓ;

(E_b) *a subplane not containing* ℓ.

(ii) *If* $N \trianglelefteq G$, *then* $F(N)$ *is of type* (B_i), *where* $i \neq 1$, *or* (E_a).

(iii) *If* $N \trianglelefteq\trianglelefteq G$, *then* $F(N)$ *is of type* (B_i), *where* $i \neq 1$, *or* (E_a), *or* $F(N) = (\overline{P}, \overline{L})$, *where* $\overline{P} \subseteq \ell$, $\ell \in \overline{L} \subseteq [P]$ *for some point* $P \in \overline{P}$, $|\overline{P}| \geq 2$ *and* $|\overline{L}| \geq 3$.

PROOF: (i) Use [4, Theorem 3.2]. Let $X \in L(G)$. If X is of type (B_i), in the notation of [4], then G leaves invariant the unique line ℓ' of X. But G fixes just one line, namely ℓ, so that $\ell' = \ell$. Types (B_1), (Bd_i) and (C_{ij}) (again, we use the notation of [4]) for $j > 0$ are impossible, as G does not fix any point. If X is of type (D_i), then $i = 2$, and no vertex of X lies on ℓ.

(ii) Clearly $\ell \in F(N) \in L(G)$.

(iii) Assume that $F(N)$ is not of type (B_i) or (E_a), and let $N = N_0 \trianglelefteq N_1 \trianglelefteq N_2 \trianglelefteq \ldots \trianglelefteq N_r = G$. Let s be the largest integer such that N_s fixes points. Then $F(N_s)$ is of type B_i, $i \geq 2$ by (ii). Let $t < s$ be the largest integer such that N_t fixes a line m different from ℓ. Then $F(N_{t+1})$ is still of type (B_i), where $i \geq 2$. If N_t fixes a point $P \in P \setminus \ell$, then N_t fixes at least two points P and Q in $T \setminus \ell$, and hence the line PQ containing the three fixed points P, Q and $PQ \cap \ell$. As N_{t+1} does not leave invariant PQ, this implies that N_t is planar, a contradiction. So all fixed lines of N_t lie in $[m \cap \ell]$, and there are at least three of them. Suppose now that N_{t-1} fixes a point $P \in P \setminus \ell$. As before, N_{t-1} fixes a second point Q in $P \setminus \ell$, and PQ is one of the fixed lines of N_t and hence contained in $[m \cap \ell]$. Connecting P and Q with a fixed point of N_{t-1} in $\ell \setminus \{m \cap \ell\}$ we obtain two different lines containing at least three fixed points so that N_{t-1} is planar.

LEMMA 2: *Let* α *and* β *be shears of an affine plane* A. *Then one of the following statements holds:*

(i) $\alpha\beta$ *fixes exactly one affine point* P *and possibly some, however not all points of the line at infinity.* P *is left invariant by* α, *and* $o(\alpha\beta) \neq 2$.

(ii) $\alpha\beta$ *is a shear.*

(iii) $\alpha\beta$ *is a translation.*

PROOF: If α or β is trivial, then $\alpha\beta$ is a shear. Assume that $\alpha, \beta \neq 1$. Let A be the centre of α, a the axis of α, B the centre of β and b the axis of β. Then A and B lie on the line at infinity ℓ_∞ of A. If $A \neq B$, then the product might fix some points on ℓ_∞, however it certainly does not fix the points A and B. Also, by [3, (2.4)] the only affine fixed point of $\alpha\beta$ is $a \cap b$. Suppose that $\alpha\beta$ is an involution. Then $\alpha\beta$ must be a perspectivity with axis ℓ_∞ by Baer's Theorem [1]. As this is impossible, we have (i). Assume now that $A = B$. Then $\alpha, \beta \in G(A,A)$, so that clearly we have (ii) or (iii).

THEOREM 1: *Let* (P,L) *be a projective plane and* G *a finite group of automorphisms of* (P,L) *which leaves invariant a line* ℓ *but no point. Assume that* G *contains a non-trivial elation, whose axis is different from* ℓ, *and let* $(\overline{P},\overline{L})$ *be the sub-*

structure generated by all centres and axes of such elations in G. *Then* $(\overline{P}, \overline{L})$ *is a subplane. Let the representation of* G *on* $(\overline{P}, \overline{L})$ *be denoted by* $-$. *If* $G(\ell, \ell) = 1$, *then* $\overline{G}$ *contains exactly one minimal normal subgroup* M, M *is non-abelian simple and* $C_{\overline{G}}M = 1$.

PROOF: Let K be the kernel of $-$. Also, for convenience, we call an elation a translation if its axis is ℓ, and a shear otherwise. As ℓ contains at least two centres of non-trivial shears, $\ell \in \overline{L}$. Thus $(\overline{P}, \overline{L})$ is a subplane by Lemma 1 (i).

Let α be a shear in G, and $x \in F(G)$. If $[\alpha, x] \neq 1$, then $[\alpha, x] = \alpha^{-1}\alpha^{x}$ is a non-trivial shear or an automorphism fixing exactly one point in $P \setminus \ell$ by Lemma 2. Hence $[\alpha, x]$ fixes exactly one point on ℓ or exactly one point in $P \setminus \ell$. But $[\alpha, x] \in F(G)$ implies that $\langle[\alpha, x]\rangle \trianglelefteq\trianglelefteq G$, so that G likewise fixes exactly one point in ℓ or exactly one point in $P \setminus \ell$. This contradiction shows that F(G) centralizes every shear and hence

PROPOSITION 1: $F(G) \leq K$.

The induced group $\overline{G}$ fulfils all hypotheses of our theorem except, possibly, that $\overline{G}(\ell, \ell)$ might be non-trivial. Suppose that there exists an element $\overline{\beta} \in \overline{G}(\ell, \ell) \setminus \{1\}$. Let B be the centre of $\overline{\beta}$. As G does not fix any point, there exists a non-trivial shear $\alpha \in G$, whose centre A is different from B. Now $\alpha^{-1}\beta^{-1}\alpha\beta = (\overline{\beta}^{-1})^{\overline{\alpha}}\,\overline{\beta}$ is a translation of $(\overline{P}, \overline{L})$ so that $\alpha^{-1}\beta^{-1}\alpha\beta = \alpha^{-1}\alpha^{\beta}$ fixes at least three points on ℓ. But also, this commutator lies in G(A,A). Hence it is a translation, and in fact $\alpha^{-1}\beta^{-1}\alpha\beta = 1$. This implies that $\overline{\beta}$ fixes the axis of α, so that $A = B$, a contradiction. Therefore, $\overline{G}(\ell, \ell) = 1$ and we can apply Proposition 1 to $\overline{G}$. It follows that $F(\overline{G}) = 1$.

Let M be a minimal normal subgroup of $\overline{G}$ and E a component of M. As $E \neq 1$, there exists a shear $\alpha \in \overline{G}$ such that $[E, \alpha] \neq 1$. We prove

PROPOSITION 2 : E *contains an element* u *fixing exactly one point* U *in* $\overline{P}\setminus \ell$.

PROOF : Assume at first that α normalises E and let x be an element in E not centralising α . Then $1 \neq [\alpha, x] \in E$. If $[\alpha, x]$ fixes exactly one affine point, we are finished. If not, then $[\alpha, x]$ is a shear with centre, say $Z \in \ell$, by Lemma 2. If all non-trivial shears in E lie in $\overline{G}(Z,Z)$, then Z is the unique fixed point of E on ℓ . As $E \trianglelefteq\trianglelefteq \overline{G}$, this implies that G fixes Z , a contradiction. So E contains a shear $\beta \in E\setminus \overline{G}(Z,Z)$ and the product $[\alpha, x]\beta$, which has exactly one fixed point in $\overline{P}\setminus \ell$.

Assume now that α does not normalise E . By the theorem of Feit and Thompson [2] , E contains an involution τ . As E and E^{α} are two different simple normal subgroups of M , they have trivial intersection and hence commute elementwise. In particular, τ and τ^{α} commute, so that $\tau^{\alpha}\tau = \alpha^{-1}\tau\alpha\tau = \alpha^{-1}\alpha^{\tau}$ is an involution. By Lemma 2, this implies that $\tau^{\alpha}\tau$ is a shear. This shear, however, does normalise E . Also $\tau^{\alpha}\tau \notin C_{\overline{G}}E$, as $\tau^{\alpha} \in C_{\overline{G}}E$. We now finish as above.

Suppose that $E \neq M$, let $E_2 , \ldots , E_r$ be the further components and $C = E_2 \ldots E_r$. Then C fixes the point U , as $[u, C] = 1$ (U and u as constructed in Proposition 2). Also, $C \trianglelefteq\trianglelefteq \overline{G}$ so that C is planar by Lemma 1 (iii). Thus $E_2 , \ldots$ and E_r are planar, while E is not, as $u \in E$. It follows that $E \trianglelefteq \overline{G}$ and $E = M$.

As $M \trianglelefteq \overline{G}$, $C_{\overline{G}}M \trianglelefteq \overline{G}$. Also, $C_{\overline{G}}M$ leaves invariant U so that $C_{\overline{G}}M$ is planar. As the subplane $F(C_{\overline{G}}M)$ of $(\overline{P},\overline{L})$ is left invariant by $\overline{G}$, it contains all centres and axes of non-trivial perspectivities in $\overline{G}$ and hence $(\overline{P},\overline{L})$. Thus $C_{\overline{G}}M = 1$, and M is the unique minimal normal subgroup of $\overline{G}$. This completes the proof of Theorem 1.

Let, in Theorem 1, α be a non-trivial shear in $\overline{G}$. Let $x \in M \setminus C_M\alpha$ and denote $H = N_M\langle[\alpha , x]\rangle$. Then $[\alpha, x] = \alpha^{-1}\alpha^{x}$

is a non-trivial shear or an element fixing exactly one point in $\overline{P}\setminus \ell$ by Lemma 2. In the first case, $<\alpha, H>$ leaves invariant the centre Z of α. If here $M \leq <\alpha, H>$, then Z is the unique fixed point of M on ℓ and hence left invariant by $\overline{G}$, a contradiction. So $M \nleq <\alpha, H>$. In the second case $<\alpha, H>$ fixes a point in $\overline{P}\setminus \ell$, and again $M \nleq <\alpha, H>$ by Lemma 1 (ii). Hence α *is an abstract perspectivity*, as defined in [5] (in fact, α has property (*1)).

Assume now that $M \cong PSL(2,q)$, q even, or $M \cong Sz(q)$. By [6, proof of Proposition 4.2 and Theorem 3.1] it follows that α is an involution. Suppose that $\alpha \notin M$. There exists an element $x \in M$ such that $[x,\alpha]$ is an involution, and in fact a shear by Lemma 2. Therefore, M itself contains involutory shears. If $M \cong PSL(2,q)$, then $(\overline{P}, \overline{L})$ is Desarguesian by [3, Theorem 2.8]. So we have

THEOREM 2 : *If in Theorem* 1 *the unique minimal normal subgroup* M *is isomorphic to* PSL(2,q), *where* $2\mid q$, *then* $(\overline{P}, \overline{L})$ *is a Desarguesian subplane of order* q.

BIBLIOGRAPHY

1. R. Baer, "Projectivities with fixed points on every line of the plane", *Bull. Amer. Math. Soc.*, 52 (1946), 273-286.
2. W. Feit and J.G. Thompson, "Solvability of groups of odd order", *Pacific J. Math.*, 13 (1963), 775-1029.
3. C. Hering, "On projective planes of type VI", *Teorie Combinatorie*, Tomo II, Rome, 1973. Atti dei Convegni Lincei 17, (Accad. Naz. Lincei, Rome, 1976), pp.29-53.
4. C. Hering, "On the structure of finite collineation groups of projective planes", *Abh. Math. Sem. Univ. Hamburg*, 49 (1979), 155-182.
5. C. Hering, "On irreducible collineation groups of projective planes which contain non-trivial perspectivities", *Summer Institute on Finite Group Theory*, Santa Cruz, California 1979 (Amer. Math. Soc., to appear).
6. C. Hering and M. Walker, "Perspectivities in irreducible collineation groups of projective planes, II", *J. Statist. Plann. Inference*, 3 (1979), 151-177.

Mathematisches Institut
Universität Tübingen
74 Tübingen 1
Auf der Morgenstelle 10
Federal Republic of Germany

ON (k,n)-ARCS AND THE FALSITY OF THE LUNELLI-SCE CONJECTURE

R. Hill* and J.R.M. Mason

1. INTRODUCTION

A *(k, n)-arc* in PG(2,q) is a set of k points, such that some n, but no n + 1, of them are collinear. The maximum value of k for which a (k,n)-arc exists in PG(2,q) will be denoted by $m(n)_{2,q}$. Clearly $m(1)_{2,q} = 1$ and $m(q+1)_{2,q} = q^2 + q + 1$, and so from now on we assume that $2 \leq n \leq q$.

The following two theorems are well-known, and proofs may be found in Chapter 12 of Hirschfeld [12].

THEOREM 1.1 :

(i) $m(n)_{2,q} \leq (n-1)q + n$.

(ii) (Cossu [8]). For $n \leq q$, equality can occur in (i) only if $n \mid q$.

A ((n-1)q + n, n)-arc is known as a *maximal arc.*

THEOREM 1.2 : There exists a maximal arc in PG(2,q) when

(i) $n = q$, for any q, a maximal arc being the complement of a line,

(ii) (Denniston [10]). $q = 2^h$, and n is any divisor of q.

It is not known whether maximal arcs exist when q is odd and $2 \leq n < q$, although it has been proved by Thas [16] that there are no (2q+3, 3)-arcs in $PG(2, 3^h)$ for $h > 1$.

For $n \nmid q$ and $n \leq q$, in which case maximal arcs cannot exist by Theorem 1.1 (ii), Lunelli and Sce [13] made the following conjecture.

L-S CONJECTURE 1.3 : For $n \le q$ and $n \nmid q$,

$$m(n)_{2,q} \le (n-1)q + 1 .$$

The conjecture is known to be true for $n = 2$ (Bose [4]), $n = 3$ (Barlotti [1]) and $n = 4$, this last result being a particular case of

THEOREM 1.4 : (Lunelli and Sce [13]):

(i) For $4 \le n \le q$ and $n \nmid q$, $m(n)_{2,q} \le (n-1)q + n - 3$.

(ii) For $9 \le n \le q$ and $n \nmid q$, $m(n)_{2,q} \le (n-1)q + n - 4$.

Lunelli and Sce [13] also made a detailed study of (k,n)-arcs in PG(2,q) for small values of q and for $q \le 7$ left just two values of $m(n)_{2,q}$ undetermined. These two cases ($q = 7$, $n = 5,6$) have since been resolved by Ciechanowicz [7], thereby confirming that L-S Conjecture 1.3 is true for all $n \le q \le 7$ (see Table 5.1 of §5).

Our aim in this paper is to improve on known lower and upper bounds on $m(n)_{2,q}$, paying particular attention to cases where n is close to q.

In §2 we give some examples of (k,n)-arcs, including two families of counterexamples to L-S Conjecture 1.3, while in §3 and §4 upper bounds on $m(n)_{2,q}$ are obtained for $n = q - 1$ and $n = q - 2$ respectively.

A survey of results concerning $m(n)_{2,q}$ for $n < q \le 9$ is given in §5, these results being summarized in Table 5.1.

It will usually be more convenient to consider complements of (k,n)-arcs in PG(2,q), and so we make the following definition.

A *(k, ≥ d)-set* K in PG(2,q) is a set of k points which meets every line of PG(2,q) in at least d points. We denote by $m(\ge d)_{2,q}$ the smallest value of k for which a $(k, \ge d)$-set exists in PG(2,q). Clearly

$$m(\ge d)_{2,q} = q^2 + q + 1 - m(q + 1 - d)_{2,q} . \qquad (1.5)$$

If K is any k-set in $PG(2,q)$, then a line ℓ is called an i-*secant* of K if $|\ell \cap K| = i$. We denote by R_i the set of i-secants of K and let $r_i = |R_i|$.

The ordered $(q+2)$-tuple $(r_0, r_1, \ldots, r_{q+1})$ is called the *secant distribution* of K. It is often convenient to give the secant distribution in the alternative form

$\left(i_1^{\,r_{i_1}}, i_2^{\,r_{i_2}}, \ldots, i_t^{\,r_{i_t}}\right)$, where $i_1, i_2, \ldots, i_t$ are the values of i, in increasing order, for which $r_i \neq 0$. The following relations among the r_i's are well-known.

LEMMA 1.6 :

(i) $$\sum_{i=0}^{q+1} r_i = q^2 + q + 1 .$$

(ii) $$\sum_{i=1}^{q+1} i\, r_i = k(q+1).$$

(iii) $$\sum_{i=2}^{q+1} \frac{i(i-1)}{2} r_i = \frac{k(k-1)}{2} .$$

PROOF :

(i) is trivial.

(ii) Count in two ways the number of ordered point-line pairs (P, ℓ) where

$P \in \ell \cap K$.

(iii) Count in two ways the number of ordered pairs $(\{P_1, P_2\}, \ell)$, where

$P_1, P_2 \in \ell \cap K$.

We call an ordered $(q+2)$-tuple $(r_0, r_1, \ldots, r_{q+1})$ of non-negative integers a *feasible secant distribution* (f.s.d.) for a (k,n)-arc (or for a $(k, \geq d)$-set) if $r_i = 0$ for $i > n$ (or for $i < d$) and if equations (i), (ii) and (iii) of Lemma 1.6 are satisfied.

2. SOME EXAMPLES OF (k,n)-ARCS

Most of the examples will be obtained as complements of $(k, \geq d)$-sets. L-S Conjecture 1.3 may be restated as

L-S CONJECTURE 2.1 : $m(\geq d)_{2,q} \geq (d+1)q$, for $(q+1-d) \nmid q$ and $d \geq 2$.

We first construct $(k, \geq d)$-sets which meet this conjectured bound for $d = 2$ and $d = 3$ and which violate it for $d = 3$ when q is even, $q \geq 8$.

EXAMPLE 2.2 : A triangle (i.e. the union of three non-concurrent lines) is a $(3q, \geq 2)$-set.

EXAMPLE 2.3 : Let ℓ_1, ℓ_2, ℓ_3 and ℓ_4 be lines, no three of which are concurrent. Then $A = \ell_1 \cup \ell_2 \cup \ell_3 \cup \ell_4$ is a $(4q-2)$-set which meets every line of PG(2,q) in at least three points with the exception of the three lines $\ell = \langle \ell_1 \cap \ell_2, \ell_3 \cap \ell_4 \rangle$, $\ell' = \langle \ell_1 \cap \ell_3, \ell_2 \cap \ell_4 \rangle$ and $\ell'' = \langle \ell_1 \cap \ell_4, \ell_2 \cap \ell_3 \rangle$, each of which A meets in just two points. If q is even, and, it is easily shown, only if q is even, the ℓ_i's may be chosen so that ℓ, ℓ' and ℓ'' are concurrent, as indicated in Figure 1. (ℓ_1, ℓ_2, ℓ_3, ℓ_4 are represented by unbroken lines and ℓ, ℓ' and ℓ'' by broken lines.)

The set $K = A \cup \{\ell \cap \ell' \cap \ell''\}$ is a $(4q-1, \geq 3)$-set having secant distribution $(3^{6q-9}, 4^{(q-4)(q-2)}, 5^{q-2}, (q+1)^4)$. The complement of K in PG(2,q) is a $(q^2 - 3q + 2, q-2)$-arc, and this shows L-S Conjecture 1.3 to be false for q even, $q \geq 8$. The smallest of these counter-examples is a (42, 6)-arc in PG(2,8).

When q is odd, the set $A \cup \{\ell \cap \ell', \ell \cap \ell''\}$ is a $(4q, \geq 3)$-set. Examples 2.2 and 2.3 give

THEOREM 2.4 :

(i) $m(\geq 2)_{2,q} \leq 3q$, for all q.

(ii) $$m(\geq 3)_{2,q} \leq \begin{cases} 4q\,, & \text{for } q \text{ odd} \\ 4q-1\,, & \text{for } q \text{ even.} \end{cases}$$

We obtained the following result for the case $m = 2$, and we are grateful to J.W.P. Hirschfeld for acquainting us with the more general result of de Finis.

THEOREM 2.5 : (de Finis [9]): When q is a square, there exists a $d(q + \sqrt{q} + 1)$-set for which the only non-zero r_i's are

$$r_{d+\sqrt{q}} = d(q + \sqrt{q} + 1)$$

and (2.6)

$$r_d = q^2 + q + 1 - d(q + \sqrt{q} + 1)$$

(The result for $d = 1$ was first observed by Tallini-Scafati [15]).

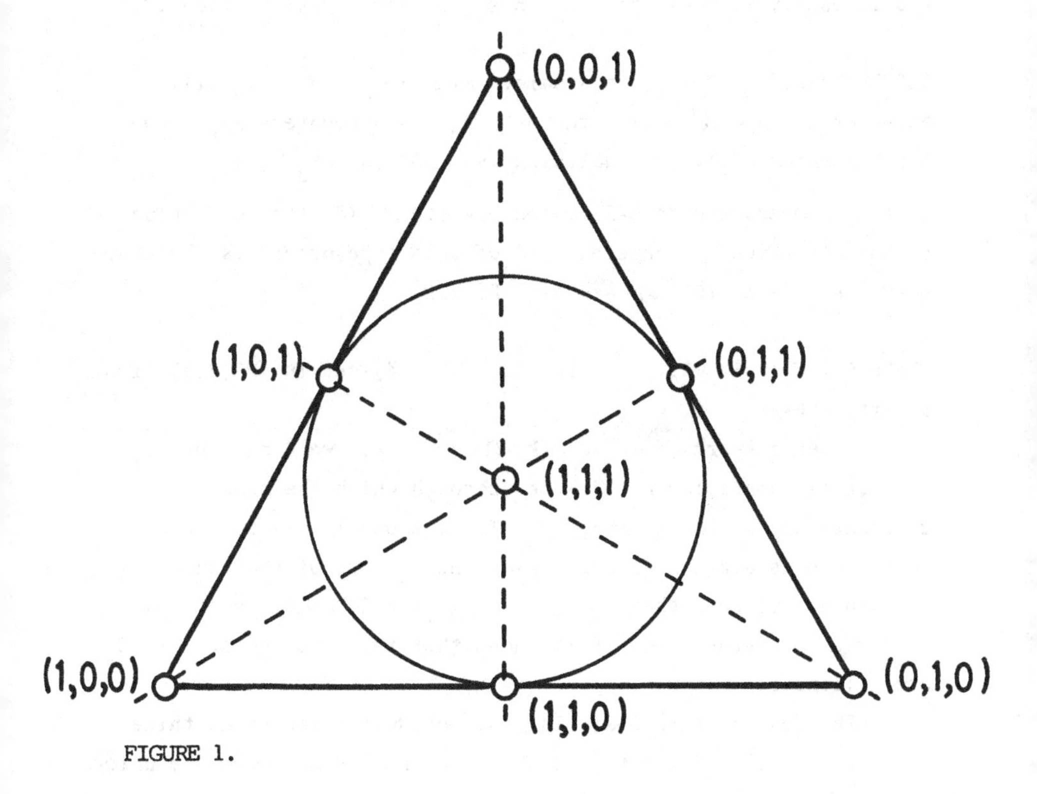

FIGURE 1.

PROOF : It was shown by Hirschfeld [11] (or see Chapter 4 of [12]) that when q is a square, $PG(2,q)$ may be partitioned into $q-\sqrt{q}+1$ Baer subplanes $PG(2,\sqrt{q})$. Let K_d denote the union of d of these disjoint Baer subplanes $(1 \le d \le q-\sqrt{q}+1)$.

Consider first the secants of K_1. Clearly there are $q+\sqrt{q}+1$ lines $(\sqrt{q}+1)$-secant to K_1, and no other line of $PG(2,q)$ can meet K_1 in more than one point. The only f.s.d. of a $(q+\sqrt{q}+1)$-set with this property is that given by (2.6) with $d=1$.

The result for $d=1$ shows that any i-secant of K_2 must have $i=2$, $\sqrt{q}+2$ or $2\sqrt{q}+2$ and it is straightforward to show that the only f.s.d. of a $2(q+\sqrt{q}+1)$-set having $r_i=0$ for all other values of i is that given by (2.6) with $d=2$.

It follows from the case $d=2$ that, given d disjoint Baer subplanes, any line meets at most one of them in $\sqrt{q}+1$ points, and it easily follows that K_d has the secant distribution (2.6).

COROLLARY 2.7 : For q a square, $m(\ge d)_{2,q} \le d(q+\sqrt{q}+1)$. Whenever $2 \le d < q/(\sqrt{q}+1)$, the set K_d is a counterexample to L-S Conjecture 2.1, or, equivalently, $PG(2,q) \setminus K_d \left(= K_{q-\sqrt{q}+1-d}\right)$ is a counterexample to L-S Conjecture 1.3. The smallest value of q for which a counterexample of this type occurs is therefore $q=9$; K_5 is a $(65,8)$-arc in $PG(2,9)$.

EXAMPLE 2.8 : We will construct a $(40,\ge 4)$-set in $PG(2,8)$ from a certain 6-arc.

Given a k-arc (i.e. a $(k,2)$-arc) K, we denote by C_i the set of points in $PG(2,q) \setminus K$ through which the number of 2-secants of K is i exactly. It is shown in §14.6 of [12] that there is a 6-arc (a 5-arc plus the nucleus of the conic containing it) such that $|C_0| = 4$, $|C_1| = 24$, $|C_2| = 36$ and $|C_3| = 3$. Furthermore, it is shown that the three points of C_3 are collinear.

The dual of this 6-arc is a set of 6 lines with no three concurrent. Let S be the 39-set of points formed from the union of these six lines. The dual of a point in C_0, C_1, C_2 or C_3 is a line meeting S in 6, 5, 4 or 3 points respectively. Furthermore, the three lines meeting S in 3 points are concurrent, in a point P, say. Thus $S \cup \{P\}$ is a $(40, \geq 4)$-set. Its secant distribution is found to be $(4^{38}, 5^{20}, 6^{9}, 9^{6})$.

For completeness, we also mention some families of (k, n)-arcs which contribute to the lower bounds on $m(n)_{2,q}$ given in Table 5.1.

EXAMPLES 2.9 (Bose [4]):

(i) A conic in $PG(2,q)$ is a $(q+1, 2)$-arc.

(ii) A conic plus its nucleus is a $(q+2, 2)$-arc, for q even.

EXAMPLES 2.10 (from §3.4 of Barlotti [2]): Let C be a conic in $PG(2,q)$, where q is odd. Let J denote the set of points of $PG(2,q) \setminus C$ which lie on no 1-secant of C. Then

(i) $J \cup \{P\}$ is a $(\frac{1}{2}q(q-1) + 1, \frac{1}{2}(q+1))$-arc, for any point P in C;

(ii) $J \cup C$ is a $(\frac{1}{2}(q^2+q+2), \frac{1}{2}(q+3))$-arc.

3. UPPER BOUNDS ON $m(q-1)_{2,q}$

Our approach to finding upper bounds on $m(q-1)_{2,q}$ is to tackle the equivalent problem of finding lower bounds on $m(\geq 2)_{2,q}$. The bound given by Theorem 1.4 (i) is

$$(m, \geq 2)_{2,q} \geq \left\{ \begin{array}{ll} 2q+5 & \text{for } q \geq 5 \\ 2q+6 & \text{for } q \geq 11 \end{array} \right\} . \qquad (3.1)$$

We will improve on this for $q = 7$, 8 and 9.

Example 2.2 shows that $m(\geq 2)_{2,q} \leq 3q$ for all q. Following a preliminary lemma, we first prove some general results about $(k, \geq 2)$-sets for which $k < 3q$.

LEMMA 3.2 : Suppose K is a $(k, \geq 2)$-set in $PG(2,q)$. Suppose $\ell_1, \ell_2, \ldots, \ell_u$ are lines concurrent in the point P. Then

$$\text{(i)} \quad \text{if } P \in K, \quad \sum_{i=1}^{u} |\ell_i \cap K| \leq k - 2(q+1-u) + q ; \qquad (3.3)$$

$$\text{(ii)} \quad \text{if } P \notin K, \quad \sum_{i=1}^{u} |\ell_i \cap K| \leq k - 2(q+1-u) . \qquad (3.4)$$

PROOF : In each case consider $\sum_{\ell} |\ell \cap K|$ summed over *all* the lines through P and use the fact that $|\ell \cap K| \geq 2$ for each line ℓ.

THEOREM 3.5 : Suppose K is a $(3q-t, \geq 2)$-set, where $t \geq 1$. Then

(i) $r_{q+1} \leq 1$;

(ii) $r_q = r_{q-1} = \ldots = r_{q+1-t} = 0$.

PROOF : (i) Suppose $r_{q+1} \geq 2$. Let ℓ_1 and ℓ_2 be $(q+1)$-secants and $P = \ell_1 \cap \ell_2$. Then (3.3) gives $2(q+1) \leq 2q + 2 - t$, contradicting $t \geq 1$.

(ii) Suppose ℓ_1 is a $(q+1-s)$-secant for some $1 \leq s \leq t$. Let $P \in \ell_1 \setminus K$. Then (3.4) gives $q + 1 - s \leq q - t$, contradicting $s \leq t$.

The next lemma is required for the proof of the following theorem.

LEMMA 3.6 : Let $GF(q)$ denote the field of q elements. Then, if $q > 2$, $\sum_{\lambda \in GF(q)} \lambda = 0$.

PROOF : Let α be a primitive root of $GF(q)$. Then

$$\sum_{\lambda \in GF(q)} \lambda = 1 + \alpha + \alpha^2 + \ldots + \alpha^{q-2} = \frac{\alpha^{q-1} - 1}{\alpha - 1} = \frac{1-1}{\alpha - 1} = 0 .$$

THEOREM 3.7 : Suppose that K is a $(3q-t, \geq 2)$-set with $t \geq 1$, and that $r_{q+1} = 1$. Then $r_{q-t} = 0$.

PROOF : Suppose ℓ_1 is the $(q+1)$-secant of K and suppose ℓ_2 is a $(q-t)$-secant. Without loss of generality, we may assume that $\ell_1 = \langle (0,0,1), (1,0,0) \rangle$, that $\ell_2 = \langle (0,0,1), (0,1,0) \rangle$ and that $(0,1,0)$ and $(0,1,1)$ are points in $\ell_2 \setminus K$.
Let S be the q-set $K \setminus (\ell_1 \cup \ell_2)$. Then each of the q lines, other than ℓ_2, through the point $(0,1,0)$ meets S in exactly one point and so also does each of the q lines, other than ℓ_2, through the point $(0,1,1)$. Also, of the $q-1$ lines, other than ℓ_1 and ℓ_2, through the point $(0,0,1)$ $(= \ell_1 \cap \ell_2)$, $q-2$ lines must meet S in exactly one point, while the remaining line meets S in exactly two points. Correspondingly, the points of S can be expressed in the following three ways. Let
$GF(q) = \{\lambda_0, \lambda_1, \ldots, \lambda_{q-1}\}$, where $\lambda_0 = 0$. Then

(a) $S = \{(1, a_i, \lambda_i): i = 0,1,\ldots, q-1\}$ for some $a_0, a_1, \ldots, a_{q-1}$ in $GF(q) \setminus \{0\}$;

(b) $S = \{(1, b_i, \lambda_i + b_i): i = 0,1,\ldots, q-1\}$ for some $b_0, b_1, \ldots, b_{q-1}$ in $GF(q) \setminus \{0\}$;

(c) $S = \{(1, \lambda_i, c_i): i = 1,2,\ldots, q-1\} \cup \{(1, \lambda_j, c'_j)\}$ for some $c_1, c_2, \ldots, c_{q-1}, c'_j$ in $GF(q), j \neq 0$.

Summing the third co-ordinates given by (a) and (b), we have

$$\sum_{i=0}^{q-1} \lambda_i = \sum_{i=0}^{q-1} (\lambda_i + b_i), \quad \text{and so} \quad \sum_{i=0}^{q-1} b_i = 0 .$$

Summing the second co-ordinates given by (b) and (c), we have

$$\lambda_j + \sum_{i=1}^{q-1} \lambda_i = \sum_{i=0}^{q-1} b_i .$$

But $\sum_{i=1}^{q-1} \lambda_i = 0$ by Lemma 3.6, and so $\lambda_j = 0$, a contradiction.

We now consider the lower bounds on $m(\geq 2)_{2,q}$ for $q = 7, 8$ and 9. We have improved the respective lower bounds of 19, 21 and 23 given by (3.1) to 21, 22 and 25. Our approach to showing the non-existence of a $(k, \geq 2)$-set in $PG(2,q)$ for a particular k and q is to eliminate all of its f.s.d.'s by various means. (A complete list of f.s.d.'s for given k and q is easily obtained either by hand or with the aid of a simple computer program.) We illustrate our methods by treating the case $q = 9$ in detail; some comments on the other cases will follow.

THEOREM 3.8 : $m(\geq 2)_{2,9} \geq 25$.

PROOF : Suppose K is a $(24, \geq 2)$-set in $PG(2,9)$ with secant distribution $(r_0 = 0,\ r_1 = 0,\ r_2,\ r_3, \ldots, r_{10})$. Then, by Theorem 3.5, $r_{10} \leq 1$ and $r_9 = r_8 = r_7 = 0$. There are just 43 f.s.d.'s satisfying these conditions and application of Theorem 3.7 $(r_{10} = 1 \Longrightarrow r_6 = 0)$ reduces the list to the 17 f.s.d.'s shown in Table 3.9. These are eliminated, as indicated, by various parts of the following lemma, thus showing that a $(24, \geq 2)$-set in $PG(2,9)$ does not exist.

r_2	r_3	r_4	r_5	r_6	r_{10}	
73	1	0	11	6	0	(i)
73	0	3	8	7	0	(i)
72	3	0	9	7	0	(i)
72	2	3	6	8	0	(i)
71	5	0	7	8	0	(iv)
72	1	6	3	9	0	(i)
71	4	3	4	9	0	(i)
70	7	0	5	9	0	(iv)
72	0	9	0	10	0	(ii)

r_2	r_3	r_4	r_5	r_6	r_{10}	
71	3	6	1	10	0	(ii)
70	6	3	2	10	0	(ii)
69	9	0	3	10	0	(iii)
69	8	3	0	11	0	(ii)
68	11	0	1	11	0	(iii)
70	1	8	11	0	1	(i)
69	4	5	12	0	1	(i)
68	7	2	13	0	1	(i)

TABLE 3.9

LEMMA 3.10 : Suppose K is a $(24, \geq 2)$-set in $PG(2,9)$. Then

(i) $4r_6 + 5r_5 + f(r_4) \leq 67$, where $f(r)$ is defined by $f(0) = 0$,

$$f(r) = \sum_{i=0}^{r-1} (6-i) \quad \text{for} \quad 1 \leq r \leq 6, \quad \text{and}$$

$$f(r) = 3r \quad \text{for} \quad r \geq 7.$$

(ii) $r_4 > 0 \implies r_6 \leq 8$.

(iii) $r_3 > 0 \implies r_6 \leq 9$.

(iv) $r_5 > 0$ and $r_5 + r_6 \geq 12 \implies r_5 + 2r_6 \leq 21$.

PROOF : First we note that (3.3) and (3.4) become: if $\ell_1, \ell_2, \ldots, \ell_u$ are concurrent in P, then

$$\sum_{i=1}^{u} |\ell_i \cap K| = \begin{cases} 13 + 2u, & \text{if } P \in K & (3.11) \\ 4 + 2u, & \text{if } P \notin K. & (3.12) \end{cases}$$

(i) By (3.12), any point of $PG(2,q) \setminus K$ lies on at most one line in $R_4 \cup R_5 \cup R_6$, except that it may lie on two, but at most two, 4-secants. Let S denote the set $\{Q \in PG(2,q) \setminus K : Q \in \ell$ for some 4-secant $\ell\}$. Then $|S| \leq 67 - 4r_6 - 5r_5$, since of the 67 points not in K, exactly $4r_6 + 5r_5$ lie on 5- and 6-secants. Also $|S| \geq f(r_4)$, since $f(r_4)$ is the least number of points not in K which can be in the union of r_4 4-secants.

The desired inequality now follows.

(ii) Let ℓ be a 4-secant. Every 6-secant meets ℓ in a point of K, for otherwise (3.12) is violated. By (3.11), through each of the four points of $\ell \cap K$ there pass at most two 6-secants. Hence $r_6 \leq 8$.

(iii) Similar to (ii); at most three 6-secants can intersect a 3-secant in the same point.

(iv) Let ℓ be a 5-secant and suppose $r_5 + r_6 \geq 12$. Let R be the line-set $R_6 \cup R_5 \setminus \{\ell\}$. For (3.12) to hold, every line in R meets ℓ in a point of K. We call a point P of

$\ell \cap K$ a v-point (with respect to R) if v lines of R pass through P. Then, by (3.11), v can be at most 3 and any 3-point must lie on at least two 5-secants other than ℓ. At least $r_5 - 1 + r_6 - 10$ of the five points of $\ell \cap K$ must be 3-points and so $r_5 - 1 \geq 2(r_5 + r_6 - 11)$, giving the desired inequality.

This concludes the proof of Lemma 3.10 and also that of Theorem 3.8.

<u>CASES $q = 7$ and $q = 8$</u> : The proof that $m(\geq 2)_{2,8} \geq 22$ is easier than that of Theorem 3.8, for only three f.s.d.'s of a $(21, \geq 2)$-set in PG(2,8) survive Theorem 3.5, and none of these three survives Theorem 3.7.

However, our proof that $m(\geq 2)_{2,7} \geq 21$ is more difficult, for, after application of Theorems 3.5 and 3.7 and the analogue of Lemma 3.10 (i), several f.s.d.'s of a $(20, \geq 2)$-set in PG(2,7) remain. Each of these remaining cases satisfies either $r_8 = 1$ and $r_5 \geq 1$ or $r_6 \geq 3$. The second author [14] has shown, by a rather lengthy case-by-case treatment, that a $(20, \geq 2)$-set cannot satisfy either of these conditions. Example 2.2 shows that this lower bound of 21 is attained. As we remarked in §1, the equivalent result that $m(6)_{2,7} = 36$, had already been obtained by Ciechanowicz [7].

4. UPPER BOUNDS ON $m(q-2)_{2,q}$

The upper bounds on $m(q-2)_{2,q}$ given by Theorem 1.4 are $q^2 - 2q - 5$ for $q \geq 7$ and $q^2 - 2q - 6$ for $q \geq 11$. We will obtain a general upper bound which improves on these for all $q \geq 7$ with the exception of $q = 7$ and $q = 11$, where the same bounds are obtained.

THEOREM 4.1 : Suppose $q \geq 7$. Then

$$m(q-2)_{2,q} \leq q^2 - 2q - 3/2 - \left(2q - 7/4\right)^{\frac{1}{2}} ,$$

or equivalently,

$$m(\geq 3)_{2,q} \geq 3q + 5/2 + \left(2q - 7/4\right)^{\frac{1}{2}} .$$

PROOF : Suppose K is a $(k, \geq 3)$-set in $PG(2,q)$ and that $q \geq 7$. We will show that

$$k \geq 3q + 5/2 + \left(2q - 7/4\right)^{\frac{1}{2}} . \qquad (4.2)$$

Suppose there are two $(q+1)$-secants of K. Then they meet in a point P of K. Considering $\sum_{\ell} |\ell \cap K|$, summed over all lines through P, we have

$$k + q \geq 2(q+1) + (q-1)\ 3 ,$$

and so $k \geq 4q - 1$

$$\geq 3q + 5/2 + \left(2q - 7/4\right)^{\frac{1}{2}} , \text{ provided } q \geq 7 .$$

So we may assume that $r_{q+1} \leq 1$.

We now put $k = 3q + t$, where we may assume $t < q + 1$, for otherwise (4.2) follows immediately.

Suppose ℓ_1 is an s-secant, where $s \leq q$, and let Q be a point of $\ell_1 \setminus K$. Then, considering $\sum_{\ell} (\ell \cap K)$ summed over all lines through Q, we have

$$3q + t \geq s + q \,.\, 3 , \quad \text{and so} \quad t \geq s .$$

Hence $r_s = 0$ for $t \leq s \leq q$.

Equations (i), (ii), (iii) of Lemma 1.6 now become

$$\text{(i)} \quad \sum_{i=3}^{t} r_i + r_{q+1} = q^2 + q + 1 ;$$

$$\text{(ii)} \quad \sum_{i=3}^{t} i\, r_i + (q+1) r_{q+1} = (3q + t)(q + 1) ;$$

$$\text{(iii)} \quad \sum_{i=3}^{t} \tfrac{1}{2} i(i-1) r_i + \tfrac{1}{2} q(q+1) r_{q+1} = \tfrac{1}{2}(3q + t)(3q + t - 1) .$$

Taking $(2+t)$ (ii) - 2 (iii) - 3t (i), thereby eliminating r_3 and r_t,

$$\sum_{i=3}^{t} (i-3)(t-i)\, r_i + (q-2)(t-q-1)\, r_{q+1} = q\,(t^2 - 4t - 3q + 9)\,.$$

Now $(i-3)(t-i) \geq 0$ for $3 \leq i \leq t$. Also, as shown above, $r_{q+1} = 0$ or 1, and since $t - q - 1 < 0$, we have in either case

$$q\,(t^2 - 4t - 3q + 9) \geq (q-2)(t-q-1)\,.$$

Hence

$$\tfrac{1}{2}t(t-5) - q + 4 \geq (1-t)/q\,,$$

and since the left-hand side is an integer, and since $-1 < \frac{1-t}{q} < 0$, this inequality holds if and only if $\frac{1}{2}t(t-5) - q + 4 \geq 0$.
Hence

$$t \geq 5/2 + \left(2q - 7/4\right)^{\frac{1}{2}}\,,$$

giving the lower bound (4.2) on k.

In particular, Theorem 4.1 gives

$$m(5)_{2,7} \leq 30\,, \quad m(6)_{2,8} \leq 42 \quad \text{and} \quad m(7)_{2,9} \leq 57\,.$$

Theorem 2.4 (ii) shows that this bound on $m(6)_{2,8}$ is attained.

5. RESULTS CONCERNING $m(n)_{2,q}$ for $n < q \leq 9$

Table 5.1 summarizes the best results known to the authors concerning the values of $m(n)_{2,q}$ for $n < q \leq 9$.

ADDENDUM

Three of the unresolved cases of Table 5.1 have now been settled by the second author. They are

$$m(5)_{2,8} = 33$$

$$m(5)_{2,9} = 37$$

and $m(6)_{2,9} = 48$.

These results will be proved in a sequel to this paper.

The last-mentioned case is of particular interest. It can be shown that the $(48, 6)$-arc in $PG(2,9)$ is projectively unique and may be described as follows.

Let T be the triangle $\ell_1 \cup \ell_2 \cup \ell_3$, where ℓ_i denotes the line $\{(x_1, x_2, x_3) : x_i = 0\}$. Let A be the 16-set

$$\{(1, x, x^3) : x \in GF(9),\ x \neq 0\} \cup \{(1, x, \alpha x^3) : x \in GF(9),\ x \neq 0\},$$

where α is a fixed primitive root of $GF(9)$. Then $T \cup A$ is a 43-set having the secant distribution $(4^{72}, 7^{16}, 10^3)$. The complement of $T \cup A$ in $PG(2,9)$ is the desired $(48, 6)$-arc.

n \ q	3	4	5	7	8	9
2	4[a]	6[a]	6[a]	8[a]	10[a]	10[a]
3		9[b]	11[b]	15[c]	15[d] - 17[e]	17[f]
4			16[b]	22[b]	28[g]	28[h]
5				29[i]	33[j] - 34[k]	37[ℓ] - 38[k]
6				36[i]	42[m]	46[ℓ] - 48[k]
7					49[n] - 51[o]	55[p] - 57[q]
8						65[r] - 66[s]

Key to Table 5.1 :

a	Bose [4] (see Examples 2.9)	k	Lunelli and Sce (Theorem 1.4)
b	Lunelli and Sce [13]	ℓ	Barlotti [2] (see Examples 2.10)
c	Bramwell [6]	m	Theorems 2.4 and 4.1
d	An ad hoc construction [14]	n	Example 2.2
e	Barlotti [1]	o	See §3
f	Barnabei et al [3]	p	Theorem 2.4
g	Cossu [8]	q	Theorem 4.1
h	Bose [5]	r	Corollary 2.7
i	Ciechanowicz [7]	s	Theorem 3.8
j	Example 2.8		

TABLE 5.1

BIBLIOGRAPHY

1. A. Barlotti, "Sui {k; n}-archi di un piano lineare finito", *Boll. Un. Mat. Ital.*, 11 (1956), 553-556.

2. A. Barlotti, *Some Topics in Finite Geometrical Structures*, Institute of Statistics mimeo series no. 439 (University of North Carolina, 1965).

3. M. Barnabei, D. Searby and C. Zucchini, "On small {k; q}-arcs in planes of order q^2", *J. Combin. Theory Ser. A*, 24 (1978), 241-246.

4. R.C. Bose, "Mathematical theory of the symmetrical factorial design", *Sankhyā*, 8 (1947), 107-166.

5. R.C. Bose, "On the application of finite projective geometry for deriving a certain series of balanced Kirkman arrangements", *Golden Jubilee Commemoration Volume* (1958-9), *Calcutta Math. Soc.*, 341-354.

6. D.L. Bramwell, Ph.D. Thesis, University of London, 1973.

7. Z. Ciechanowicz, Ph.D. Thesis, University of London, 1980.

8. A. Cossu, "Su alcune proprieta dei {k; n}-archi di un piano proiettivo sopra un corpo finito", *Rend. Mat. e Appl.*, 20 (1961), 271-277.

9. M. De Finis, "Sui k-insieme a due caratteri in un piano di Galois", Seminario di Geometrie Combinatorie 13 (Universita di Roma, 1979), cf. these proceedings.

10. R.H.F. Denniston, "Some maximal arcs in finite projective planes", *J. Combin. Theory Ser. A*, 6 (1969), 317-319.

11. J.W.P. Hirschfeld, "Cyclic projectivities in PG(n,q)", *Teorie Combinatorie* (Rome, 1973), vol. I (Accad. Naz. dei Lincei, 1976), 201-211.

12. J.W.P. Hirschfeld, *Projective Geometries over Finite Fields* (Clarendon Press, Oxford, 1979).

13. L. Lunelli and M. Sce, "Considerazione arithmetiche e risultati sperimentali sui $\{K; n\}_q$-archi", *Ist. Lombardo Accad. Sci. Rend. A*, 98 (1964), 3-52.

14. J.R.M. Mason, Ph.D. Thesis, University of Salford, 1980.

15. M. Tallini-Scafati, "{k,n}-archi di un piano grafico finito, con particulare riguardo a quelli con due caratteri", *Atti Accad. Naz. Lincei Rend.*, 40 (1966), 812-818; 1020-1025.

16. J.A. Thas, "Some results concerning {(q+1)(n-1);n}-arcs and {(q+1)(n-1)+1; n}-arcs in finite projective planes of order q", *J. Combin. Theory Ser. A*, 19 (1975), 228-232.

Department of Mathematics
University of Salford
Salford M5 4WT
U.K.

CUBIC SURFACES WHOSE POINTS ALL LIE ON THEIR 27 LINES

J.W.P. Hirschfeld

Let F be a cubic surface with 27 lines in $PG(3,q)$. Theorem 30.1 in Manin [2] states that, if $q > 34$, then there exists a point of F on none of its lines. There is, however, sufficient information in [1] to work out the precise list of cubic surfaces with no such point.

When the cubic curves of $PG(2,q)$ are mapped by their coefficients to the points of $PG(9,q)$, then the set of triple lines is mapped to the *Del Pezzo* surface V_2^9. Successive projections, always from points of themselves, are also called Del Pezzo surfaces by Manin. A cubic surface with 27 lines in $PG(3,q)$ is in this sense a Del Pezzo surface of order three, whose lines are its *exceptional curves*. In what follows, F is always such a surface. The 27 lines of F lie by threes in 45 *tritangent* planes, in e of which the three lines are concurrent at an *Eckardt* point. All the subsequent lemmas come from [1].

LEMMA 1 : F *exists over all fields except* $GF(q)$ *with* $q = 2, 3$ *or* 5.

LEMMA 2 : *The number of points on* F *is* $q^2 + 7q + 1$.

LEMMA 3 : *The number of points on the* 27 *lines of* F *is* $27(q-4) + e$.

LEMMA 4 : *For* q *odd,* $e \leq 18$; *for* q *even,* $e \leq 45$.

THEOREM 1 : *If all the points of* F *lie on the lines, then* $e = (q-10)^2 + 9$, *whence the possibilities are*

q	4	7	8	9	11	13	16
e	45	18	13	10	10	18	45 .

PROOF : This follows from Lemmas 1-4 .

Let

$$E = V(x_0^3 + x_1^3 + x_2^3 + x_3^3),$$

$$\mathcal{D} = \begin{cases} V(x_0^3 + x_1^3 + x_2^3 + x_3^3 + x_4^3, \sum x_i) & p \neq 3 \\ V(\sum_{i<j<k} x_i x_j x_k, \sum x_i) & p = 3, \end{cases}$$

$$C = V(x_0 x_1 (x_0 + x_1) + x_2 x_3 (x_0 + x_2 + x_3)) :$$

the summations are all from 0 to 4 .

E is the *equianharmonic* surface, $\mathcal{D}$ is the *diagonal* surface and C the *cyclic, non-equianharmonic* surface. When $q = 4$, E is also the *Hermitian* surface. We write $F \sim G$ when F and G are projectively equivalent.

LEMMA 5 : *If* F *has exactly* 18 *Eckardt points, then* $F \sim E$. *This occurs when* $x^2 + x + 1$ *has two roots in* GF(q), q *odd; that is, when* $q \equiv 1 \pmod 6$.

LEMMA 6 : *If* F *has exactly* 10 *Eckardt points, then* $F \sim \mathcal{D}$. *This occurs when* $x^2 - x - 1$ *has two roots in* GF(q), q *odd; that is, when* $q \equiv \pm 1 \pmod{10}$.

LEMMA 7 : *If* F *has exactly* 45 *Eckardt points, then* $F \sim E \sim \mathcal{D}$. *This occurs for* $q = 4^m$.

LEMMA 8 : *Over* GF(4), $F \sim \mathcal{D} \sim E$ *and* F *is Hermitian.*

LEMMA 9 : *Over* GF(7), $F \sim E$.

LEMMA 10 : *Over* GF(8), $F \sim C$.

THEOREM 2 : *There are seven fields over which all points of a cubic surface* F *lie on its lines, and in each case* F *is projectively unique. The table below gives the list of surfaces* F, *their number of points* $|F|$, *the number of points* $e = e_3, e_2, e_1$ *on three, two and one line of* F, *the order* $|G|$ *of the projective group* G *of* F, *and the order* $|\Gamma|$ *of the collineation group* Γ *of* F.

q	F	|F|	e_3	e_2	e_1	|G|	|Γ|
4	E, D	45	45	0	0	25,920	51,840
7	E	99	18	81	0	648	648
8	C	121	13	96	12	192	576
9	D	145	10	105	30	120	240
11	D	179	10	105	64	120	120
13	E	261	18	81	162	648	648
16	E, D	396	45	0	324	25,920	103,680

PROOF : Theorem 1 gives the possible values for q and the corresponding e. In fact, each case exists and is unique. For $q = 4$ and 16, Lemma 7 gives the result. For $q = 7$ and 13, Lemma 5 gives the result. For $q = 9$ and 11, Lemma 6 gives the result. Lemmas 8-10 show that, for $q = 4$, 7, and 8, the only cubic surfaces with 27 lines are the ones whose points are all on the lines.

BIBLIOGRAPHY

1. J.W.P. Hirschfeld, "Classical configurations over finite fields : I. The double-six and the cubic surface with 27 lines", *Rend. Mat. e Appl.*, 26 (1967), 115-152.
2. J.I. Manin, *Cubic Forms* (North Holland, Amsterdam, 1974).

Mathematics Division
University of Sussex
Brighton BN1 9QH
U.K.

EXISTENCE RESULTS FOR TRANSLATION NETS

Dieter Jungnickel

We consider $(s, r; \mu)$-nets admitting a group G of automorphisms acting regularly on the point set and fixing each parallel class; such nets will be called translation nets (due to Sprague for $\mu = 1$). Our main interest is in deriving bounds on the maximum possible value of r (given s and μ) subject to certain restrictions on G (e.g. for abelian groups). Translation nets are equivalent to a generalization of the congruence partitions defined by André. We prove a decomposition theorem in the case of nilpotent groups and show that here the problem may be reduced to finding the maximum value of r for $(p^i, r; p^j)$-translation nets with elementary abelian translation group; this is related to partial t-spreads. Using a result of Schulz, we show that every translation affine design has the parameters of an affine space or desarguesian affine plane and has an elementary abelian translation group. A similar result holds for symmetric translation nets with a nilpotent translation group.

1. INTRODUCTION AND PRELIMINARY KNOWLEDGE

In this paper we consider a generalization of the well-known nets of Bruck [4], [5] where non-parallel blocks intersect μ times (μ not necessarily equal to 1); we are interested in such structures admitting a translation group G (i.e. G acts regularly on the point set and fixes each parallel class). The aim of the paper is to obtain existence results for translation nets; in particular, we study bounds on the possible number of parallel classes (subject to various restrictions on G, which may be required to be nilpotent or abelian etc.). Translation nets with $\mu = 1$ have been introduced by Sprague [25].

In the general case, the translation nets coincide with the "group constructible nets" of Drake and Freeman [8]. Still earlier, they have been studied in disguise in the investigation of regular Klingenberg structures by Drake and the present author (see [9], [15] and [16]): they are equivalent to the "pre-uniform c-Klingenberg-matrices" of [9]. Not surprisingly, then, some of our results are not new but only slight generalizations of known results.

We begin by collecting the necessary preliminary knowledge. Our notation agrees with that of Dembowski [6]; in particular, points are denoted by lower case and blocks by upper case letters. Also, $[x,y]$ is the number of blocks joining the points x and y. (All structures considered are finite). $[X,Y]$ is defined dually. We write $x \parallel y$ (resp. $X \parallel Y$) if $[x,y] = 0$ or $x = y$ (resp. $[X,Y] = 0$ or $X = Y$) and call x,y (resp. X,Y) *parallel.*

The general notion of a net was introduced by Drake and Jungnickel [9]; but these structures have already been studied for a long time in the form of orthogonal arrays, transversal designs or affine 1-designs. We recall the precise definition:

DEFINITION 1.1: Let Σ be an incidence structure satisfying

$\parallel$ is an equivalence relation on the block set; (1.1.a)

each parallel class partitions the point set; (1.1.b)

$[X,Y] = \mu$ whenever $X \nparallel Y$; (1.1.c)

there are $r \geq 3$ parallel classes and some parallel class has exactly $s \geq 2$ elements. (1.1.d)

Then Σ is called an $(s, r; \mu)$-*net*. The dual structure Σ^* of an $(s, r; \mu)$-net Σ is called a *transversal design* (briefly, an $(s, r; \mu)$-TD).

A simple proof of the following counting result may be found in Drake and Jungnickel [9, Section 5].

PROPOSITION 1.2 : *Let* Σ *be an* $(s, r; \mu)$*-net. Then every parallel class has* s *elements, each block has* $s\mu$ *points and the total number of points is* $s^2\mu$ *(i.e.* Σ *is an affine* $1-(s^2\mu, \mu, r)$*-design. Also*

$$r \le (s^2\mu - 1)/(s - 1) = f(s, \mu) \tag{1.2.a}$$

with equality iff Σ *is an (affine) 2-design.*

In this case, Σ will also be called a *complete* (s, μ)*-net*. An $(s, r; \mu)$-net is called *maximal*, if it cannot be embedded in an $(s, r'; \mu)$-net with $r' > r$. Obviously a complete (s, μ)-net can only exist if $s - 1$ divides $\mu - 1$. If this is not the case, (1.2.a) may be improved as follows.

THEOREM 1.3 : (Bose and Bush [3]) *Assume the existence of an* $(s, r; \mu)$*-net where* $s - 1$ *does not divide* $\mu - 1$. *Put* $\mu - 1 = b(s - 1) + d$ *with* $0 < d < s - 1$ *and define* θ *by* $2\theta = (1 + 4s(s - 1 - d))^{\frac{1}{2}} - (2s - 2d - 1)$. *Then*

$$r \le [(s^2\mu - 1)/(s - 1)] - [\theta] - 1 \tag{1.3.a}$$

where as usual $[x]$ *denotes the greatest integer not exceeding* x.

We will also consider a second special class of nets (besides the affine designs). In this connection, the following result is interesting.

THEOREM 1.4 : (Hine and Mavron [12]) *Let* Σ *be an* $(s, r; \mu)$*-net which possesses distinct parallel points. Then*

$$r \le s\mu . \tag{1.4.a}$$

If $r = s\mu$, *then parallelism is an equivalence relation on the point set too, and each parallel class of points has exactly* s *elements iff* Σ^* *is also an* $(s, s\mu; \mu)$*-net.*

The last part of Theorem 1.4 has also been shown in [14] in the dual setting. We introduce a name for these special nets.

DEFINITION 1.5: A *symmetric* (s, μ)*-net* is an $(s, s\mu; \mu)$-net Σ such that Σ^* is also an $(s, s\mu; \mu)$-net. Another common name is "symmetric transversal design".

Symmetric nets have found considerable interest lately; we refer the reader to Mavron [23] for a survey (Mavron uses the term "hypernet" instead). Nets, respectively transversal designs, admitting certain nice types of automorphism groups have been studied lately, in particular in the symmetric case. We will need the following special case later.

DEFINITION 1.6: (Jungnickel [14], [17]) An $(s, r; \mu)$-TD Σ is called *class regular* with respect to the group G if G acts an an automorphism group of Σ which is regular on each point class.

It can be shown that G then is semi-regular on the block set of Σ, which was originally a second condition in [14]. By results of [14], Σ is equivalent to an "$(s, r; \mu)$-difference matrix over G". By a result of Drake [7], which generalizes the theorem on complete mappings of Hall and Paige [11], we can conclude a certain existence criterion which will here only be needed for $\mu = 1$ in a special case. Then an elementary counting proof is available in [18]. We state only this special case.

THEOREM 1.7: *Let* $s \equiv 2 \bmod 4$. *Then there is no class regular* $(s, r; 1)$*-TD whenever* $r \geq 3$.

Class regular transversal designs have been studied in [14]. In the symmetric case, the author has also studied nets with a Singer group (a group acting regularly on the point set and the block set) in [17] and nets with higher degrees of transitivity in [19]. Here we will consider translation nets to be defined in the next section.

2. TRANSLATION NETS AND PARTIAL CONGRUENCE PARTITIONS

DEFINITION 2.1 : Let Σ be an $(s, r; \mu)$-net admitting an automorphism group G acting regularly on the point set and fixing each parallel class of Σ. Then Σ is called a *translation net* and G the *translation group* of Σ.

This generalizes a definition of Sprague [25] who has considered the case $\mu = 1$. As we will see shortly, the translation nets coincide with the "group constructible nets" of Drake and Freeman [8] and are equivalent to the "uniform c-Klingenberg matrices" of Drake and Jungnickel [9]. Their group theoretic equivalent generalizes the notion of a "congruence partition" defined by André [1]. This explains the terminology in the following definition.

DEFINITION 2.2 : Let G be a group of order $s^2\mu$ and $U_1, \ldots, U_r$ subgroups of G such that

$$|U_i| = s\mu \quad \text{for all } i ; \qquad (2.2.a)$$

$$|U_i \cap U_j| = \mu \quad \text{whenever } i \neq j . \qquad (2.2.b)$$

Then $\mathcal{U} = \{U_1, \ldots, U_r\}$ is called a *partial congruence partition* with parameters s, r, μ (briefly, an $(s, r; \mu)$-*pcp*); the U_i are the *components* of $\mathcal{U}$.

PROPOSITION 2.3 : *The existence of an* $(s, r; \mu)$-*translation net* Σ *with translation group* G *is equivalent to that of an* $(s, r; \mu)$-*pcp* $\mathcal{U}$ *in* G.

PROOF : Given Σ, choose a point p and label the blocks through p as $B_1, \ldots, B_r$. Let $U_i = G_{B_i}$ be the stabilizer of B_i and put $\mathcal{U} = \{U_1, \ldots, U_r\}$ to obtain the desired pcp. Given a pcp $\mathcal{U} = \{U_1, \ldots, U_r\}$ choose the elements of G as points and all cosets of the U_i as blocks with containment as incidence to obtain Σ.

Note that the $(q, q+1; 1)$-pcp's are precisely the congruence partitions of André [1]. We introduce some notation:

NOTATION 2.4 : A translation net (resp. a pcp) in G will be called *nilpotent (or abelian* or *elementary abelian)* iff G has this property. We denote by $Tn(s, \mu)$, $Np(s, \mu)$, $Ab(s, \mu)$ and $Ea(s, \mu)$ the maximum value of r, for which a translation net (resp. a nilpotent, abelian, elementary abelian translation net) exists. Here we understand $EA(s^2 \mu)$ to be the abelian group of order $s^2\mu$ that is the direct product of elementary abelian p-groups; by abuse of language we call this group *elementary abelian* again. Finally, denote by $N(s, \mu)$ the maximum of r for which an $(s, r; \mu)$-net exists at all.

Using (1.2.a), we trivially have

$$Ea(s, \mu) \leq Ab(s, \mu) \leq Np(s, \mu) \leq Tn(s, \mu) \leq N(s, \mu) \leq f(s, \mu) . \tag{2.4.a}$$

We will show later that the first two inequalities in (2.4.a) are in fact equalities. A lower bound on $Ea(s, \mu)$ will be given in Section 3. We conclude this section with a few remarks.

REMARK 2.5 : The elements of the translation group G of a complete translation net as defined above are indeed translations as defined in Dembowski [6] or Schulz [23] for 2-designs, i.e. fixed point free automorphisms α satisfying

$$x, x^{\alpha}\, I\, B \implies B^{\alpha} = B . \tag{2.5.a}$$

Automorphisms with (2.5.a) are usually called *dilatations*. Of course, one might study the analogue of this notion in an arbitrary net. But the usual results do not in general carry over; e.g. a dilatation might have more than one fixed point. (For an example, take G as the direct product of two cyclic groups of order 4 with generators a, b; then the net belonging to the pcp $\{\langle a\rangle, \langle b\rangle, \langle ab\rangle\}$ has $\alpha : x \mapsto x^{-1}$ as a dilatation, and α has 4 fixed points.) On the other hand, any 2-design with a transitive translation group G (in the usual sense) has a parallelism (though $X \parallel Y$ does not necessarily follow from $[X,Y] = 0$) and G is regular on points and fixes each parallel class (Schulz [23]). Thus the terminology of Definition 2.1 makes sense and we may apply the known results on transitive translation groups in the case of complete nets.

REMARK 2.6 : In the case of affine planes, any translation has a centre (in the projective extension); i.e. if a translation τ fixes one line of a parallel class, then it fixes every line of this class. This is not necessarily true in the general case. In fact, let B be a block and G_B the stabilizer of B; then every element of G_B fixes every block in the parallel class of B iff G_B is a normal subgroup of G (the proof of Sprague [25, Lemma 5.1] for $\mu = 1$ holds in the general case too).

REMARK 2.7 : Drake and Jungnickel [9] define a "c-(t,k)-K-matrix" $A = (a_{ij})$ $(i = 0,\dots,k;\ j = 1,\dots,t)$ over a group G of order t^2/c by the following condition:

The multiset $a_{ie} - a_{jm}$ of differences belonging to rows i and j of A contains each element of G exactly c times $(i \neq j)$. (2.7.a)

A is "pre-uniform" iff each row of A is a coset of a subgroup of G. Then clearly each row may be assumed to be a subgroup itself (otherwise add $-a_{i1}$ to the i-th row) and (2.7.a) means that any two of these subgroups of order t of G intersect in c elements. Hence c divides t^2/c and $s = t/c$ is an integer. This shows that a pre-uniform c-(t,k)-K-matrix is nothing but a $(t/c, k+1; c)$-pcp. Thus we will be able to use the results of [9] in the sequel; in this case proofs will only be sketched.

3. LOWER BOUNDS

In this section, we give lower bounds on $Ea(s, \mu)$ by some recursive constructions from starting material obtained from affine spaces. The following main recursive construction is Proposition 9 of Drake and Freeman [8], but the method of proof is already due to Drake and Jungnickel [9, Proposition 8.7]. For $\mu = 1$, the result is also in Sprague [25].

PROPOSITION 3.1 : (Drake, Freeman, Jungnickel) *Assume the existence of an* $(s, r; \mu)$*-pcp in* G *and of an* $(s', r'; \mu')$*-pcp in* H. *Then there also exists an* $(ss', r; \mu\mu')$*-pcp in* $G \oplus H$.

PROOF : Let $\mathcal{U} = \{U_1, \dots, U_r\}$ and $\mathcal{V} = \{V_1, \dots, V_r\}$ be the given pcp's and put $\mathcal{W} = \{W_1, \dots, W_r\}$ with $W_i = U_i \oplus V_i$ for $i = 1, \dots, r$.

Thus we look for large pcp's in p-groups.

PROPOSITION 3.2 : (Drake and Jungnickel [9, 8.3]) *Let* q *be a prime power and* d *a non-negative integer. Then there exists a* $(q, q^{d+1} + \dots + q + 1; q^d)$*-pcp in* $EA(q^{d+2})$. *Thus*

$$EA(q, q^d) = N(q, q^d) = q^{d+1} + \dots + q + 1 . \qquad (3.2.a)$$

PROOF : The points and hyperplanes of the affine space $AG(d+2, q)$ form a (complete) translation net with these parameters.

Thus the maximum conceivable value of r (according to (1.5)) is indeed possible for pcp's with $s = p^i$, $\mu = p^j$ and $i \mid j$ (p prime). If $i \nmid j$, we need other recursive methods to obtain large pcp's. We first note a corollary of Propositions 3.1 and 3.2.

COROLLARY 3.3 : (Drake and Jungnickel [9, 8.8]) *Let* $s = q_1 \cdot \ldots \cdot q_r$ *be the prime power factorization of* s *and put* $q = \min\{q_i : i = 1, \dots, r\}$. *Then there exist* $(s, q+1; \mu)$ *pcp's for all* μ *in* $G = EA(s^2) \oplus H$, *where* H *is any group of order* μ.

PROOF: This is immediate for $\mu = 1$. If $\mathcal{U} = \{U_1, \ldots, U_r\}$ is a pcp in this case, then take $\mathcal{V} = \{V_1, \ldots, V_r\}$ with $V_i = U_i \oplus H$ in the general case. We might call $\{H, \ldots, H\}$ a trivial $(1, r; \mu)$-pcp in H.

PROPOSITION 3.4: *Let* $\mathcal{U} = \{U_1, \ldots, U_r\}$ *be an* $(s, r; \mu)$*-pcp in* G *and let* H *be a normal subgroup of order* t *of* G *such that* $H \cap U_i = 1$ *for all* i. *Then there exists an* $(s/t, t; \mu t)$*-pcp in* G/H.

PROOF: Define V_i by $V_i = U_i H / H$ for $i = 1, \ldots, r$. Then it is easily checked that $\{V_1, \ldots, V_r\}$ is the desired pcp in G/H. Note that all V_i have order $s\mu = (s/t)(\mu t)$ as $H \cap U_i = 1$.

COROLLARY 3.5: (Jungnickel [16]) *Let* p *be a prime,* i *a positive and* j *a non-negative integer. Then*

$$Ea(p^i, p^j) \geq p^{i+j} . \tag{3.5.a}$$

PROOF: Let $k = i + j$ and consider the pcp $\{U_1, \ldots, U_{p^k+1}\}$ belonging to any given translation plane of order p^k. Removing U_{p^k+1} leaves a $(p^k, p^k; 1)$-pcp in $EA(p^{2k})$ which in fact corresponds to a symmetric $(p^k, 1)$-net. Now choose a subgroup H of U_{p^k+1} of order p^j and apply Proposition 3.4. Note that the resulting pcp yields a symmetric (p^i, p^j)-net.

The next proposition glues together pcp's. It is similar to the construction part of Theorem 3.3 of Jungnickel and Sane [20]; but this construction idea really goes back to Shrikhande [24] who used it in the language of orthogonal arrays.

PROPOSITION 3.6: *Let* $\mathcal{U} = \{U_1, \ldots, U_r\}$ *be an* $(s, r; \mu)$*-pcp in* G *with* $s \mid \mu$, *let* H *be a normal subgroup of order* s *of* G *with* $H \cap U_i = 1$ *for all* i, *and assume the existence of an* $(s, r'; \mu/s)$*-pcp in* G/H. *Then* $\mathcal{U}$ *may be extended to an* $(s, r+r'; \mu)$*-pcp in* G.

PROOF: Let $\{V_1,\dots,V_{r'}\}$ be the given pcp in G/H and define $U_{r+i} = \phi^{-1}(V_i)$ for $i=1,\dots,r'$, where ϕ is the canonical epimorphism from G onto G/H. Then each U_{r+i} has order $|H|\,s(\mu/s) = s\mu$ and clearly $|U_{r+i} \cap U_{r+j}| = |H|\,|V_i \cap V_j| = s(\mu/s) = \mu$. Also, $|U_i \cap U_{r+j}| \geq \mu$, as both U_i and U_{r+j} have order $s\mu$, whereas G has order $s^2\mu$. But for each $x \in U_{r+j}$ with $x \in U_i$, there are $s-1$ elements $x' \equiv x \bmod H$ (but $x' \neq x$) in U_{r+j} which cannot be in U_i as $U_i \cap H = 1$. Thus also $|U_i \cap U_{r+j}| \leq s\mu/s$ which proves the assertion.

When $s \nmid \mu$, we may still sometimes adjoin one component to a given pcp.

PROPOSITION 3.7: *Let* $\mathcal{U} = \{U_1,\dots,U_r\}$ *be an* $(s,r;\mu)$*-pcp in* G *and let* H *be a normal subgroup of order* s *of* G *with* $H \cap U_i = 1$ *for all* i. *Then* $\mathcal{U}$ *may be extended to an* $(s,r+1;\mu)$*-pcp.*

PROOF: Choose any subgroup V of order μ of G with $V \cap H = 1$, e.g. $V = U_1 \cap U_2$, and put $U_{r+1} = VH$. The rest of the argument is as in the last part of Proposition 3.6.

COROLLARY 3.8: *Let* p *be a prime,* i *a positive and* j *a non-negative integer. Put* $j = ai+b$ *with* $0 \leq b < i$. *Then*

$$\mathrm{Ea}(p^i, p^j) \geq p^{(a+1)i+b} + p^{ai+b} + \dots + p^{i+b} + 1. \tag{3.8.a}$$

PROOF: Use Corollary 3.5, and Propositions 3.6 and 3.7 with induction on a. Also observe that, for the pcp $\{V_1,\dots,V_{p^k}\}$ constructed in Corollary 3.5, G/H has a subgroup K of order p^i with $K \cap V_i = 1$ for all i, namely $K = U_{p^k+1}/H$. Then $G/K \cong \mathrm{Ea}(p^{i+j})$.

The pcp's just constructed are maximal by the results of Jungnickel and Sane [20]. Note that Corollary 3.8 provides an alternative proof of Proposition 3.2 for $d > 0$. Now Corollary 3.8

and Proposition 3.1 yield the following result (using again a trivial $(1, r; \mu')$-pcp if necessary).

THEOREM 3.9 : *Let* s *and* μ *be natural numbers,* P *the set of primes dividing* s *and let* $p_i^{e_i} \| s$, $p_i^{f_i} \| \mu$ *for* $p_i \in P$. *(Here* $p^a \| x$ *means* $p^a \mid x$ *and* $p^{a+1} \nmid x$.*) Then*

$$Ea(s, \mu) \geq \min \{ Ea(p_i^{e_i}, p_i^{f_i}) : p_i \in P\} \geq \min \{ p_i^{e_i+f_i} + 1 : p \in P\} , \tag{3.9.a}$$

where in fact the last bound may be improved as in (3.8.a).

We conclude this section by mentioning the connection with the partial t-spreads considered in Beutelspacher [2] ; this is really due to Drake and Freeman [8, Proposition 6], at least the first part of the assertion (and the converse is very similar). Recall that a *partial t-spread* in $PG(n-1, q)$ (the projective space of dimension $n-1$ over $GF(q)$) is a collection of pairwise disjoint t-dimensional subspaces.

PROPOSITION 3.10 : (Drake and Freeman) *The existence of a partial* $(t-1)$*-spread with* r *components in* $PG(n-1, q)$ *implies that of an elementary abelian* $(q^t, r; q^{n-2t})$*-pcp. The converse holds for primes* q .

Then Corollary 3.8 may be obtained from Beutelspacher's results. On the other hand, Corollary 3.8 yields Beutelspacher's lower bound [2, Theorem 4.2] for primes p.

4. UPPER BOUNDS

In this section we start with the derivation of upper bounds. One main tool is a splitting of pcp's dually to the direct product construction in Proposition 3.1 . For this to work, we need G to be the direct product of its Sylow subgroups, i.e. G should be nilpotent. The following result is due to Drake and Jungnickel [9, Proposition 8.10] in the abelian case, but the proof carries over to the nilpotent case.

THEOREM 4.1 : *Let* $\mathcal{W} = \{W_1, \ldots, W_r\}$ *be an* $(s, r; \mu)$*-pcp in the nilpotent group* G. *If* $s = s's''$ *and* $\mu = \mu'\mu''$ *with* $(s'\mu', s''\mu'') = 1$, *then there exists an* $(s', r; \mu')$*-pcp* $\mathcal{U} = \{U_1, \ldots, U_r\}$ *and an* $(s'', r; \mu'')$*-pcp* $\mathcal{V} = \{V_1, \ldots, V_r\}$ *such that* $\mathcal{W}$ *is constructed from* $\mathcal{U}$ *and* $\mathcal{V}$ *as in the proof of Proposition 3.1 .*

SKETCH OF PROOF : As G is nilpotent, we have $G = H \oplus K$ with $|H| = s'^2\mu'$ and $|K| = s''^2\mu''$. Put $U_i = W_i \cap H$ and $V_i = W_i \cap K$.

COROLLARY 4.2 : *With the notation of Theorem 3.9, we have*

$$Np(s, \mu) = \min\{ Np(p_i^{e_i}, p_i^{f_i}) : p_i \in P \}; \tag{4.2.a}$$

$$Ab(s, \mu) = \min\{ Ab(p_i^{e_i}, p_i^{f_i}) : p_i \in P \}; \tag{4.2.b}$$

$$Ea(s, \mu) = \min\{ Ea(p_i^{e_i}, p_i^{f_i}) : p_i \in P \}. \tag{4.2.c}$$

We will see later that in fact $Np(s, \mu) = Ea(s, \mu)$. By Corollary 4.2, we are back to the problem of finding bounds on pcp's in p-groups. The result of Bose and Bush (see Theorem 1.3) here yields the following bound.

PROPOSITION 4.3 : (Bose and Bush) *Let* p *be a prime,* d *a positive integer and* $j = di + b$ *with* $0 < b < i$. *Then*

$$N(p^i, p^j) \le p^{(d+1)i+b} + p^{di+b} + \ldots + p^{i+b} + p^b - [\theta] - 1, \tag{4.3.a}$$

where $2\theta = (1 + 4p^i(p^i - p^b))^{\frac{1}{2}} - (2p^i - 2p^b - 1)$. *For example, for* $b = 1$ *(4.3.a) implies*

$$N(p^i, p^{di+1}) \le p^{(d+1)i+1} + \ldots + p^{i+1} + (p+3)/2. \tag{4.3.b}$$

In this case, we have a better bound in the elementary abelian case. Applying Proposition 3.10 to Theorem 4.1 of Beutelspacher [2], we get as an upper bound for $Ea(p^i, p^{di+1})$ the lower bound of (3.8.a). Hence we have

THEOREM 4.4 : *Let* p *be a prime,* i *a positive and* d *a non-negative integer. Then*

$$Ea(p^i, p^{di+1}) = p^{(d+1)i+1} + p^{di+1} + \ldots + p^{i+1} + 1 . \quad (4.4.a)$$

It should be mentioned that no $(p^i, r; p^{di+1})$-net with r exceeding the bound of (4.4.a) is known. It is thus conceivable that the Bose-Bush bound (which only yields (4.3.b)) can be improved. In the remainder of this section we show that the largest pcp's in p-groups do arise in the elementary abelian ones. We first consider a special case, i.e. the case $s = p$.

LEMMA 4.5 : *Assume the existence of a* $(p, r; p^d)$*-pcp* $\mathcal{U}$ *in a group* G *with* $|\Phi(G)| = p^e$, *where* $\Phi(G)$ *is the Frattini group of* G. *Then*

$$r \leq p^{d+1-e} + \ldots + p + 1 . \quad (4.5.a)$$

PROOF : Note that in this case all components of the pcp are maximal subgroups of G. Also, any two maximal subgroups of G clearly intersect in p^d elements. Thus the maximum possible r arises if we take $\mathcal{U}$ as the set of all maximal subgroups of G. We do this and consider $\Phi = \Phi(G)$, i.e. the intersection of all maximal subgroups of G. Thus $\Phi \subset U$ for every $U \in \mathcal{U}$; hence $\overline{\mathcal{U}} = \{U/\Phi : U \in \mathcal{U}\}$ will be a pcp in $H = G/\Phi$ provided $|H| \geq p^2$. More precisely, if $e \leq d$, then $\overline{\mathcal{U}}$ is a $(p, r; p^{d-e})$-pcp in H. But clearly $e \neq d+2$, as G has maximal subgroups, and $e \neq d+1$, as otherwise there is only one maximal subgroup and (4.5.a) holds trivially. Now the assertion follows from (1.2.a), applied to the pcp in H.

COROLLARY 4.6 : *Let* $\mathcal{U}$ *be a* $(p, r; p^d)$*-pcp in a group* $G \ncong EA(p^{d+2})$. *Then*

$$r \le p^d + \ldots + p + 1 \; ; \qquad (4.6.a)$$

whereas we recall that

$$Ea(p, p^d) = p^{d+1} + \ldots + p + 1 \; . \qquad (4.6.b)$$

PROOF: We have to show that $\Phi(G) \neq 1$ in this case. But this is clear, as $\Phi(G)$ is the smallest normal subgroup N of G such that G/N is elementary abelian (see e.g. Huppert [13, III.3.16]).

We will consider examples of equality in (4.6.a) below. Here, we wish to use Corollary 4.6 as the basis of an inductive proof that $r \le p^{i+j-1} + \ldots + p + 1$ for any non-elementary abelian $(p^i, r; p^j)$-pcp. This requires taking factor groups and considering the induced structure of a pcp; as this is in general not a pcp, we need a more general concept which will only be used to prove the stated result.

DEFINITION 4.7: Let G be a group of order p^m $(m \ge 2)$ and let $\mathcal{U} = \{U_1, \ldots, U_r\}$ be a set of proper subgroups of G satisfying $r \ge 2$ and

$$U_i U_j = G \quad \text{whenever} \quad i \neq j \; . \qquad (4.7.a)$$

Then $\mathcal{U}$ is called a *generalized partial congruence partition* with *parameter* $a = \max \{ i : \exists\, U \in \mathcal{U} \text{ with } |U| = p^i \}$ or for short a *gpcp(a)*. The U_i are called the *components* of $\mathcal{U}$.

Note that we do consider the complex product of U_i and U_j, *not* the subgroup generated by U_i and U_j. The following lemma is obvious.

LEMMA 4.8: *Any* $(p^i, r; p^j)$*-pcp is a gpcp* $(i+j)$. *Next let* $\mathcal{U}$ *be a gpcp(a) in the group* G *of order* p^m. *Then*

(*i*) $m \le 2a$ *with equality iff* $\mathcal{U}$ *is a* $(p^a, r; 1)$*-pcp;*

(ii) *each maximal subgroup of* G *contains at most one component of* $\mathcal{U}$.

We also need to know that no gpcp(i + j) can have too many components in the elementary abelian case.

LEMMA 4.9 : *Let* $\mathcal{U}$ *be a gpcp(a) in* $EA(p^m)$. *Then*

$$|\mathcal{U}| \leq p^a + \ldots + p + 1 . \qquad (4.9.a)$$

PROOF : If $a = m - 1$, we use the fact that $|\mathcal{U}|$ is bounded by the number of maximal subgroups of G, i.e. by $Ea(p, p^{a-1})$, because of Lemma 4.8. But this implies (4.9.a) because of (4.6.b). Now let $a < m - 1$ and choose any subgroup P of G with $|P| = p$. Put $H = G/P$ and $\overline{\mathcal{U}} = \{ UP/P : U \in \mathcal{U} \}$. Then it is easily seen that $\overline{\mathcal{U}}$ is a gpcp in H with parameter at most a and thus we obtain (4.9.a) by induction on m. (Note that the case $m = 2$ implies $a = 1$, which is covered by the first argument.)

THEOREM 4.10 : *Let* $\mathcal{U}$ *be a gpcp(a) in a group* G *of order* p^m *with* $G \ncong EA(p^m)$. *Then*

$$|\mathcal{U}| \leq p^{a-1} + \ldots + p + 1 . \qquad (4.10.a)$$

PROOF : Assume first that $a = m - 1$. Then again $|\mathcal{U}|$ is bounded above by the number of maximal subgroups of G and the assertion follows from (4.6.a). This also takes care of the case $m = 2$. Now assume that $a < m - 1$ and choose any normal subgroup P of G with $|P| = p$ (this is possible, as the centre Z(G) is not trivial). Put $H = G/P$ and $\mathcal{U}^* = \{ UP/P : U \in \mathcal{U} \}$. Clearly $\mathcal{U}^*$ is a gpcp with parameter at most a ; unless H is elementary abelian, induction on m yields the assertion. But if H is elementary abelian, then $P = \Phi(G)$, as $\Phi(G)$ is the smallest normal subgroup N of G such that G/N is elementary abelian and as G is not elementary abelian itself. In this case, we may also assume that Z(G) is cyclic, as otherwise we may choose another normal subgroup P_1 of G of order p and apply the above argument to P_1 instead of P; clearly G/P_1 cannot be elementary abelian, too. Thus $Z = Z(G)$ is cyclic and $K = G/Z(G)$ elementary abelian. Note that $|Z| \leq p^2$ as $\Phi(G) = G'G^p$; see, for example, Huppert [13, III. 3.14]. Therefore, G is not abelian. Put $\overline{U} = UZ/Z$ and $\overline{\mathcal{U}} = \{\overline{U} : U \in \mathcal{U}\}$. Clearly $\overline{\mathcal{U}}$ is a gpcp in K. We want to show that in this case $\overline{\mathcal{U}}$ has parameter $a-1$; then the assertion follows by Lemma 4.9. Now, if $|U| \leq p^{a-1}$ for some $U \in \mathcal{U}$, then clearly $|\overline{U}| \leq p^{a-1}$. It suffices to show that $P \leq U$ for all those $U \in \mathcal{U}$ with $|U| = p^a$. But $\Phi(U) \leq \Phi(G) = P$ (see e.g. Huppert [13, III. 3.14]), i.e. $P = \Phi(U) \leq U$ or $\Phi(U) = 1$, and in the latter case U is elementary abelian. We will show that this is impossible. By our assumptions on G, $K = G/Z(G)$ has square order p^{2b} and is in fact a symplectic space over $GF(p)$ with respect to the form defined by $f(xZ, yZ) = c$ iff $[x, y] = z^c$ where z is a fixed generator of G' (see Huppert [13, III. 13.7]). As U is abelian by our assumption, $\overline{U}$ will be a totally isotropic subspace of K and thus $\dim \overline{U} \leq b$, i.e. $|U| \leq p^b$. But also by assumption $|U| = p^a$ and $2b+1 \leq m \leq 2a$ by Lemma 4.8, which is the desired contradiction.

COROLLARY 4.11: *Assume the existence of a* $(p^i, r; p^j)$*-pcp in* G, *where* G *is not elementary abelian. Then*

$$r \leq p^{i+j-1} + \ldots + p + 1. \tag{4.11.a}$$

As $p^{i+j-1} + \ldots + p + 1 < p^{i+j}$ and as $Ea(p^i, p^j) \geq p^{i+j} + 1$ by Corollary 3.8, we see that $Tn(p^i, p^j) = Np(p^i, p^j) = Ab(p^i, p^j) = Ea(p^i, p^j)$. Hence, using Corollary 4.2, we have

THEOREM 4.12: *One always has (using the notation of Theorem 3.9)*

$$Np(s, \mu) = Ab(s, \mu) = Ea(s, \mu) = \min\{Ea(p_i^{e_i}, p_i^{f_i}) : p_i \in P\}, \tag{4.12.a}$$

and the unique group G *of order* $s^2\mu$ *for which the maximum value is realized is* $EA(s^2\mu)$. *A lower bound for (4.12.a) is given by (3.8.a), an upper bound by (3.2.a) and (4.3.a).*

Our results show that the maximum size of a $(p^i, r; p^j)$-pcp in any other group but $EA(p^{2i+j})$ is roughly at most $1/p$ of that in $EA(p^{2i+j})$. Two more applications of this fact will be given in Section 6. We conclude this section with some examples where the bounds of (4.6.a) resp. (4.11.a) are met.

EXAMPLES 4.13 :

(a) Let G be an extra special group of order p^{2m+1}, e.g. in Huppert [13, p.355]. Then $H = G/\Phi(G) \cong EA(p^{2m})$ and thus H has a $(p, p^{2m-1} + \ldots + 1; p^{2m-2})$ - pcp $\mathcal{U}$. If Φ is the canonical epimorphism from G onto H, then $\Phi^{-1}(\mathcal{U})$ is a $(p, p^{2m-1} + \ldots + 1; p^{2m-1})$ - pcp in G. Here we have equality in (4.6.a) for $d = 2m - 1$, m any positive integer.

(b) Similarly, let G be the abelian group of order p^{d+2} and type $(2, 1, \ldots, 1)$, let P be the subgroup of order p of a cyclic subgroup $\cong \mathbb{Z}_{p^2}$ of G and put $H = G/P$. Then H is elementary abelian and has a $(p, p^d + \ldots + 1; p^{d-1})$ - pcp $\mathcal{U}$; again $\Phi^{-1}(\mathcal{U})$ is a $(p, p^d + \ldots + 1; p^d)$ - pcp in G meeting the bound in (4.6.a) (for any positive integer d).

(c) Let G be the metacyclic group of order p^4 and exponent p^2, say with generators a and b satisfying $a^{p^2} = b^{p^2} = 1$ and $a^b = a^{p+1}$ (see Huppert [13, III.12.6]). An easy induction proof shows that

$$(ba^i)^m = b^m a^{im+pim(m-1)/2} \tag{4.13.a}$$

in G. From (4.13.a), one shows that $(ba^i)^m = (ba^j)^n$ iff $m \equiv n \equiv 0 \pmod{p^2}$, provided $i \not\equiv j \bmod p$. Hence the subgroups $U_i = \langle ba^i \rangle$ $(i = 1, \ldots, p)$ and $U_{p+1} = \langle a \rangle$ form a $(p^2, p+1; 1)$-pcp in G which meets the bound in (4.11.a). (Note that, on the other hand, for $i \equiv j \pmod p$, also $m \equiv n \equiv p \pmod{p^2}$ implies $(ba^i)^m = (ba^i)^n$, so that the exhibited pcp really is maximal, as it should be.)

5. TRANSLATION NETS WITH $\mu = 1$

In this section, we consider the special case $\mu = 1$ only. Here normality of the components has interesting consequences as Sprague has shown in [25, 2.8 and 5.2].

PROPOSITION 5.1 : (Sprague) *Let* $\mathcal{U}$ *be an* $(s, r; 1)$*-pcp with two normal components* U, V *and* $r \geq 3$. *Then* $U \cong V$ *and* $G = U \oplus V$. *If* $\mathcal{U}$ *has three normal components, then* G *is abelian.*

Sprague has also given examples of translation nets with $\mu = 1$ and $r \geq 3$ and exactly two normal components. Note that 5.1 does not hold for $\mu > 1$; e.g. all $p+1$ components of Example 4.13.a for $m = 1$ are normal, as they are maximal subgroups of the non-abelian p-group of order p^3. Sprague has also shown in [25, Theorem 5.5], that a translation net with $\mu = 1$ and two normal parallel classes is elementary abelian if $r \geq s/3 + 2$. We will improve this considerably for abelian groups and also give a related result for arbitrary groups. The following result is essentially a combination of an idea of S.S. Sane with a lemma communicated to the author by P.J. Cameron.

PROPOSITION 5.2 : *Let* G *be any non-abelian group of order* s^2 *and assume the existence of an* $(s, r; 1)$*-pcp* $\mathcal{U}$ *in* G. *Then*

$$r \leq \left[(2s - 2)/(p + 1)\right] + 2, \tag{5.2.a}$$

where p *is the smallest prime divisor of* s.

PROOF : As G is non-abelian, at most two components of $\mathcal{U}$ are normal. Also note that $U \cap V^g = 1$ for any two distinct components U, V and each $g \in G$ (this is due to Sprague [25, Lemma 2.2]). Thus we want to have a lower bound on the number of elements in the union $\bigcup_{g \in G} U^g$, where U is not normal in G. As p is the smallest prime dividing s, the normalizer $N_G(U)$ of U has index at least p in G; hence U has at least p conjugates, say $U = U_1, \ldots, U_p$. For any two conjugates, their intersection $U_i \cap U_j$ has at least index p in U_i. Then it is clear that

$$|\bigcup_{i=1} U_i| \geq |U| (1 + (1-1/p) + (1-2/p)+\ldots+ (1-(p-1)/p)) = |U|(p+1)/2 .$$

Now counting only the non-identity elements of G and even assuming two subgroups to be normal, we obtain

$$(r-2)((p+1)/2)(s-1) + 2(s-1) \leq s^2 - 1 ,$$

which is (5.2.a). (If less than two subgroups are normal, this can be improved slightly.)

We next consider abelian p-groups again.

PROPOSITION 5.3 : *Let* G *be an abelian group of order* p^{2n} *and assume the existence of a* $(p^n, r; 1)$*-pcp* $\mathcal{U}$ *in* G *with* $r \geq 3$. *Then all components of* $\mathcal{U}$ *are isomorphic, say to* U *of order* p^n, *and* $G = U \oplus V$, *where* $U, V \in \mathcal{U}$. *Let the type of* U *be* $(a_1, \ldots, a_m)$ *with* $a_1 \geq \ldots \geq a_m$ *and let* h *be the index with* $a_1 = \ldots = a_h > a_{h+1}$ *resp.* $h = m$ *if no such index exists. Then*

$$r \leq p^h + 1 . \qquad (5.3.a)$$

PROOF : Let W be any other component of G but U and V and assume $(u,v) \in W$ $(u \in U, v \in V)$. Then $o(u) = o(v)$, as otherwise some power $\neq 1$ of (u,v) would be in either U or V, contradicting the definition of a pcp for $\mu = 1$. Also, no u can occur twice as a first coordinate of an element of W, as this would yield a pair $(1,v)$ with $v \neq 1$ in W. Hence every u occurs precisely once as a first coordinate in W, and similarly for every $v \in V$. Now consider the element $z = (x_1, 1, \ldots, 1; y_1, \ldots, y_h, y_{h+1}, \ldots, y_m)$ of W where x_1 is a fixed element of order p^{a_1}. If we put

$x = x_1^{p^{a_1-1}}$, then $z^{p^{a_1-1}} = (x, 1, \ldots, 1; y_1', \ldots, y_h', 1, \ldots, 1)$ is in W; here $(y_1', \ldots, y_h', 1, \ldots, 1)$ has order p, as $(y_1, \ldots, y_m)$ has order p^{a_1} by the remarks above. But V contains exactly $p^h - 1$ elements $(y_1', \ldots, y_h'; 1, \ldots, 1)$ of order

p and none of these can be used in more than one of the remaining components W together with the first coordinate $(x, 1, \ldots, 1)$. Hence there are at most $p^h - 1$ choices for W, which proves the assertion.

COROLLARY 5.4 : *Let* G *be an abelian group of order* p^{2n} *and exponent* p^a. *Assume the existence of a* $(p^n, r; 1)$*-pcp in* G. *Then*

$$r \le p^{[n/a]} + 1 . \tag{5.4.a}$$

In particular, if G *is not elementary abelian, then*

$$r \le p^{n/2} + 1 . \tag{5.4.b}$$

This is enough to prove

THEOREM 5.5 : *Assume the existence of an* $(s, r; 1)$*-pcp in* G, *where* G *is abelian. Then* G *is elementary abelian if* $r > \sqrt{s} + 1$.

PROOF : This follows from (5.4.b) if s is a prime power. But if s is composite, then the smallest factor in its prime power factorization is at most $\sqrt{s}$ and so we would have $r \le \sqrt{s} + 1$ by (4.2.b) and (1.2.a).

Note that Theorem 5.5 does not hold in the non-abelian case; Sprague [25] has exhibited a non-abelian $(8, 4; 1)$-pcp. Corollary 5.4 and Theorem 5.5 improve the results in Section 4 for $\mu = 1$. If we combine Theorem 5.5 and Proposition 5.2, we get

COROLLARY 5.6 : *Assume the existence of an* $(s, r; 1)$*-pcp in* G, *and let* p *be the smallest prime dividing* s. *Then* G *is elementary abelian if* $r > [(2s - 1)/(p + 1)] + 2$.

Note that this bound is inferior to Sprague's result mentioned above if $p = 2$ or 3 (though the assumption of the existence of

2 normal components has been removed). We give one more example showing that Proposition 5.3 is best possible.

EXAMPLE 5.7 : Take $G = \mathbb{Z}_{p^2} \oplus \mathbb{Z}_{p^2}$. Then $U_1 = \langle(1,0)\rangle$, $U_2 = \langle(0,1)\rangle$ and $U_i = \langle(1, i-2)\rangle$ for $i = 3,\dots, p+1$ form a $(p^2, p+1; 1)$-pcp. As there is a $(p, p+1; 1)$-pcp in $\mathbb{Z}_p \oplus \mathbb{Z}_p$, we may obtain a $(p^{a+2}, p+1; 1)$-pcp in $\mathbb{Z}_{p^2} \oplus \mathbb{Z}_{p^2} \oplus \mathbb{Z}_p \oplus \dots \oplus \mathbb{Z}_p$ by Proposition 3.1 ; hence (5.3.a) is best possible for $h = 1$ and all types $(2,1,\dots,1)$.

Next, we improve a non-existence result of Sprague who has given the following result for prime powers q in [25, Corollary 2.7].

THEOREM 5.8 : *Let* q *be any odd number. Then there is no* $(2q, 4; 1)$*-pcp* .

PROOF : Let $\mathcal{U}$ be a $(2q, 4; 1)$-pcp in some group G of order $4q^2$, say $\mathcal{U} = \{U_1,\dots, U_4\}$. Now U_4 is a system of coset representatives for each of U_1, U_2, U_3 ; in other words, U_4 is an automorphism group of the $(2q, 3; 1)$-translation net Σ belonging to $\{U_1, U_2, U_3\}$ and U_4 acts regularly on each parallel class of Σ . Hence the dual Σ^* of Σ will be a class regular $(2q, 3; 1)$-TD with respect to U_4 , contradicting Theorem 1.7 .

Finally, we will briefly examine a construction for certain difference sets essentially due to Hall [10] . He gave a particular example for the following result.

PROPOSITION 5.9 : *The existence of a* $(2n, n; 1)$*-pcp in* G *implies that of a* $(4n^2, 2n^2\text{-}n, n^2\text{-}n)$*-difference set in* G .

PROOF : If $\mathcal{U} = \{U_1,\dots, U_n\}$ is the given pcp, put $D = \bigsqcup_{i=1}^{n} U_i \setminus \{0\}$ (where G is written additively). It is easily seen that D is the desired difference set.

This result seems to be folklore and is also proved somewhat indirectly in Sprague [25] who also has determined all $(6,3;1)$- and all $(8,4;1)$-pcp's.

6. SYMMETRIC AND COMPLETE TRANSLATION NETS

We finally use the results of Section 4 together with known results to obtain some more information about symmetric or complete translation nets.

THEOREM 6.1 : *A complete* (s,μ)*-translation net* Σ *with translation group* G *exists iff*

(i) s *is a prime power;*

(ii) $\mu = s^d$ *for some non-negative integer* d*;*

(iii) G *is elementary abelian.*

PROOF : Note that Σ is an affine 2-design and that the translation group of G consists of "translations" in the sense of Schulz [23], cf. Remark 2.5 . Hence the main result of Schulz implies that G is a p-group of exponent p . This proves (i); then the divisibility condition for completeness given by (1.2.a) shows that (ii) holds (cf. the proof of Jungnickel [15, Theorem 5.4]). But then G is elementary abelian by Corollary 4.11 .

This result was previously known when Σ also admits a non trivial central dilatation, cf. Dembowski [6, 2.3.24b]. In fact, the central dilatations of Σ fixing G (G written additively) correspond to those automorphisms of G fixing all components of the pcp $\mathcal{U}$ corresponding to Σ (Schulz [23, 4.5]). This yields

COROLLARY 6.2 : *Let* Σ *be a complete* (s,μ)*-translation net,* s *odd,* G *the translation group and* D *the group of dilatations of* Σ *(as in [6] and [23]). Then* $G \lneq D$ *; if* s *is a power of the prime* p*, then* D_0 *contains a cyclic subgroup of order* $p-1$.

PROOF : Clearly multiplication with an element of the multiplicative group $\mathbf{Z}_p^*$ of the prime field on p elements leaves each subgroup of order p of G invariant, hence each component of $\mathcal{U}$.

We remark that there are complete translation nets *not* isomorphic to the system of points and hyperplanes of some affine space, see e.g. Mavron [21]. If G is abelian, then a simple proof of Theorem 6.1 may be found in [15, Theorem 5.4]. For nilpotent G, the result follows directly from our Theorem 4.12. But in the general case, it seems necessary to quote the rather deep result of Schulz. As no analogue of his result is known for symmetric nets, we are only able to show

THEOREM 6.3 : *A symmetric translation* (s,μ)*-net with nilpotent translation group* G *exists iff*

(i) s *and* μ *are powers of the same prime* p;

(ii) G *is elementary abelian.*

PROOF : That these conditions are necessary follows immediately from Theorem 4.12. On the other hand, we have constructed examples in Corollary 4.5 (really due to [16]).

7. CONCLUDING REMARKS

We want finally to point out some of the problems that remain open.

(a) The method of Section 4 does not seem to yield any better information on the precise value of $Ea(p^i, p^j)$. This is the most important open problem and is, as we have seen, equivalent to the analogous problem for partial t-spreads over $GF(p)$. If it could be solved, we would have the exact upper bound for the nilpotent case and all values of s and μ. Currently, we only have this knowledge for $(s,\mu) = 1$ and a few other cases.

(b) In this connection, the author conjectures that the precise value of $Ea(p^i, p^j)$ is always given by the lower bound in (3.8.a). It also remains possible that this is even true for nets in general.

(c) The condition that G is nilpotent in Theorem 6.3 should certainly be unnecessary. In fact, it should be possible to prove that G has to be a p-group in a manner similar to the proof of Schulz' theorem using the classification of all finite groups with a partition.

(d) In general, one would expect that non-nilpotent groups yield inferior bounds than in the nilpotent case.

ACKNOWLEDGEMENT

The author would like to thank the Deutsche Forschungsgemeinschaft for its support via a Heisenberg grant during the time of this research.

BIBLIOGRAPHY

1. J. André, "Über nicht-desarguesche Ebenen mit transitiver Translationsgruppe", *Math. Z.*, 60 (1954), 156-186.
2. A. Beutelspacher, "Partial spreads in finite projective spaces and partial designs", *Math. Z.*, 145 (1975), 211-229.
3. R.C. Bose and K.A. Bush, "Orthogonal arrays of strength two and three", *Ann. Math. Statistics*, 23 (1952), 508-524.
4. R.H. Bruck, "Finite nets. I. Numerical invariants", *Canadian J. Math.*, 3 (1951), 94-107.
5. R.H. Bruck, "Finite nets. II. Uniqueness and imbedding", *Pacific J. Math.*, 13 (1963), 421-457.
6. P. Dembowski, *Finite Geometries* (Springer-Verlag, Berlin, 1968).
7. D.A. Drake, "Partial λ-geometries and generalized Hadamard matrices over groups", *Canad. J. Math.*, 31 (1979), 617-627.
8. D.A. Drake and J.W. Freeman, "Partial t-spreads and group constructible (s,r,μ)-nets", *J. Geom.*, 13 (1979), 210-216.
9. D.A. Drake and D. Jungnickel, "Klingenberg structures and partial designs II. Regularity and uniformity", *Pacific J. Math.*, 77 (1978), 389-415.

10. M. Hall Jr., "Difference sets", *Combinatorics*, Tract 57 (Math. Centrum, Amsterdam, 1975).
11. M. Hall Jr. and L.J. Paige, "Complete mappings of finite groups", *Pacific J. Math.*, 5 (1955), 541-549.
12. T.C. Hine and V.C. Mavron, "Embeddable transversal designs", *Discrete Math.*, 29 (1980), 191-200.
13. B. Huppert, *Endliche Gruppen I.* (Springer-Verlag, Berlin, 1967).
14. D. Jungnickel, "On difference matrices, resolvable transversal designs and generalized Hadamard matrices", *Math. Z.*, 167 (1979), 49-60.
15. D. Jungnickel, "Some new combinatorial results on finite Klingenberg structures", *Utilitas Math.*, 16 (1979), 249-269.
16. D. Jungnickel, "A class of uniform Klingenberg matrices", *Ars. Combin.*, to appear.
17. D. Jungnickel, "On automorphism groups of divisible designs", *Canad. J. Math.*, submitted.
18. D. Jungnickel, "On difference matrices and regular latin squares", *Abh. Math. Sem. Univ. Hamburg*, to appear.
19. D. Jungnickel, "Transitive symmetric nets", *Arch. Math. (Basel)*, to appear.
20. D. Jungnickel and S.S. Sane, "On extensions of nets", to appear.
21. V.C. Mavron, "Translations and parallel classes of lines in affine designs", *J. Combin. Theory Ser. A.*, 22 (1977), 322-330.
22. V.C. Mavron, "Resolvable designs, affine designs and hypernets", to appear.
23. R.H. Schulz, "Über Blockpläne mit transitiver Dilatationsgruppe", *Math. Z.*, 98 (1967), 60-82.
24. S.S. Shrikhande, "Generalized Hadamard matrices and orthogonal arrays of strength two", *Canad. J. Math.*, 16 (1964), 736-740.
25. A.P. Sprague, "Nets with Singer group fixing each parallel class", to appear.

Mathematisches Institut
Justus-Liebig-Universität
Arndtstr. 2
6300 Giessen
Federal Republic of Germany

TRANSLATION PLANES HAVING PSL(2,w) OR SL(3,w) AS A COLLINEATION GROUP

Michael J. Kallaher

A recurring theme in the theory of affine and projective planes is the interplay between a plane and a subgroup of the plane's collineation group. One particular aspect of this theme is the investigation of planes having a particular type of group as a collineation group. In this talk we report on joint work with V. Jha; we investigated finite translation planes with either $PSL(2,w)$ or $SL(3,w)$, when w is a prime power, as a collineation group. We will limit ourselves to giving the main results and outlining some proofs; complete proofs and related results will appear elsewhere. (See Jha and Kallaher [4].)

1. NOTATION

All planes discussed will be finite (affine) translation planes. To save time and space we adopt the following conventions which will be used for the remainder of the talk:

Π	:	finite translation plane
p	:	characteristic of Π (i.e. Π has order p^r where p is a prime and r a positive integer)
ℓ_∞	:	line at infinity of Π
$(Q, +, \cdot)$	:	a quasi-field coordinatizing Π
$K = K(\Pi)$	:	kernel of Π (i.e. $K = \{k \mid k \in Q,\ (ka)b = k(ab)$ and $k(a+b) = ka + kb$ for all $a, b \in Q\}$)
$q = p^k$	:	order of K
d	:	dimension of Π over K (i.e. $q^d = p^r$, Q has dimension d and Π dimension $2d$ as vector spaces over K)
$G = G(\Pi)$	:	collineation group of Π
$T = T(\Pi)$	:	translation group of Π

$C = C(\Pi)$: translation complement of Π (i.e. $C = G_O$ where O is the origin of the coordinate system). This is a subgroup of $\Gamma L(2d,q)$.

$LC = LC(\Pi)$: linear translation complement of Π (i.e. $LC = C \cap GL(2d,q)$)

component : line through the origin O. This is a subspace of Π of dimension d over K.

2. THE LORIMER-RAHILLY AND JOHNSON-WALKER TRANSLATION PLANES.

Besides certain translation planes of characteristic two possessing $PSL(2,2^s) = SL(2,2^s)$, where $s \geq 2$, as collineation groups, there are two other known translation planes, both of order 16, having a collineation group isomorphic to $PSL(2,w)$ for some prime power w. These are the Lorimer-Rahilly and Johnson-Walker translation planes which both possess the group $PSL(2,7)$ as a subgroup of the linear translation complement. These planes also are unique in that they have a collineation group isomorphic to $SL(3,w)$ for some prime power w. (Recall that $PSL(2,7) = SL(3,2)$.)

The action of $PSL(2,7) = SL(3,2)$ on these two translation planes is as follows. In both planes the group $PSL(2,7)$ is a subgroup of the group LC, the involutions in $PSL(2,7)$ are Baer, and on ℓ_∞ the group $PSL(2,7)$ has three fixed points and an orbit of length 14. In the Lorimer-Rahilly plane the group $PSL(2,7)$ fixes pointwise a subplane of order two with respect to which the group is tangentially transitive. In the Johnson-Walker plane each Sylow 7-subgroup of $PSL(2,7)$ fixes a subplane of order two pointwise but these eight subplanes are distinct subplanes of a 3-net in the plane.

Moreover, both the Lorimer-Rahilly translation plane and the Johnson-Walker translation plane are derivable from the translation plane coordinatized by the unique semifield of order 16 with kernel $K = GF(4)$. Thus one can be obtained from the other by means of a suitable net replacement. (See Johnson and Ostrom [5].)

3. QUESTIONS.

The above examples lead naturally to the following two questions:

A. What finite translation planes have a collineation group $H = PSL(2,w)$ for some prime power w ?

B. What finite translation planes have a collineation group $H = SL(3,w)$ for some prime power w ?

The second question was originally due to Lorimer. (In his version, the prime power w is q, the order of the kernel.)

It is not difficult to show that if a translation plane Π has a collineation group $H = PSL(2,w)$, then $w > 3$ and $p > 3$ imply the linear translation complement $LC(\Pi)$ has $PSL(2,w)$ as a subgroup. Note that $PSL(2,2) = S_3$ and $PSL(2,3) = A_4$. Similarly, if Π has a collineation group $H = SL(3,w)$ then $LC(\Pi)$ has $SL(3,w)$ as a subgroup (without restriction on w or p). It follows that there is almost no restriction, assuming in questions A and B that the groups $PSL(2,w)$ and $SL(3,w)$ are in $LC(\Pi)$.

In our investigation the following interesting result, a special case of a more general theorem due to Ostrom [6], plays an important role in limiting the possibilities for the translation plane Π.

THEOREM 1 : *Let the characteristic* p *of the finite translation plane* Π *be odd. If the linear translation complement* $LC(\Pi)$ *contains a Klein-four group* H *in which all the involutions are mutually conjugate (within* $LC(\Pi)$*), then* 4 *divides the dimension* d *of* Π *and the involutions of* H *are Baer.*

PROOF : (Outline) If one of the involutions in H is a homology then all three involutions of H are homologies. Since they are conjugate and H has no nontrivial elation, these involutory homologies have the same center $P \in \ell_\infty$ and the same axis ℓ, a component of Π. But this contradicts the fact that H is a Frobenius complement on the component OP.

Hence the three involutions $\alpha_1, \alpha_2, \alpha_3$ of H must be Baer and $2 \mid d$. For each i, let Π_i be the Baer subplane fixed pointwise by α_i. By a result of Foulser [1] the planes Π_1, Π_2, Π_3 are distinct. Note that α_i fixes each Π_j. If α_i induces an involutory homology in Π_j then an application of Maschke's theorem yields that $\alpha_i \alpha_j = \alpha_k$ is a homology, a contradiction. Thus α_i induces a Baer involution in Π_j and hence $2 \mid d/2$, or $4 \mid d$.

REMARK 1: Note that Theorem 1 applies to many classical groups including S_n, A_n, PSL(2,w), and SL(n,w).

4. THE GROUP PSL(2,w) AS A COLLINEATION GROUP.

Our first major result is the following theorem.

THEOREM 2: *If the finite translation plane* Π *has a collineation group* $H = PSL(2,w)$ *with* $w \geq 3$, *then one of the following statements holds:*

(i) *The characteristic* p *is odd and* 4 *divides the dimension* d.

(ii) *The characteristic* $p = 2$, *the prime power* w *is odd and at least* 7, *and* 2 *divides the dimension* d.

(iii) *The characteristic* $p = 2$ *and* $w = 3, 5$ *or* 2^s *with* $s \geq 2$.

PROOF: If p is odd then an application of Theorem 1 gives statement (1). If $p = 2$ and w is odd with $w \geq 7$, an application of the Hering-Ostrom theorem (Hering [3]) on groups generated by affine elations gives $2 \mid d$.

REMARK 2: The restrictions on w in Theorem 2 are necessary as the following examples show.

(a) The group $PSL(2,2) = SL(2,2) = S_3$ occurs as the collineation group of many translation planes, including nearfield planes, Hall planes and generalized André planes.

(b) The group $PSL(2,3) = A_4$ occurs in translation planes of odd dimension and even order; examples include semifield planes of characteristic two.

(c) The group $PSL(2,5) = SL(2,4)$ occurs in desarguesian planes or characteristic two (and these have dimension 1 over their kernel) as well as certain Hall planes of characteristic two.

(d) The groups $PSL(2,2^s) = SL(2,2^s)$ with $s \geq 2$ occur in both desarguesian and Hall planes of characteristic two.

Looking at the orbit structure of the group $PSL(2,7)$ on the line ℓ_∞ in both the Lorimer-Rahilly and the Johnson-Walker plane, it appears reasonable to add the hypothesis that $PSL(2,w)$ fixes pointwise a nontrivial subset Δ of ℓ_∞ and is transitive on $\ell_\infty - \Delta$. The following can then be obtained.

THEOREM 3 : *Let* Π *be a finite translation plane of characteristic* p *having a subgroup* $H = PSL(2,p^s)$, *where* $s \geq 1$, *in its linear translation complement* $LC(\Pi)$. *If* H *fixes pointwise a subset* Δ *of* ℓ_∞ *with* $|\Delta| = p^t + 1$ *for some integer* $t \geq 1$ *and* H *is transitive on* $\ell_\infty - \Delta$, *then* $p = 2$.

PROOF : (outline) Assume p is odd. Using (by now) standard arguments involving a prime p-primitive divisor (of $p^{2s} - 1$), we obtain $s \leq r/2$ where p^r is the order of Π. Using the transitivity of H on $\ell_\infty - \Delta$, we obtain $t \leq s$ and $(r-t)|2s$. It follows that $t = s = r/2$. Thus the p-elements of H are Baer p-elements. Using results of Foulser [2] on Baer p-elements, we find $p = 3$, $s = 1$, and $r = 2$. This contradicts (i) of Theorem 2.

5. THE GROUP SL(3,w) AS A COLLINEATION GROUP

For the group $SL(3,w)$ we get the following results similar to Theorem 2.

THEOREM 4 : *If the finite translation plane* Π *has the group* $H = SL(3,w)$ *as a collineation group then one of the following statements holds:*

(i) *The characteristic* p *is odd and* 4 *divides the dimension* d .

(ii) *The characteristic* $p = 2$ *and the dimension* d *is even and at least* 4 .

PROOF : (Outline) If p is odd Theorem 1 gives statement (i). If $p = 2$ an application of the Hering-Ostrom theorem on affine elations shows that H contains Baer Involutions. This yields statement (ii).

Analogous to Theorem 3 we have the following.

THEOREM 5 : *Let* Π *be a finite translation plane of characteristic* p *having a subgroup* $H = SL(3,p^s)$, *where* s *is a positive divisor of* k, *in its linear translation complement* $LC(\Pi)$. *If* H *leaves invariant a subset* Δ *of* ℓ_∞ *with* $|\Delta| = p^s + 1$ *and* H *is transitive on* $\ell_\infty - \Delta$, *then the following statements hold:*

(i) *The characteristic* $p = 2$, *the dimension* $d = 4$, *and* $2^s = q = 2^k$.

(ii) *The group* H *fixes the set* Δ *pointwise.*

PROOF : (Outline) By Theorem 4 the dimension d is at least 4 . Using the fact that H is transitive on $\ell_\infty - \Delta$ and the fact that $s \mid k$, a number-theoretic argument shows that $d \leq 4$. Hence $d = 4$. It then follows that $s = k$, or $p^s = q = p^k$. Let u be a prime divisor of $q^2 + q + 1$ not dividing $q^4 - 1$ and let $\sigma \in H$ be an element of order u . Standard arguments show that σ must fix pointwise the set Δ and act semi-regularly on $\ell_\infty - \Delta$. Since H is generated by its elements of order u , statement (ii) holds.

To prove $p = 2$, assume instead that it is odd. Then $H = SL(3,q)$ has an involution τ whose centralizer contains

$H_0 = SL(2,q)$. The involution τ is a Baer involution and the group H_0 leaves invariant the associated Baer subplane Π_0. By Foulser [1] the group $\overline{H}_0$ of collineations induced by H_0 on Π_0 must be $PSL(2,q)$. Theorem 1 then says that Π_0 has dimension 4 over the kernel $K = GF(q)$. Since Π itself has dimension 4 over K, this gives a contradiction. Hence $p = 2$.

Using the work of Johnson and Ostrom [5] on translation planes of order 16 we have the following immediate corollary

THEOREM 6: *Let* Π *be a finite translation plane of characteristic* p *having a collineation group* $H = SL(3,p)$. *If* H *leaves invariant a subset* Δ *of* ℓ_∞ *with* $|\Delta| = p+1$ *and is transitive on* $\ell_\infty = \Delta$, *then* Π *is either the Lorimer-Rahilly translation plane or the Johnson-Walker translation plane.*

REMARK 3: Obviously, the above results do not answer completely questions A and B stated in Section 3. But they do give strong evidence that the Lorimer-Rahilly and Johnson-Walker translation planes are the only answers to both questions.

Vikram Jha has just recently investigated translation planes of characteristic p having a subgroup H of the linear translation complement with $H = PSL(2,w)$ and leaving invariant a subset Δ of ℓ_∞ with $|\Delta| = p+1$ while being transitive on $\ell_\infty - \Delta$. He has shown that the only such planes are the Lorimer-Rahilly and Johnson-Walker translation planes.

BIBLIOGRAPHY

1. D.A. Foulser, "Subplanes of partial spreads in translation planes", *Bull. London Math. Soc.*, 4 (1972), 32-38.
2. D.A. Foulser, "Baer p-elements in translation planes", *J. Algebra*, 31 (1974), 354-366.
3. C. Hering, "On shears of translation planes", *Abh. Math. Sem. Univ. Hamburg*, 37 (1972), 258-268.
4. V. Jha and M.J. Kallaher, "On the Lorimer-Rahilly and Johnson-Walker translation planes", to appear.
5. L. Johnson and T.G. Ostrom, "Tangentially transitive planes of order 16", *J. Geom.*, 10 (1977), 146-163.

6. T.G. Ostrom, "Elementary abelian 2-groups in finite translation planes", to appear.

Department of Mathematics
Washington State University
Pullman
Washington 99164
U.S.A.

SEQUENCEABLE GROUPS : A SURVEY

A.D. Keedwell

A finite group $(G,\cdot)$ of order n is said to be *sequenceable* if its elements can be arranged in a sequence $a_0, a_1, a_2, \ldots, a_{n-1}$ in such a way that the partial products $b_0 = a_0$, $b_1 = a_0a_1$, $b_2 = a_0a_1a_2, \ldots,$ $b_{n-1} = a_0a_1 \ldots a_{n-1}$ are all distinct and so are again the elements of G. It is immediately evident that for this to be possible, the first element a_0 must be equal to the identity element e of G.

Sequenceable groups arise in connection with the construction of so-called complete latin squares. A latin square on n symbols is called *row complete* if each of the $n(n-1)$ ordered pairs of distinct symbols occurs in adjacent positions (cells) in exactly one row of the latin square. Since there are $n-1$ pairs of adjacent cells in each row of the square, we get an exact match between the ordered pairs and the places in which they may occur. An example is given in Figure 1. There is an analogous definition of *column completeness*. A latin square which is both row complete and column complete is called *complete*. The square given in Figure 1 is a complete latin square.

0	1	3	2
1	2	0	3
3	0	2	1
2	3	1	0

FIGURE 1

A practical application of row complete latin squares of small size is to the statistical design of sequential experiments in which several treatments are to be administered in succession to a number of different subjects. When referred to in this connection, these squares are usually called *balanced crossover designs* or *balanced changeover designs*. It was discovered very early on by statisticians that it is easy to obtain row complete (and complete) latin squares of all even orders (see, for example, [14] and [3]) but that to construct such squares of odd order cannot easily be accomplished. It has been shown more recently by several authors that row complete latin squares of orders 3, 5 and 7 do not exist. Quite recently, Mertz and Tillson have independently obtained row complete latin squares of order 9 by means of computer searches but these squares are not based on groups and it is not known whether they can be made column complete as well as row complete. (A latin square which is the multiplication table of a group is said to be *based on* that group.)

We return to the question of sequenceable groups. By a theorem of Paige [12], the product of all the elements of an abelian group is the identity element e unless the group contains a unique element of order 2. In the latter case, the product is equal to the element of order 2. To see this, we arrange the elements in a sequence $e\,g_1g_1^{-1}g_2g_2^{-1}\ldots g_rg_r^{-1}h_1h_2\ldots h_s$ where each non-identity element g_i whose order is not two is immediately followed by its inverse, and the elements $h_1,h_2,\ldots,h_s$ are the elements of order two. If there is only one element of order two, it is obvious that the product is equal to this element. If there are no elements of order two, the product is e. Finally, suppose that there is more than one element of order two. In that case, the elements of order two, together with the identity element e, form a subgroup H of order $s+1=2^t$ for some integer $t>1$ with generating elements $u_1,u_2,\ldots,u_t$, say. Each h_i is expressible in the form $u_1^{\varepsilon_1}u_2^{\varepsilon_2}\ldots u_t^{\varepsilon_t}$, where $\varepsilon_i = 0$ or 1

$(i = 1,2,\ldots, t)$, and each expression of this form represents one of the elements of order two. Consequently, 2^{t-1} of the elements of order two involve the generator u_i. So, in the product of all the elements of order two, each generator u_i occurs an even number of times. Since $u_i^2 = e$, the product is equal to e.

It follows immediately from Paige's theorem that an abelian group cannot be sequenceable unless it has one and only one element of order two. For, the partial product b_0 is equal to a_0 $(= e)$ and so we require that $b_{n-1} \neq e$.

Gordon [7] showed in 1961 that the above necessary condition for an abelian group to be sequencesable is also sufficient. He proved: "A finite abelian group is sequenceable if and only if it is a direct product $A \times B$, where A is cyclic of order 2^k $(k > 0)$ and B is of odd order". Gordon also showed that, if $(G,\cdot)$ is a sequenceable group with partial products $b_0, b_1, \ldots, b_{n-1}$ then the latin square whose (i, j)th entry is $b_i^{-1} b_j$ is a complete latin square. Since, in particular, all cyclic groups of even order are sequenceable, this leads to an easy construction of row complete (or complete) latin squares based on such groups. A sequencing for the cyclic group Z_{2n} written additively as the group of residue classes modulo n is $0, 2n-1, 2, 2n-3, 4, 2n-5, 6, \ldots, 2n-2, 2n-(2n-1)$, and this leads to the following construction for a row complete latin square of order $2n$. In the first row, write $0, 2n-1, 1, 2n-2, 2, 2n-3, 3, \ldots, n-1, n$. To obtain the ith row add $i-1$ to each entry of the first row. To make the resulting square column complete as well as row complete, reorder the rows in such a way that the entries of the first column of the square are the same as is given above for the first row.

It has been pointed out by the present author that the multiplication table of a finite group can be written in the form of a row complete latin square if *and only if* the group is sequenceable and that, in consequence, a row complete latin square which is based on a group can always be made column complete as well as row complete by a suitable reordering of its rows; see [9]. We shall give the proof.

Suppose first that the row complete latin square $L = (g_{ij})$ is the multiplication table of a group $(G,\cdot)$ with elements $g_0, g_1, \ldots, g_{n-1}$, so that $g_{ij} = g_i g_j$. In that case $g_{i,j-1}^{-1} g_{ij} = g_{j-1}^{-1} g_i^{-1} g_i g_j = g_{j-1}^{-1} g_j = a_j$ say, for $i = 0,1,\ldots, n-1$. Suppose that $a_j = a_{j'}$ for $j' \neq j$. Then, because $g_{i',j'-1} = g_{i,j-1}$ for some value of i' (each element of G occurs exactly once in each column of L), we have $g_{i'j'} = g_{i',j'-1} a_{j'} = g_{i,j-1} a_j = g_{ij}$. However, this contradicts the row completeness of L. Thus, $a_{j'} \neq a_j$ unless $j' = j$. Consider now the first row of L. Its j th element is

$$g_{0j} = g_{00}(g_{00}^{-1} g_{01})(g_{01}^{-1} g_{02}) \ldots (g_{0,j-1}^{-1} g_{0j}) = g_{00} a_1 a_2 \ldots a_j .$$

Since the elements of the first row of L are all different, it follows that the partial products $\prod_{k=1}^{j} a_k$ for $j = 1,2,\ldots, n-1$ are all distinct, where the elements $a_1, a_2, \ldots, a_{n-1}$ are the non-identity elements of G. That is, the elements $e, a_1, a_2, \ldots, a_{n-1}$ form a sequencing for G.

Suppose now that L is the latin square whose (i,j)th element is $b_i^{-1} b_j$ where $b_k = e a_1 a_2 \ldots a_k$ for $k = 0,1,\ldots, n-1$ and where $e, a_0, a_1, \ldots, a_{n-1}$ is a sequencing of the group $(G,\cdot)$. We require to show that the ordered pair of elements u, v of G occur consecutively in some row of L. That is, we require to find integers i, j such that $b_i^{-1} b_{j-1} = u$ and $b_i^{-1} b_j = v$. From these two equations, $u a_j = v$. This determines j. Then $b_i = b_{j-1} u^{-1}$ and this fixes i. Thus, every pair of elements of G occurs exactly once and L is row complete. An entirely similar argument shows that L is also column complete.

After the work of Paige and Gordon which, taken together, determined the necessary and sufficient condition that an abelian group should be sequenceable, only very minor contributions to the question "Which finite groups are sequenceable?" were achieved until quite recently.

Gordon himself discovered [7] that the quaternion group of order 8 and the dihedral groups D_3 and D_4 of orders 6 and 8 are not sequenceable but that the dihedral group D_5 of order 10 is sequenceable. These facts were confirmed by Dénes and Török [4] in 1968. These authors showed further that the dihedral groups D_6, D_7 and D_8 are sequenceable by obtaining computer-generated sequencings for them. They were also able to generate sequencings for the non-abelian group of order 21. (Note that 21 is the smallest odd order for which a non-abelian group exists.) In the same year, Mendelsohn [11] independently showed that the latter group is sequenceable. He formed partial sequencings of the group by hand and then completed them by computer. In 1973, the present author [8] obtained by hand a sequencing for the non-abelian group of order 27 and exponent 9. Also in 1973, Wang [13] used a computer to obtain sequencings of the non-abelian groups of orders 39, 55 and 57.

In 1975 and 1976 some rather more far-reaching results concerning the sequenceability of dihedral groups were obtained. Anderson [1] indicated a way of reducing the question of whether certain of the dihedral groups are sequenceable or not to a simpler question about their derived groups (which are cyclic). He employed the concept of a patterned starter (which had earlier been used for the construction of Room squares) and the type of dihedral group he was able to deal with was a dihedral group D_p of order $2p$, where p is a prime which has a primitive root r such that $3r \equiv -1 \bmod p$. He was able to obtain sequencings for D_5, D_{11}, D_{17} and D_{23}. Probably the most important feature of his technique was his use of an idea subsequently made more explicit by Friedlander and called by the latter author a *quotient sequencing*. Anderson used the term *feasible string*.

For the purposes of the present survey paper, we shall make the following definition:

If α is any sequence $u_1, u_2, \ldots, u_n$ of elements of a group G, we shall denote by $P(\alpha)$ its sequence of partial products $v_1 = u_1$, $v_2 = u_1u_2$, $v_3 = u_1u_2u_3$, ..., $v_n = u_1u_2 \ldots u_n$. Let G be a non-abelian group of order pq, where p is a prime less than the integer q, and suppose that G has a normal subgroup H of order q. Let $G/H = gp\{x_1, x_2, \ldots, x_p\}$. A sequence α of length pq consisting of elements of G/H is called a *quotient sequencing* of G if each x_i, $1 \le i \le p$, occurs q times in both α and $P(\alpha)$.

In the case of the dihedral group D_p, we may take H as the unique normal subgroup of order p and write $D_p / H = gp\{1, x : x^2 = 1\}$. Then the quotient sequencing (feasible string) used by Anderson is the sequence comprising $\frac{1}{2}(p+1)$ ones followed by x repeated p times, followed by $\frac{1}{2}(p-1)$ ones. This has as its sequence of partial products: $\frac{1}{2}(p+1)$ ones, followed by the sequence x, one repeated $\frac{1}{2}(p-1)$ times, followed by x repeated $\frac{1}{2}(p+1)$ times.

In the year following publication of Anderson's paper [1], Friedlander [6] published a paper in which, with the aid of the same quotient sequencing as that used by Anderson, he established that, if p is a prime congruent to 1 modulo 4, then D_p is sequenceable. He also stated that he had found sequencings for D_7, D_{11}, D_{19} and D_{23}. We shall give an outline of Friedlander's proof.

We suppose that the quotient sequencing given above arises from a sequencing $a^{\alpha_1}, a^{\alpha_2}, \ldots, a^{\alpha_{k+1}}, ba^{\beta_1}, ba^{\beta_2}, \ldots, ba^{\beta_{2k+1}}$, $a^{\gamma_1}, a^{\gamma_2}, \ldots, a^{\gamma_k}$, where $p = 2k+1$ (prime) and $ab = ba^{-1}$. In that case the corresponding partial products are $a^{\alpha_1}, a^{\alpha_1+\alpha_2}, \ldots$, $a^{S_\alpha}, ba^{\beta_1-S_\alpha}, a^{\beta_2-\beta_1+S_\alpha}, ba^{\beta_3-\beta_2+\beta_1-S_\alpha}, a^{\beta_4-\beta_3+\beta_2-\beta_1+S_\alpha}, \ldots$, $ba^{S_\beta-S_\alpha}, ba^{S_\beta-S_\alpha+\gamma_1}, ba^{S_\beta-S_\alpha+\gamma_1+\gamma_2}, \ldots, ba^{S_\beta-S_\alpha+S_\gamma}$, where

$S_\alpha = \alpha_1 + \alpha_2 + \ldots + \alpha_{k+1}$, $S_\beta = \beta_{2k+1} - \beta_{2k} + \beta_{2k-1} - \ldots + \beta_3 - \beta_2 + \beta_1$, and $S_\gamma = \gamma_1 + \gamma_2 + \ldots + \gamma_k$.

If the above sequence is indeed to be a sequencing of D_p, then the following four conditions must be satisfied.

(i) The integers $\alpha_1, \alpha_2, \ldots, \alpha_{k+1}, \gamma_1, \gamma_2, \ldots, \gamma_k$ must be all different, modulo p ;

(ii) the integers $\beta_1, \beta_2, \ldots, \beta_{2k+1}$ must be all different modulo p ;

(iii) $\alpha_1, \alpha_1+\alpha_2, \ldots, S_\alpha, \beta_2-\beta_1+S_\alpha, \beta_4-\beta_3+\beta_2-\beta_1+S_\alpha, \ldots,$ $\beta_{2k}-\beta_{2k-1}+\ldots+\beta_2-\beta_1+S_\alpha$ must be distinct, modulo p ; and

(iv) $\beta_1-S_\alpha, \beta_3-\beta_2+\beta_1-S_\alpha, \ldots, S_\beta-S_\alpha, S_\beta-S_\alpha+\gamma_1,$ $S_\beta-S_\alpha+\gamma_1+\gamma_2, \ldots, S_\beta-S_\alpha+S_\gamma$ must be distinct, modulo p .

If we write $\delta_i = \beta_{2i} - \beta_{2i-1}$ and $\varepsilon_i = \beta_{2i+1} - \beta_{2i}$ for $1 \le i \le k$ and $\varepsilon_0 = \beta_1 - S_\alpha$ then conditions (iii) and (iv) can be re-written as

(iii)' the partial sums of the sequence $\alpha_1, \alpha_2, \ldots, \alpha_{k+1},$ $\delta_1, \delta_2, \ldots, \delta_k$ must be distinct, modulo p ; and

(iv)' the partial sums of the sequence $\varepsilon_0, \varepsilon_1, \ldots, \varepsilon_k,$ $\gamma_1, \gamma_2, \ldots, \gamma_k$ must be distinct, modulo p .

Also, condition (ii) is equivalent to

(ii)' the partial sums of the sequence $\varepsilon_0, \delta_1, \varepsilon_1, \delta_2, \varepsilon_2,$ $\ldots, \delta_k, \varepsilon_k$ must be distinct, modulo p .

Friedlander showed that, for $p \equiv 1 \mod 4$, the conditions (i), (ii)', (iii)', (iv)' are all satisfied when we choose the sequences $\alpha, \gamma, \delta, \varepsilon$ to be as follows:

$$\{\alpha_1, \alpha_2, \ldots, \alpha_{k+1}\} = \{0, 2, 4, \ldots, 2k\}$$
$$\{\gamma_1, \gamma_2, \ldots, \gamma_k\} = \{1, 3, 5, \ldots, 2k-1\}$$
$$\{\delta_1, \delta_2, \ldots, \delta_k\} = \{-f, -3f, -5f, \ldots, -(2k-1)f\}$$
$$\{\varepsilon_0, \varepsilon_1, \ldots, \varepsilon_k\} = \{0, 2f, 4f, \ldots, 2kf\}$$

where f is an integer which is a non-square modulo p .

Friedlander also conjectured that, for all primes p congruent to -1 modulo 4, D_p is sequenceable. Recent work of the present author reinforces this conjecture. Indeed, he further conjectures that D_n is sequenceable for all odd integers n and has obtained sequencings for all such dihedral groups up to and including D_{37}. An undergraduate student (G.B. Hoghton of the University of Surrey) has proved that if p is an odd prime not equal to 5 such that $\frac{1}{2}(p-1)$ is a primitive root of p, then D_p is sequenceable. Also, he has shown that the necessary and sufficient condition that $\frac{1}{2}(p-1)$ should be a primitive root of p is that either (i) $p \equiv 5 \mod 8$ and 2 is a primitive root of p or (ii) $p \equiv 7 \mod 8$ and 2 belongs to the exponent $\frac{1}{2}(p-1)$. The case (i) is embraced by Friedlander's result. The case (ii) shows that, in particular, the dihedral groups D_{47}, D_{71} and D_{79} are sequenceable. Anderson [2] has proved that there are non-abelian sequenceable groups of arbitrarily large even order. Also, Eynden [5] has proved that every countably infinite group is sequenceable.

Apart from the above-mentioned results, the only other significant result known to the author is the proof, obtained earlier this year, that if p and q are distinct odd primes with $q = 2ph+1$ for some integer h (which is the necessary and sufficient condition that a non-abelian group of order pq exists) and if p has 2 as a primitive root, then the non-abelian group of order pq is sequenceable; see [10]. We shall end our survey by giving an outline of the method used to prove this result.

Once again the concept of quotient sequencing was used. Suppose that p is an odd prime which has 2 as a primitive root and let σ satisfy the congruence $2\sigma \equiv 1 \mod p$. Then σ also is a primitive root of p since $\sigma^h \equiv 1 \mod p$ implies $2^h \equiv 1 \mod p$. If G is a non-abelian group of order pq then $q = 2ph+1$ for some integer h and G has a unique normal subgroup H of order q. Let $G/H = \mathrm{gp}\{x : x^p = 1\}$. Then the following is a quotient sequencing of G : a sequence of $2ph$ ones, followed by x, followed by a sequence of $2ph-1$ $(p-1)$-tuples $x^{\sigma-1}, x^{\sigma^2-\sigma}, x^{\sigma^3-\sigma^2}, \ldots, x^{\sigma^{p-2}-\sigma^{p-3}}, x^{1-\sigma^{p-2}}$, where indices are computed modulo p, followed by a sequence $x^{\sigma-1}, x^{\sigma^2-\sigma}, \ldots, x^{\sigma^{p-2}-\sigma^{p-3}}, 1, x^2, x^4, x^8, \ldots, x^{2^{p-2}}, x^{1-\sigma^{p-2}}$. We leave the reader to check that the partial products of this quotient sequencing are $2ph$ ones followed by the sequence $x, x^{\sigma}, x^{\sigma^2}, \ldots, x^{\sigma^{p-2}}$ repeated $2ph-1$ times, and then the sequence $x, x^{\sigma}, x^{\sigma^2}, \ldots, x^{\sigma^{p-2}}, x^{\sigma^{p-2}}, x^{\sigma^{p-3}}, x^{\sigma^{p-4}}, \ldots, x^{\sigma^2}, x^{\sigma}, x, 1$. We try now to construct a sequencing of our group G of order pq of which the above is the quotient sequencing under the mapping $\phi : G \to G/H$.

Let r be a primitive element of $GF[q]$. Then, we observe that, since $q = 2ph+1$ is prime, the cyclic subgroup $H = gp\{a : a^q = e\}$ of the group $G = gp\{a, b : a^q = e,\ b^p = e,\ ab = ba^s,\ \text{where } s^p \equiv 1 \bmod q\}$ has a partial sequencing $e, a^{r^h-r^{h-1}}, a^{r^{h+1}-r^h}, \ldots, a^{r^{q-2}-r^{q-3}}, a^{1-r^{q-2}}, a^{r-1}, \ldots, a^{r^{h-2}-r^{h-3}}$ with partial products $e, a^{r^h-r^{h-1}}, a^{r^{h+1}-r^{h-1}}, \ldots, a^{r^{q-2}-r^{h-1}}, a^{1-r^{h-1}}, a^{r-r^{h-1}}, \ldots, a^{r^{h-2}-r^{h-1}}$. This may be used to replace the first $2ph(=q-1)$ ones in the quotient sequencing. The element of H which does not appear in this partial sequencing is $a^{r^{h-1}-r^{h-2}}$ and so this element must replace the last 1 in the quotient sequencing. The element of H which does not appear in the sequence of partial products is $a^{-r^{h-1}}$ and so this element must replace the last 1 in the partial products of the quotient sequencing : that is, $a^{-r^{h-1}}$ must be the product of all the elements of G when they have been ordered in the required manner.

The remainder of the proof follows the style of Friedlander's argument for the dihedral group D_p, although it is considerably more complicated. To take the case $p = 3$ as an example, we suppose that the above quotient sequencing arises from a sequencing of G of the form $e, a^{r^h-r^{h-1}}, a^{r^{h+1}-r^h}, \ldots, a^{r^{h-2}-r^{h-3}}$, $ba^{\alpha}, ba^{\alpha_1}, b^2a^{\beta_1}, ba^{\alpha_2}, b^2a^{\beta_2}, \ldots, ba^{\alpha_{q-2}}, b^2a^{\beta_{q-2}}, ba^{\alpha_{q-1}}$, $a^{r^{h-1}-r^{h-2}}, b^2a^{\beta_{q-1}}, b^2a^{\beta_q}$. There are then four conditions to be met by the indices α_i and β_i which are somewhat similar to those arising in Friedlander's argument, together with a fifth condition to express the fact that the product of all the elements of the sequencing is $a^{-r^{h-1}}$. It is shown in [10] that the various conditions can be met, though the proof is somewhat lengthy. Sequencings for the non-abelian groups of orders 21, 39 and 55 are given as illustrative examples.

BIBLIOGRAPHY

1. B.A. Anderson, "Sequencings of certain dihedral groups", *Proc. Sixth S.E. Conf. on Combinatorics, Graph Theory and Computing,* Florida Atlantic Univ., 1975, (Utilitas Math., Winnipeg, 1975), pp. 65-76.
2. B.A. Anderson, "Sequencings and starters", *Pacific J. Math.,* 64 (1976), 17-24.
3. J.V. Bradley, "Complete counterbalancing of immediate sequential effects in a latin square design", *J. Amer. Statist. Assoc.,* 53 (1958), 525-528.
4. J. Dénes and E. Török, "Groups and Graphs", *Combinatorial Theory and its Applications* (North Holland, Amsterdam, 1970), pp.257-289.
5. C.V. Eynden, "Countable sequenceable groups", *Discrete Math.,* 23 (1978), 317-318.
6. R. Friedlander, "Sequences in non-abelian groups with distinct partial products", *Aequationes Math.,* 14 (1976), 59-66.
7. B. Gordon, "Sequences in groups with distinct partial products", *Pacific J. Math.,* 11 (1961), 1039-1313.
8. A.D. Keedwell, "Some problems concerning complete latin squares", *Proc. British Combinatorial Conf.,* Aberystwyth, 1973, London Math. Soc. Lecture Note Series 13 (Cambridge Univ. Press, Cambridge 1974), pp. 89-96.
9. A.D. Keedwell, "Latin squares, P-quasigroups and graph decompositions", (Symposium on Quasigroups and Functional Equations, Belgrade, 1974), *Zbornik Rad. Mat. Inst. Beograd (N.S.)* 1(9), (1976), 41-48.

10. A.D. Keedwell, "On the sequenceability of non-abelian groups of order pq", submitted to *Discrete Math.*

11. N.S. Mendelsohn, "Hamiltonian decomposition of the complete directed n-graph", *Theory of Graphs,* Proc. Colloq. Tihany, 1966 (Academic Press, New York, 1968), pp. 237-241.

12. L.J. Paige, "A note on finite abelian groups", *Bull. Amer. Math. Soc.,* 53 (1947), 590-593.

13. L.L. Wang, "A test for the sequencing of a class of finite groups with two generators", *Amer. Math. Soc. Notices* 20 (1973), 73T-A275.

14. E.J. Williams, "Experimental designs balanced for the estimation of residual effects of treatments", *Australian J. Sci. Research, Ser.A,* 2 (1949), 149-168.

Department of Mathematics
University of Surrey
Guildford GU2 5XH
U.K.

POLAR SPACES EMBEDDED IN A PROJECTIVE SPACE

Christiane Lefèvre-Percsy

This paper is a survey of the results related to polar spaces "embedded" into a projective space. This involves polar spaces embedded in a polarity (Veldkamp and Tits), fully embedded polar spaces (Buekenhout, Lefèvre and Dienst) and weakly embedded polar spaces (Lefèvre).

1 . INTRODUCTION

The study of orthogonal, hermitian and symplectic quadrics led Veldkamp [14] to a notion of polar space. A simpler set of axioms was stated by Tits in order to classify buildings [13] . In 1974, Buekenhout and Shult [5] simplified Tits' axioms. They defined a *polar space* as an incidence structure, with non void set of points P, non void set of lines L and incidence relation I , such that, for each line L and each point p not incident with L , one of the following occurs:

(i) there exists exactly one point p' incident with L and a line L' incident to both p and p' ;

(ii) for each point p' incident with L , there is a line L' incident to both p and p' .

We also suppose that there is no point p of P such that each point of P is incident to a line incident with p .

A polar space (P, L, I) such that (ii) does not occur is a *generalized quadrangle*.

Tits [13] has classified all polar spaces which are not generalized quadrangles. (It is presently hopeless to classify the latter.) Up to a few known exceptions, they are isomorphic to the polar space associated with a pseudo-quadratic form. However, this classification does not determine all possible "representations" or "embeddings" of a polar space into a projective space.

We summarize all results related to the classification of polar spaces (including generalized quadrangles) which are "embedded" in a projective space. Since the results on "weakly embedded" spaces are still unpublished, we give the main steps of their proof.

2 . POLAR SPACES EMBEDDED IN A POLARITY

Veldkamp and Tits have studied polar spaces which are embedded in a polarity. Tits called them, in [13], "embeddable polar spaces".

DEFINITION : A polar space $(\mathcal{P}, \mathcal{L}, \mathrm{I})$ is *embedded in a polarity* π of a projective space P if $\mathcal{P}$ is a set of absolute points of π, spanning P and disjoint from the kernel of π, if $\mathcal{L}$ is a set of totally isotropic lines of π such that $\mathcal{P} = \bigcup \mathcal{L}$ and if I is the incidence relation of P between the members of $\mathcal{P}$ and $\mathcal{L}$.

The *kernel* of π is the set of all points p of P such that $\pi(p) = \mathrm{P}$.

Examples of polar spaces which are embedded in a polarity are provided by the semi-quadrics defined in [2].

Tits [13] proved the following theorem, which is an improvement of a result of Veldkamp [14].

THEOREM 1 : *If* $(\mathcal{P}, \mathcal{L}, \mathrm{I})$ *is a polar space embedded in a polarity, then* $(\mathcal{P}, \mathcal{L}, \mathrm{I})$ *is a semi-quadric.*

3 . FULLY EMBEDDED POLAR SPACES

Fully embedded polar spaces have been called projective generalized quadrangles or projective Shult spaces in [3] and [4].

DEFINITION : A polar space $(\mathcal{P}, \mathcal{L}, \mathrm{I})$ is *fully embedded* in a projective space P if $\mathcal{P}$ is a set of points of P spanning P, if $\mathcal{L}$ is a set of lines of P such that $\mathcal{P} = \bigcup \mathcal{L}$ and if I is the incidence relation of P between $\mathcal{P}$ and $\mathcal{L}$.

The following theorem has been proved in several steps (Buekenhout [1]); Buekenhout and Lefèvre [3], [4] : Lefèvre-Percsy [8], Dienst [6]. New proofs are given in [10] and [12]).

THEOREM 2 : *If* $(\mathcal{P}, \mathcal{L}, I)$ *is a polar space, fully embedded in a finite dimensional projective space, then* $(\mathcal{P}, \mathcal{L}, I)$ *is a semi-quadric.*

Except in [1] and [12], the proof of this theorem consists in building a polarity of P in which $(\mathcal{P}, \mathcal{L}, I)$ is embedded, and so Theorem 1 applies.

4. WEAKLY EMBEDDED POLAR SPACES

Polar spaces may be "embedded" in a somewhat weaker sense : we consider embeddings which are isometric in the terminology of Percsy [11], who enlarged Kantor's one [7].

DEFINITION : A polar space $(\mathcal{P}, \mathcal{L}, I)$ is *weakly embedded* in a projective space P if $\mathcal{P}$ is a set of points of P spanning P, if $\mathcal{L}$ is a set of lines of P and if I is the incidence relation of P between $\mathcal{P}$ and $\mathcal{L}$, such that:

(1) each line of $\mathcal{L}$ is incident with at least two points of $\mathcal{P}$;

(2) for each point $p \in \mathcal{P}$, the linear variety of P generated by the set of lines of $\mathcal{L}$ incident to p, intersects $\mathcal{P}$ exactly in these lines;

(3) if $L_1, L_2 \in \mathcal{L}$ and if L_1 and L_2 intersect in P, then $L_1 \cap L_2$ is a point of $\mathcal{P}$.

Examples of weakly embedded polar spaces in P are provided by semi-quadrics in a projective subspace of P (i.e. in a projective space over a subfield of the ground field of P).

We prove that weakly embedded polar spaces in a finite projective space are essentially of this kind. A precise statement of this result needs the following terminology.

DEFINITION : A line L of P is a *secant line* to $(\mathcal{P}, \mathcal{L}, I)$ if $L \notin \mathcal{L}$ and L meets $\mathcal{P}$ in at least two points. It is possible to show that all secant lines intersect $\mathcal{P}$ in the same number of points. Let $t+1$ be this number.

THEOREM 3 : *Let* $(\mathcal{P}, \mathcal{L}, \mathrm{I})$ *be a polar space, weakly embedded in a finite projective space* P. *If* $t > 1$, *then* $(\mathcal{P}, \mathcal{L}, \mathrm{I})$ *is a semi-quadric in a subspace (over a subfield) of* P.

The idea of the proof is the following : from $(\mathcal{P}, \mathcal{L}, \mathrm{I})$, one builds a subset $\overline{\mathcal{P}}$ of points of P and shows that $\overline{\mathcal{P}}$ is a projective subspace of P, in which $(\mathcal{P}, \mathcal{L}, \mathrm{I})$ is fully embedded. Then the result follows from Theorem 2.

In fact, this proof is rather long and is based on various preliminary properties of weakly embedded polar spaces. So, we only give here the main steps of the proof. A detailed version of it will be published separately in [9].

DEFINITION : Let α be a plane of P meeting $(\mathcal{P}, \mathcal{L}, \mathrm{I})$ in $t+1$ lines through a point a. Let $\mathcal{P}_\alpha$ be the set of all points of α belonging to at least two secant lines of α, together with the point a itself. Consider the union $\overline{\mathcal{P}} = \bigcup_\alpha \mathcal{P}_\alpha$, for all such planes α.

We show that $\overline{\mathcal{P}}$ is a projective subspace (over a subfield) of P, in two steps.

1 : The 3-dimensional case.

If P has dimension 3, then $(\mathcal{P}, \mathcal{L}, \mathrm{I})$ must be a generalized quadrangle. Using known combinatorial properties of the latter structures and constructing projectivities of P leaving $\mathcal{P}$ invariant, it is possible to prove that $\overline{\mathcal{P}}$ is a 3-dimensional projective subspace of P.

2 : The higher dimensional case.

The result is obtained using an embedding theorem of Kantor [7]. Consider the linear geometry G of all sections of $\mathcal{P}$ by subspaces of P. Using an induction assumption, it is possible to show that G is locally affino-projective. Then, by Kantor's theorem, G is embeddable in a projective space and this one must coincide with the above set $\overline{\mathcal{P}}$. And so we can conclude.

BIBLIOGRAPHY

1. F. Buekenhout, "Ensembles quadratiques des espaces projectifs", *Math. Z.*, 110 (1969), 306-318.
2. F. Buekenhout, "Characterizations of semi-quadrics. A survey", *Teorie Combinatorie*, Tomo I, Rome, 1973. Atti dei Convegni Lincei 17 (Accad. Naz. Lincei, Rome, 1976), pp.393-421.
3. F. Buekenhout and C. Lefèvre, "Generalized quadrangles in projective spaces", *Arch. Math.*, 25 (1974), 540-552.
4. F. Buekenhout and C. Lefèvre, "Semi-quadratic sets in projective spaces", *J. Geom.*, 7 (1976), 17-42.
5. F. Buekenhout and E. Shult, "On the foundations of polar geometry", *Geom. Dedicata*, 3 (1974), 155-170.
6. K.J. Dienst, "Verallgemeinerte Vierecke in projektiven Räumen", Preprint 491 (Technische Höchschule Darmstadt, 1979).
7. W.M. Kantor, "Envelopes of geometric lattices", *J. Combin. Theory Ser. A*, 18 (1975), 12-26.
8. C. Lefèvre-Percsy, "Sur les semi-quadriques en tant qu'espaces de Shult projectifs", *Acad. Roy. Belg. Bull. Cl. Sc.*, 63 (1977), 160-164.
9. C. Lefèvre-Percsy, "Weakly embedded polar spaces", In preparation.
10. C. Lefèvre-Percsy, "A note on fully embedded polar spaces", in preparation.
11. N. Percsy, "Plongement de géométries", Thèse de doctorat, Université Libre de Bruxelles, 1980.
12. J.A. Thas and P. De Winne, "Generalized quadrangles in finite projective spaces", *J. Geom.*, 10 (1977), 126-137.
13. J. Tits, *Buildings of Spherical Type and Finite BN-pairs*, Lecture Notes in Math. 386 (Springer-Verlag, Berlin,1974).
14. F.D. Veldkamp, "Polar geometry I-V", *Proc. Koning. Ned. Akad. Wet. A*, 62 (1959), 512-551; 63 (1959), 207-212 *(Indag. Math.* 21,22).

Département de Mathématique
Université Libre de Bruxelles
Campus Plaine C.P. 216
Boulevard du Triomphe
1050 Bruxelles
Belgium

ON RELATIONS AMONG THE PROJECTIVE GEOMETRY CODES

R.A. Liebler

Let V_i denote the set of i-dimensional subspaces of $PG(n-1, q)$ and let R be a commutative ring with identity. Then the R-modules RV_i are intertwined by the incidence maps:

$$I_{ij} : RV_i \longrightarrow RV_j$$
$$x \longmapsto \sum y \quad \text{with} \quad y \subseteq x \quad \text{or} \quad x \subseteq y .$$

In case R is a field of characteristic zero, these maps are fundamental in the analysis of the $PGL(n, q)$-representation afforded by RV_i and are well understood.

In case $R = GF(p)$ with $p \mid q$, the images and kernels of these maps define the various projective geometry codes. However, much less seems to be known about how they fit together.

In spite of this, there seem to be general statements independent of q that one can make in this context. The purpose of this note is to use generalized intersection matrices to show:

THEOREM: *Suppose* $n = 4$ *and* R *is a field of characteristic dividing* q. *Then*

$$\text{Im } I_{31} \supseteq \text{Im } I_{21} \cap \text{Ker } I_{12} .$$

It is hard for me to believe that inclusions of this type are not already known. If the reader can point them out in the literature, I would be most grateful. In any case, I hope the method of proof is of some interest in its own right.

PRELIMINARIES: Let $G \le GL(4, q)$ be a cyclic group of order $(q^3 - 1)$ that fixes $<1000>$ and acts as a Singer group on $<0 *** >$. Let $K = Q_p(\zeta)$ where ζ is a primitive $(q^3 - 1)$ root of unity and Q_p is the quotient field of the p-adic integers. If O_K is the ring of algebraic integers in K, then both K and $k = O_K / p O_K \cong GF(q^3)$ are splitting fields for G.

Take R to be one of the fields K or k. Then Maschke's theorem provides the decomposition

$$R V_i = \sum_{\lambda \in G^*} \oplus R_i(\lambda) ,$$

where the summation ranges over the set G^* of irreducible representations of G. Each of the maps I_{ij} is a G-homomorphism, so we have, for example,

$$\text{Im } I_{ij} \cap \text{Ker } I_{ik} = \sum_{\lambda \in G^*} \oplus \left\{ \text{Im}(I_{ij}|_{R_i(\lambda)}) \cap \text{Ker}(I_{ik}|_{R_i(\lambda)}) \right\} .$$

Let T_i be a transversal for the G-orbits on V_i and set

$$t_R(\lambda, X) = \sum_{g \in G} \lambda(g^{-1}) x^g , \quad x \in T_i , \ \lambda \in G^* .$$

Then $t_R(\lambda, x) \ne 0$ just in the case that order of λ divides $|x^G|$ and the non-zero elements of $\{ t_R(\lambda, x) \ \ x \in T_i \}$ form a basis for $R_i(\lambda)$.

Now define the *generalized intersection matrices* $M_\lambda(i,j)$ as follows:

$$I_{ij} (t_K(\lambda, x)) = \sum (M_\lambda(i, j))_{x,y} \, t_K(\lambda, y) .$$

Observe that the entries of $M_\lambda(i,j)$ are in O_K and that

$$I_{ij} (t_k(\lambda, x)) \equiv \sum (M_\lambda(i, j))_{x,y} \, t_k(\lambda, y) \pmod p .$$

The theorem is proved by showing that

$$\text{Im } I_{31} |_{R_3(\lambda)} \supseteq \text{Im}(I_{21} |_{R_2(\lambda)}) \cap \text{Ker}(I_{12} |_{R_1(\lambda)}) \qquad (*)$$

for each $\lambda \in G^*$, and this in turn is done by comparing the appropriate generalized intersection matrices.

PROOF : Let

$$T_1 = \{\langle 1000\rangle, \langle 0001\rangle, \langle 1001\rangle\},$$

$$T_2 = \left\{\begin{pmatrix}1000\\0001\end{pmatrix}, \begin{pmatrix}0001\\0100\end{pmatrix}, \begin{pmatrix}1001\\0100\end{pmatrix}, \begin{pmatrix}1001\\1\alpha 10\end{pmatrix} \middle| \alpha\in GF(q)\right\},$$

$$T_3 = \left\{\begin{pmatrix}1000\\0010\\0001\end{pmatrix}, \begin{pmatrix}0100\\0010\\0001\end{pmatrix}, \begin{pmatrix}1001\\0100\\0010\end{pmatrix}\right\}.$$

The matrices $M_1(i,j)$ are just the familiar intersection matrices:

$$M_1(1,2) = \begin{pmatrix} 1 & 0 & 0 & \dots & 0 \\ 1 & q+1 & 1 & \dots & 0 \\ q-1 & 0 & q & \dots & q \end{pmatrix}, \quad M_1(2,1) = \begin{pmatrix} q^2+q+1 & 1 & 1 \\ 0 & q+1 & 0 \\ 0 & q-1 & q \\ \vdots & \vdots & \vdots \\ 0 & q-1 & q \end{pmatrix},$$

$$M_1(3,1) = \begin{pmatrix} 0 & 1 & 0 \\ q^2+q+1 & 1 & q+1 \\ 0 & q^2-1 & q^2 \end{pmatrix}.$$

Now reduce these matrices modulo p obtaining $\overline{M_1(i,j)}$. Observe that

$$\text{Im } M_1(3,1) = \left\langle \begin{matrix} 0 & 1 & 2 \\ 1 & 1 & 1 \end{matrix}\right\rangle \supseteq \langle 1 \quad 0 \quad 1\rangle = \text{Im } \overline{M_1(2,1)} \cap \text{Ker } \overline{M_4(1,2)};$$

consequently,

$$\text{Im } (I_{31} \mid R_3(1)) \supseteq \text{Im}(I_{21} \mid R_2(1)) \cap \text{Ker}(I_{12} \mid R_1(1)) .$$

Suppose next that $\lambda \neq 1$ has order divisible by q^2+q+1. Then

$$M_\lambda(1,2) = \begin{pmatrix} 0 & 0 & 0 & \dots & 0 \\ 1 & x & 1 & \dots & 1 \\ q-1 & 0 & x_1 & \dots & x_{q+1} \end{pmatrix}, \quad M_\lambda(2,1) = \begin{pmatrix} 0 & 1 & 1 \\ 0 & y & 0 \\ 0 & q-1 & y_1 \\ \vdots & \vdots & \vdots \\ 0 & q-1 & y_{q+1} \end{pmatrix},$$

$$M_\lambda(3,1) = \begin{pmatrix} 0 & 0 & 0 \\ 0 & 1 & 1 \\ 0 & q-1 & z \end{pmatrix},$$

for some $x, y, z, x_i, y_i \in O_K$. This meagre amount of information is actually sufficient for our purpose. Indeed

$$1 \leq \operatorname{rank} \overline{M_\lambda(3,1)} = \dim \operatorname{Im}(I_{31} \big| k_3(2)) ,$$

while

$$1 \geq \dim (\operatorname{Im} \overline{M_\lambda(2,1)} \cap \operatorname{Ker} \overline{M_\lambda(1,2)}) .$$

Finally, suppose the order of λ is not divisible by q^2+q+1. Then

$$M_\lambda(1,2) = \begin{pmatrix} 0 & 0 & 0 & \dots & 0 \\ 0 & 0 & 0 & \dots & 0 \\ 0 & 0 & x_1 & \dots & x_{q+1} \end{pmatrix}, \quad M_\lambda(2,1) = \begin{pmatrix} 0 & 0 & 0 \\ 0 & 0 & 0 \\ 0 & 0 & y_1 \\ \vdots & \vdots & \vdots \\ 0 & 0 & y_{q+1} \end{pmatrix},$$

$$M_\lambda(3,1) = \begin{pmatrix} 0 & 0 & 0 \\ 0 & 0 & 0 \\ 0 & 0 & z \end{pmatrix} .$$

For such a λ, (*) can only fail if

$$0 \equiv z \equiv x \quad \dots \quad x_{q+1} \pmod p$$

but $\quad y_i \not\equiv 0 \pmod p$ for some i.

Then it suffices to show that x_i and y_i are algebraic conjugates.

Let $r = \langle 1001 \rangle$ and let $\ell \in T_2$ be the representative of the appropriate line orbit. Define a $(q^3-1) \times (q^3-1)$ matrix $A = (a_{ij})$ by $a_{ij} = 1$ if r^{g^i} and ℓ^{g^j} are incident and zero otherwise. Then A is a circulant matrix and

$$I_{12}(t(\lambda,r)) = I_{12}(\textstyle\sum \lambda(g^{-i})\, r^{g^i}) = \sum_i \lambda(g^{-i})(\sum a_{ij}\, \ell^{g^i})$$

$$= \sum_j \left[\sum \lambda(g^{j-i})\, a_{ij}\right] \lambda(g^{-j})\, \ell^{g^i} = \sum_j \left[\sum \lambda(g^k)\, a_{1,k+1}\right] \lambda(g^{-j})\, \ell^{g^i} ,$$

since $a_{1,k+1} = a_{i,i+k}$. It follows that $x_i = \sum \lambda(g^k)\, a_{1,k+1}$

and similarly

$$y_i = \sum \lambda(g^k) a_{k+1,1} = \sum \lambda(g^{-k}) a_{1,k+1} = \bar{x}_i .$$

This completes the proof of the theorem.

Mathematics Department
Colorado State University
Fort Collins
Colorado 80524
U.S.A.

PARTITION LOOPS AND AFFINE GEOMETRIES

Mario Marchi

A *partition group* (G, Π) is a group G together with a family Π of subgroups such that every element of G, other than the identity, is in exactly one subgroup of Π. Partition groups play an important role in the study of *translation affine parallel structures* (for short *translation a.p. structures*) and, in particular, of *translation Sperner spaces* (see André [1]). The notion of partition group has been generalised in many ways. Havel (see [5], [6], and many other papers) has introduced the notion of *parallelisable partition of groupoids* and of *partition* in many other algebraic structures and has studied the correspondent incidence structures. Wähling, generalising the concept of incidence-group introduced by Ellers and Karzel, has studied the structure of *incidence loops* and of *incidence groupoids* (see [13], [14]). In the present Note we consider a suitable class of *loops* with a *partition* Π (by means of non-trivial subgroups) such that all Sperner spaces and many a.p. structures can be represented by loops of this class (see also [2] and [7]). In §§ 1,2, we construct suitable partition loops in order to represent various types of a.p. structures in the "nicest" way. The case of the existence of a group of translations is particularly considered. Conversely, in §§ 3,4, we initiate the study of a.p. structures represented by a given partition loop. The geometrical meaning of left, middle and right nuclei is introduced.

1.

As is well known, an *affine parallel structure* (for short, *a.p. structure*) in the sense of André [1] is an incidence structure $\Sigma = (P, R, \parallel, I)$ such that the following conditions hold:

(S 1) for every two distinct elements $\underline{P}$, $\underline{Q}$ in P (P = set of *points*) there is exactly one element $\underline{r}$ in R (R = set of *lines*) such that $\underline{P}\, I\, \underline{r}$, $\underline{Q}\, I\, \underline{r}$;

(S 2) in R an equivalence relation $\|$ *(parallelism)* is given such that *Euclid's axiom* holds : for all $\underline{P}$ in P and $\underline{r}$ in R, there exists a unique $\underline{s}$ in R such that $\underline{s}\, I\, \underline{P}$, $\underline{s} \| \underline{r}$;

(S 3) $|R| \geq 2$ and $|\underline{r}| \geq 2$, $\forall \underline{r}$ in R.

An a.p. structure is a *Sperner space* (*S-space,* for short) if, instead of (S 3) we have

(S 3') $|R| \geq 2$ and $|\underline{r}| = n \geq 2$, $\forall \underline{r}$ in R.

An a.p. structure will be called *regular* if the additional condition holds:

(S 4) $\forall \underline{r}, \underline{s}$ in R : $\underline{r} \| \underline{s} \Longrightarrow |\underline{r}| = |\underline{s}|$.

Of course, S-spaces are regular a.p. structures. Henceforth all a.p. structures we shall consider will be regular.

For a bibliography on S-spaces and on a.p. structures see [8].

By a *partition loop* we shall mean a loop L with a family Π of non-trivial subloops L_ω (where ω varies in a suitable set Ω) such that for all elements ℓ in L, $\ell \neq \sigma$ (σ being the neutral element in L) there exists exactly one subloop L_ω in Π such that $\ell \in L_\omega$. Henceforth we shall write (L, Π) when dealing with such a partition loop. We shall now be concerned with a particular class of partition loops (L, Π) for which every subloop of Π is a subgroup.

THEOREM 1 : *Let* (L, Π) *be a partition loop such that*

(L 1) *all subloops of* Π *are subgroups ;*

(L 2) $\forall \ell \in L$, $\forall M \in \Pi$, $\forall m_1, m_2 \in M$,

$$\ell + (m_1 + m_2) = (\ell + m_1) + m_2 .$$

Then the incidence structure $\Sigma = (P, R, \|, I)$ *where*

(i) $\mathcal{P} = L$, $\mathcal{R} = \{\ell + M \mid \forall \ell \in L, \forall M \in \Pi\}$,

(ii) $(\ell_1 + M_1) \parallel (\ell_2 + M_2) \Leftrightarrow M_1 = M_2$,

(iii) *incidence is set-theoretic inclusion,*

is a regular a.p. structure.

Conversely, every regular a.p. structure can be represented by a partition loop (L, Π) *satisfying (L 1) and (L 2).*

PROOF: Let (L, Π) be a partition loop as stated. Every element p in L may be thought of as a point $\underline{P}$ in $\mathcal{P}$ of the structure Σ : we shall write $\underline{P} \equiv (p)$. Let $\underline{P}_1 \equiv (p_1)$, $\underline{P}_2 \equiv (p_2)$ be two distinct points of Σ. The line $\underline{P}_1\underline{P}_2$ is defined in the following way. Let x in L be such that $p_1 + x = p_2$; then by definition we shall write $\underline{P}_1\underline{P}_2 \equiv (p_1 + M)$, where M in Π is such that $x \in M$. We have to show the uniqueness of the line $\underline{P}_1\underline{P}_2$. Let us consider two distinct points of $\underline{P}_1\underline{P}_2$, namely $\underline{Q}_i \equiv (q_i)$ where $q_i = p_1 + m_i$ $(i = 1,2)$; then the line $\underline{Q}_1\underline{Q}_2$ is, by definition, $\underline{Q}_1\underline{Q}_2 \equiv (q_1 + M')$, where M' is such that $q_1 + y = q_2 \Longrightarrow y \in M'$. On the other hand, by (L 2) and (L 1),

$$q_1 + (-m_1 + m_2) = (p_1 + m_1) + (-m_1 + m_2) = p_1 + [m_1 + (-m_1 + m_2)] =$$
$$= p_1 + m_2 = q_2 .$$

So $y = -m_1 + m_2$ and $M' = M$. Again, by (L 2), we have then that $\underline{Q}_1\underline{Q}_2 = \underline{P}_1\underline{P}_2$. Let $\underline{P} \equiv (p)$ be a point and $\underline{r} \equiv (\ell + M)$ be a line. Then the line $\underline{r}' \parallel \underline{r}$, $\underline{r}' \text{ I } \underline{P}$ is, by definition, $\underline{r}' = (p + M)$. We have to show the uniqueness of the line $\underline{r}'$; that is

$$\underline{r} \parallel \underline{r}', \quad \underline{r} \cap \underline{r}' \neq \phi \Longrightarrow \underline{r} = \underline{r}' .$$

Let $\underline{r} \cap \underline{r}' = \{q\}$; then there exist m_1, m_2 such that $q = \ell + m_1 = p + m_2$, hence, in view of (L 2), we have

$$\ell = (p + m_2) - m_1 = p + (m_2 - m_1),$$

whence

$$\ell + M = [p + (m_2 - m_1)] + M = p + [(m_2 - m_1) + M] = p + M .$$

It is obvious that parallel lines have the same cardinality. Conversely, let Σ be a regular a.p. structure and $\underline{O}$ be a distinguished point. Moreover, let $\underline{r}_\omega$ be any line incident with $\underline{O}$, while ω varies in a suitable set Ω. Let $R_\omega(+)$ $(\omega \in \Omega)$ be a group, with neutral element O_{R_ω} and let $\Psi_{\underline{r}_\omega}$ be a one-to-one correspondence $\Psi_{\underline{r}_\omega} : \underline{r}_\omega \to R_\omega$ such that $\Psi_{\underline{r}_\omega}(\underline{O}) = O_{R_\omega}$. In this way each point $\underline{P}$ in P, $\underline{P} \neq \underline{O}$, is represented by one element p of the set $L = \bigsqcup_{\omega\in\Omega} R_\omega$, where $p = \Psi_{\underline{OP}}(\underline{P})$. We shall write $\underline{P} \equiv (p)$, and if the line $\underline{OP}$ is represented by the group R we shall write, then, $\underline{OP} \equiv (R)$. In the set L we identify all elements O_{R_ω} $(\forall\omega \in \Omega)$ and then write

$$O_{R_\omega} = \sigma \quad (\forall \omega \in \Omega).$$

In view of this identification a one-to-one correspondence $\lambda : P \to L$ between the set L and the set of points P is given such that $\underline{P} \equiv (\lambda(\underline{P}))$, $\forall \underline{P} \in P$, $\underline{P} \neq \underline{O}$ and $\underline{O} \equiv (\lambda(\underline{O})) = (\sigma)$. We now have to introduce in L a binary operation (addition). By definition $\forall\omega \in \Omega$ we have that if $r_1, r_2 \in R_\omega$ the operation

$$r_1 + r_2 = r_3 \in R_\omega \tag{1}$$

is defined for all ω in Ω, since R_ω is a group.

Now let R, S be any two different groups of the set $\Pi = \{R_\omega \mid \forall\omega \in \Omega\}$. In order to define the addition $r + s$, where $r \in R$, $s \in S$, we may introduce a one-to-one correspondence $\Psi_{r,S}$ between the set of points of the line $\underline{s} \equiv (S)$ and the set of points of the line $\underline{s}' \parallel \underline{s}$, $s' \, I \, \underline{R}$, $\underline{R} \equiv (r)$, such that

$$\Psi_{r,S}(\underline{O}) = \underline{R}, \qquad \Psi_{\sigma,S} = \text{identity}. \tag{2}$$

This construction is a generalization of the one of the Sperner's *quasi-module* : see e.g. [2], [12]. The partition loop (L, Π) is then a generalization of the Arnold's *Vektorloop*.

For simplicity we shall identify every point $\underline{P}$ in P with the corresponding element $p = \lambda(\underline{P})$ and every line $\underline{r}_\omega I \underline{O}$ with the corresponding group R_ω. Then we can write conditions (2) in the form $(\forall r \in L, \forall S \in \Pi)$:

$$\Psi_{r,S}(\sigma) = r, \quad \Psi_{\sigma,S}(s) = s \quad (\forall s \in S). \tag{2'}$$

By definition we may now write

$$r + s = \Psi_{r,S}(s). \tag{3}$$

We now verify that L is a loop, with respect to addition defined as in (1) and (3), and σ is the neutral element.

In fact, from (2'), $\forall r, s \in L$, we have

$$r + \sigma = r, \quad \sigma + s = s. \tag{4}$$

Moreover, $\forall r, t \in L$ (r, t not in the same subgroup of the family Π), there exists a unique x in L such that

$$r + x = t; \tag{4'}$$

for, if $\underline{R} \equiv (r)$, $\underline{T} \equiv (t)$ are points and if $\underline{s} \equiv (S)$ is the line such that $\underline{s}\ I\ \underline{O}$ and $\underline{s} \parallel \underline{R}\,\underline{T}$, then x is defined by $\Psi_{r,S}(x) = t$. Again, $\forall s, t \in L$ (s, t not in the same subgroup of the family Π), there exists a unique y in L such that

$$y + s = t, \tag{4''}$$

because the function $\Psi_{r,S}$ with S fixed and for all r in L gives rise to a decomposition of L into mutually disjoint sets. If $r, t \in L$ or $s, t \in L$ are in the same subgroup of Π, the existence of x and y in (4') and (4'') is certain. In this way we see that L is a loop with the partition Π into subgroups; then the condition (L 1) is satisfied. To prove the additional condition (L 2) we have to choose functions $\Psi_{r,S}$ $(\forall r \in L, \forall S \in \Pi)$ not in a completely arbitrary way. If we choose in every left coset $(r + S)$ of S a distinguished element $\overline{r}$, we may define for all $\overline{s} \in S$ the new function $\Psi'_{r,S}$:

$$\Psi'_{\overline{r}+\overline{s},S}(s) = \Psi_{\overline{r},S}(\overline{s} + s) \quad (\forall s \in S). \tag{2''}$$

Then, according to definition (3), we have

$$(\overline{r} + \overline{s}) + s = \overline{r} + (\overline{s} + s) \qquad (\forall s, \overline{s} \in S).$$

To prove (L 2) for all r in L, let $r = \overline{r} + s_0$, then

$$(r + s_1) + s_2 = [(\overline{r} + s_0) + s_1] + s_2 = [\overline{r} + (s_0 + s_1)] + s_2 = \overline{r} + [s_0 + (s_1 + s_2)]$$
$$= (\overline{r} + s_0) + (s_1 + s_2) = r + (s_1 + s_2). \quad \square$$

Now using Theorem 1, we may examine two different aspects of the same problem. We may study a given a.p. structure Σ and try to find the "best" representation for it by means of a partition loop. Conversely, if a partition loop (L, Π) is given, we may study what geometrical properties the associated a.p. structure has. We now investigate the first point of view.

Let us suppose that an a.p. structure Σ is endowed with a group G of translations: a *translation* of Σ is the identity or a fixed-point-free collineation which maps lines onto parallel lines. It is easy to see that G is sharply transitive on each set $G(\underline{P}) \subseteq P$ $(\forall \underline{P} \in P)$. In this case it may be of interest to make the arbitrary choices of the groups of the partition Π and of the functions $\Psi_{r,S}$ $(\forall r \in L, \forall S \in \Pi)$ in such a way as to agree with the group G. The loop L would be in this case a sort of "canonical representation" of Σ with the group G. We shall say that L represents the pair (Σ, G).

Henceforth, on the basis of Theorem 1, we shall no longer distinguish the elements of the loop L from the points of the a.p. structure Σ. We recall that (see [4]) the *left nucleus* of a loop L is:

$$N_L = \{\ell \in L \mid \ell + (m + n) = (\ell + m) + n, \ \forall m, n \in L\}.$$

THEOREM 2 : *Let* Σ *be a regular a.p. structure and* G *be a translation group on* Σ. *It is possible to define a partition loop* (L, Π) *such that* $G = \{x' = \ell + x, (x, x' \in L) \mid \forall \ell \in N_L\}$ *where* N_L *is the left nucleus of the loop* L.

PROOF : The mapping

$$\tau \; : \; x' = \ell + x \tag{5}$$

is a translation if $\forall\, m, n \in L$ we have

$$\ell + (m+n) = (\ell+m) + n . \tag{6}$$

Then if $\ell \in N_L$, (5) is a translation.

Conversely, let τ_ℓ be a translation of the group G such that $\tau_\ell(\sigma) = \ell$. If we choose the functions $\Psi_{\ell,S}$ $(\forall S \in \Pi)$ such that

$$\Psi_{\ell,S}(s) = \tau_\ell(s) \qquad (\forall s \in S)$$

then, by definition (3), we may write

$$\tau_\ell \; : \; x' = \ell + x .$$

We now have to show that $\ell \in N_L$. If $m \in G(\sigma)$ (that is there exists a translation $\tau_m : x' = m + x$), we have, $\forall n \in L$,

$$\ell + (m+n) = \tau_\ell[\tau_m(n)] = \tau_{\ell+m}(n) = (\ell+m) + n .$$

So (6) holds. Let now $m \notin G(\sigma)$; in order to satisfy (6) for all m, n in L we have to choose the functions $\Psi_{m,S}$, $\Psi_{\ell+m,S}$ such that

$$\tau_\ell[\Psi_{m,S}(s)] = \Psi_{\ell+m,S}(s) \qquad (\forall s \in S,\ \forall S \in \Pi). \tag{7}$$

It must be shown that this is always possible according to (2").

If G is not point-transitive on Σ, G partitions L into mutually disjoint sets $G(m)$ $(\forall m \in L)$. If we take one representative element $\bar{m}$ for each distinct set $G(m)$, we may arbitrarily choose the functions $\Psi_{\bar{m},S}$ $(\forall S \in \Pi)$; then $\Psi_{\ell+\bar{m},S}$ $(\forall \ell \in G(\sigma),\ \forall S \in \Pi)$ are chosen according to (7) (where $m = \bar{m}$). Hence we have that $\ell \in G(\sigma) \Longrightarrow \ell \in N_L$. □

We can now investigate the geometrical meaning of the left nucleus $N_L \subset L$. We recall that a *sub-a.p. structure* $\Sigma' = (P', R', \parallel', I)$ of a given a.p. structure $\Sigma = (P, R, \parallel, I)$ is again an a.p. structure such that

(T 1) $P' \subseteq P$;

(T 2) for all $\underline{r}'$ in R' there is a unique $\underline{r}$ in R such that $\underline{r}' = \underline{r} \cap P'$;

(T 3) $\underline{r}' \parallel \underline{s}' \Leftrightarrow \underline{r} \parallel \underline{s} \quad (\forall \underline{r}' = \underline{r} \cap P', \forall \underline{s}' = \underline{s} \cap P')$.

PROPOSITION 1 : *Let* Σ *be a regular a.p. structure and* G *a translation group on* Σ. *Let* (L, Π) *be a partition loop associated with* Σ *as in Theorems 1, 2. Then the left nucleus* N_L *of* L *is a translation a.p. structure, sub-a.p. structure of* Σ.

PROOF : It is well known that N_L is a group; moreover the partition Π of L subordinates a partition $\Pi_L = \{S \cap N_L \mid \forall S \in \Pi\}$ in N_L. Then N_L is an a.p. structure with a transitive group of translations, the group G, which is isomorphic to N_L itself. □

The translation a.p. structure N_L is exactly the so-called *stem* of the a.p. structure Σ (see André [1], Teil III, §2) if G is abelian.

The *transitivity sets* of the group G are

$$G(p) = N_L + p \qquad (\forall p \in L);$$

they are not generally sub-a.p. structures of Σ.

The *cosets* of N_L are

$$C_L(m) = m + N_L \qquad (\forall m \in L);$$

they partition L in a similar way to the transitivity sets. They are also not generally sub-a.p. structures of Σ, and are mapped onto each other by translations of G. In fact for all ℓ in N_L, we have

$$\ell + (m + N_L) = (\ell + m) + N_L .$$

By a *finite a.p. structure* we shall mean an a.p. structure $\Sigma = (P, R, \parallel, I)$ for which P is a finite set. In this case we can say more about *all* translations and dilatations of Σ.

We have the following results, in addition to Theorem 2.

THEOREM 3 : *Let* Σ *be a finite regular a.p. structure and let* $\mathcal{D} \neq \{1\}$ *the dilatation group of* Σ. *It is always possible to define a partition loop* (L, Π), *as in Theorem 1, with left nucleus* N_L *such that*

(i) *all translations of* Σ *can be written in the form*

$$x' = \ell + x \qquad (\forall \ell \in N_L)\ ;$$

(ii) *all dilatations of* Σ *can be written in the form*

$$x' = \ell + f(x) \qquad (\forall \ell \in N_L)\ , \tag{8}$$

where $f(x)$ *is an automorphism of* N_L *and the condition* $f(m + S) = f(m) + S \quad (\forall m \in L\ , \ \forall S \in \Pi)$ *holds.*

PROOF : From results of André ([1] , Teil I, §5) we have that $\mathcal{D} \neq \{1\}$ implies that *all* translations of Σ form a group G, which is a normal subgroup of $\mathcal{D}$. We may then apply Theorem 2, and part (i) is proved. From the same results of André, we have that if there exists one point $\underline{Q}$ in P such that the stabilizer $\mathcal{D}_Q \subseteq \mathcal{D}$ is $\mathcal{D}_Q \neq \{1\}$, then for all points $\underline{P}$ in P we have : $\mathcal{D}_P \neq \{1\} \Leftrightarrow \underline{P} \in \mathcal{D}(\underline{Q})$, where $\mathcal{D}(\underline{Q})$ is the orbit of $\underline{Q}$ under the action of the group $\mathcal{D}$. Moreover, G is transitive on $\mathcal{D}(\underline{Q})$, all stabilizers $\mathcal{D}_P \neq \{1\}$ are isomorphic and $\mathcal{D} = G\mathcal{D}_Q$. To prove part (ii) we have to choose as point $\underline{0} = (\sigma)$ in Σ one point such that $\mathcal{D}_0 \neq \{1\}$; then $\mathcal{D}(\underline{0}) = G(\underline{0}) = N_L$. We may now state that all dilatations of Σ with fixed point $\underline{0} = (\sigma)$ are mappings $x' = f(x)$ of L onto itself such that : $f(\sigma) = \sigma$, $f(S) = S \quad (\forall S \in \Pi)$, $f(m + S) = f(m) + S \quad (\forall m \in L\ , \forall S \in \Pi)$; the proof is immediate.

The restriction of each $f(x)$ to the translation a.p. structure N_L is an equivalence (in the sense of Seier [11]) of the same translation structure; moreover each $f(x)$ is an automorphism of N_L because G is a normal subgroup of $\mathcal{D}$ (Seier

[11] , Satz 2.4) . Hence, in view of the relation $\mathcal{D} = G\mathcal{D}_0$, the proof is now complete. □

From a result of Biliotti ([3] , Teorema 2.3) , we can remark that, if N_L is not contained in one line, the set of all mappings (8) is a subgroup of the dilatation group of the translation a.p. structure N_L .

2.

Instead of looking at the existence of one translation group for the a.p. structure Σ , as in Theorem 2, we may give a "canonical representation" of Σ with respect to suitable incidence conditions. This is done in following theorems. We recall that $\underline{0}$ is the point corresponding to the neutral element σ in L .

THEOREM 4 : *Let Σ be a regular a.p. structure and $\underline{P} \equiv (p)$, $\underline{Q} \equiv (q)$ be two different points in Σ, such that $\underline{0} \not I \underline{P}\,\underline{Q}$. It is possible to define a partition loop (L, Π), as pointed out in Theorem 1, such that the line $\underline{r}$, through $\underline{P}$ and parallel to $\underline{0}\,\underline{Q}$, and the line $\underline{s}$, through $\underline{Q}$ and parallel to $\underline{0}\,\underline{P}$, intersect if and only if $p + q = q + p$.*

PROOF : Let us suppose that $p \in P \in \Pi$ and $q \in Q \in \Pi$; then we have $\underline{r} \equiv (p + Q)$, $\underline{s} \equiv (q + P)$. We see immediately that

$$p + q = q + p \Longrightarrow \underline{r} \cap \underline{s} = \{p + q\} \neq \emptyset .$$

Conversely, let us assume $\underline{r} \cap \underline{s} \neq \emptyset$. If the functions $\Psi_{r,S}$ $(\forall r \in L, \forall S \in \Pi)$ are chosen in an arbitrary way we have only that there exist some $p_1 \in P$ and $q_1 \in Q$ such that $p + q_1 = q + p_1$. On the contrary, we may state that the functions $\Psi_{r,S}$ are by definition such that

$$\underline{r} \cap \underline{s} = \{t\} \Longrightarrow \Psi_{p,Q}(q) = t , \quad \Psi_{q,P}(p) = t ;$$

then by (3), we have

$$p + q = q + p = t . \quad \square$$

We recall that a translation τ of an a.p. structure Σ is *central* if and only if $\tau(\underline{r}) = \underline{r} \Longrightarrow \forall \underline{s} \parallel \underline{r} : \tau(\underline{s}) = \underline{s}$.

The definition of the loop (L, Π) in Theorem 4 agrees with the one of Theorem 2 if and only if the translations of G are central.

As a consequence of Theorem 4 we have an interesting proposition.

A *main line* (Hauptgerade) in an a.p. structure Σ is a line $\underline{r}$ such that $\forall \underline{s} \nparallel \underline{r}$ we have $\underline{s} \cap \underline{r} \neq \emptyset$ (see André [1]).

PROPOSITION 2 : *Let* Σ *be a regular a.p. structure with one parallelism class* R_0 *of main lines. It is possible to define a partition loop* (L, Π), *as pointed out in Theorems 1 and 4, such that there exists a subgroup* A *in* Π *for which*

$$R_0 = \{\ell + A \mid \forall \ell \in L\}$$

and $$a + \ell = \ell + a, \qquad \forall a \in A, \quad \forall \ell \in L.$$

PROOF : In the arbitrary choice of subgroups of Π we choose as a group corresponding to the line $\underline{r}_0 \in R_0$ through $\underline{O}$, a commutative group A. Then $\forall \underline{r} \in R_0$, $\forall \underline{s} \in R \setminus R_0$ we have, by the definition of main line, that $\underline{r} \cap \underline{s} \neq \emptyset$. If we assume, according to Theorem 4, $\underline{Q} \equiv (a)$, for all $a \in A \setminus \{\sigma\}$, $\underline{P} \equiv (\ell)$ for all $\ell \in L \setminus A$, then the line $\underline{r}$ through $\underline{P}$ and parallel to $\underline{OQ}$ is a main line, hence incident with the line $\underline{s}$ through $\underline{Q}$ and parallel to $\underline{OP}$. Then by Theorem 4, and by the assumption that A is commutative, we have the result. □

We can also characterize the class of all a.p. structures Σ satisfying the following condition (see Permutti [10]).

(S 5) Let $S = \{R_1, R_2, \ldots R_n\}$ be a set of $n \geq 3$ parallelism classes of lines in Σ. If $\underline{r}_1 \in R_1$ and $\underline{r}_n \in R_n$ are fixed, by an *S-chain connecting* $\underline{r}_1$ *with* $\underline{r}_n$ we shall mean any set $\{\underline{r}_1, \underline{r}_2, \ldots, \underline{r}_n\}$ such that $\underline{r}_i \in R_i$ and

$\underline{r}_i \cap \underline{r}_{i+1} \neq \emptyset \quad \forall i \in \{1,\ldots,n-1\}$. Then we assume Σ *to have a set* S *of parallelism classes of lines such that for all lines* $\underline{r}_1 \in R_1$ *and* $\underline{r}_n \in R_n$ *there exists one and only one* S*-chain connecting* $\underline{r}_1$ *with* $\underline{r}_n$.

If $n=2$ this condition means that every line in R_1 intersects every line in R_2.

THEOREM 5 : *Let* Σ *be a regular a.p. structure and* (L, Π) *be the partition loop representing* Σ *according to Theorem 1. Then* Σ *satisfies condition (S 5) if and only if there exists one set* $\Pi^* = \{A_1, A_2, \ldots, A_n\}$ *of subgroups* $A_j \in \Pi$ $(j=1,\ldots,n)$ *such that*

$$[(\ldots((A_1 + A_2) + A_3) + \ldots) + A_i] \cap A_{i+1} = \{\sigma\}, \quad \forall i \in \{1,\ldots,n-1\}, \tag{9}$$

and

$$L = [\ldots((\ell + A_1) + A_2) + \ldots] + A_n, \qquad \forall \ell \in L. \tag{10}$$

PROOF : If condition (S 5) holds, we consider the line $\bar{\underline{r}}_1$ incident with the point $\underline{L} \equiv (\ell)$. Then for every point $\underline{P} \equiv (p)$ let $\underline{r}_n(\underline{P})$ be the line in R_n incident with $\underline{P}$. By $\{\bar{\underline{r}}_1, \underline{r}_2(\underline{P}), \ldots, \underline{r}_n(\underline{P})\}$ we shall indicate the (unique) S-chain connecting $\bar{\underline{r}}_1$ with $\underline{r}_n(P)$. Let us assume now that $A_1, A_2, \ldots, A_n$ are the subgroups of the partition Π corresponding to the parallelism classes $R_1, R_2, \ldots, R_n$ respectively. In view of the uniqueness of the S-chain connecting the given lines, relation (9) holds. Moreover, we have

$$\bar{\underline{r}}_1 \equiv (\ell + A_1), \qquad \underline{r}_2(\underline{P}) \equiv ((\ell + a_1) + A_2),$$

$$\underline{r}_3(\underline{P}) \equiv [((\ell + a_1) + a_2) + A_3], \ldots,$$

$$\underline{r}_n(\underline{P}) \equiv [\ldots(((\ell + a_1) + a_2) + a_3) + \ldots] + A_n$$

for suitable $a_i \in A_i$ $(i=1,\ldots,n-1)$; hence

$$p = [\ldots((\ell + a_1) + a_2) + \ldots] + a_n, \tag{11}$$

and (10) is proved too.

Conversely, if relations (9), (10) hold, for every point $\underline{P} \equiv (p)$ there are suitable $a_j \in A_j$ $(j = 1, \ldots, n)$ such that (11) is satisfied, then there exists a S-chain connecting $\bar{\underline{r}}_1$ with $\underline{r}_n(P)$.

To finish we have to prove that such a chain is unique. It is enough to prove that the decomposition (11) is unique. If there were two different decompositions such that

$$p = [\ldots((\ell + \bar{a}_1) + \bar{a}_2) + \ldots] + \bar{a}_n = [\ldots((\ell + a_1) + a_2) + \ldots] + a_n ,$$

we would have

$$((\ell + \bar{a}_1) + \bar{a}_2) + \ldots = [\ldots((\ell + a_1) + a_2) + \ldots] + (a_n - \bar{a}_n);$$

then, in view of (9),

$$a_n - \bar{a}_n = \sigma .$$

In this way we can also prove

$$a_{n-1} - \bar{a}_{n-1} = \sigma , \ldots, a_2 - \bar{a}_2 = \sigma , \quad a_1 - \bar{a}_1 = \sigma ,$$

and hence for all given ℓ, p the decomposition (11) is unique. □

If the a.p. structure Σ satisfies (S 5) with $n = 2$, then Theorem 5 takes the following form.

THEOREM 6 : *Let* Σ *be a regular a.p. structure with two parallelism classes of lines* R_1, R_2, *such that every line in* R_1 *intersects every line in* R_2. *It is possible to define a partition loop* (L, Π), *as in Theorems 1 and 4 such that there exist two subgroups* $A_1, A_2 \in \Pi$ *for which*

$$L = A_1 + A_2 = A_2 + A_1 , \quad a_1 + a_2 = a_2 + a_1$$
$$(\forall a_1 \in A_1 , \ \forall a_2 \in A_2).$$

The proof is a consequence of Theorems 4 and 5. □

3.

Let us now return to the second point of view discussed above, just after Theorem 1. We investigate some geometrical properties of the a.p. structure Σ in connexion with different algebraic conditions on L. The geometrical meaning of the left nucleus was already pointed out in §1. We shall now consider the mapping (5) in the general case.

PROPOSITION 3 : *Let* (L, Π) *be a partition loop, as in Theorem 1. The mapping*

$$\tau \;:\; x' = m + x \tag{5}$$

is a translation of the a.p. structure Σ *if and only if* m *fulfils the condition*

$$m + (p + Q) = (m + p) + Q \qquad (\forall p \in L,\ \forall Q \in \Pi). \tag{12}$$

PROOF : The mapping (5) is one-to-one and onto; if (12) holds, lines are mapped to parallel lines. Moreover, (5) is a fixed-point-free map, and so is a translation. Conversely, let (5) be a translation; then one must have

$$m + (p + Q) = p' + Q \qquad (\forall p \in L,\ \forall Q \in \Pi)$$

where p' depends on m and p but not on elements in Q. Hence there will be a q in Q such that

$$m + p = p' + q.$$

Then

$$p' = (m + p) - q,$$

and (12) is proved. □

By T we shall denote the set of all translations (5); moreover, we shall denote

$$\mathcal{Q}_L = \{ m \in L \mid m + (p + Q) = (m + p) + Q,\ \forall p \in L,\ \forall Q \in \Pi \}.$$

In the general case, T *is not closed with respect to composition.* Let

$$\tau_i \;:\; x' = m_i + x \qquad (m_i \in \mathcal{Q}_L),\quad i = 1,2, \tag{13}$$

be translations; composition gives rise to the mapping

$$\tau_1 \circ \tau_2 \;:\; x' = m_1 + (m_2 + x) = (m_1 + m_2) + \tilde{x}, \tag{14}$$

where $\tilde{x}$ is a function of x which depends on m_1 and m_2. Of course (14) may have some fixed point; in any case (14) maps lines onto parallel lines:

$$m_1 + [m_2 + (p + Q)] = m_1 + [(m_2 + p) + Q] = [m_1 + (m_2 + p)] + Q =$$
$$= [(m_1 + m_2) + \tilde{p}] + Q .$$

Nevertheless this does *not* yield, in the general case, the result $(m_1 + m_2) \in \mathcal{Q}_L$.

It may be interesting to study both the group $\langle T \rangle$ of dilatations generated by T and the algebraic structure of $\mathcal{Q}_L$.

PROPOSITION 4 : *In (14), let* $\tilde{x} = x$ *for all* $x \in L$; *then* $(m_1 + m_2) \in \mathcal{Q}_L$ *and* $\tau_1 \circ \tau_2 \in T$.

PROOF : By hypothesis we have, $\forall x \in L$,

$$m_1 + (m_2 + x) = (m_1 + m_2) + x . \tag{15}$$

Then $\forall p \in L$, $\forall Q \in \Pi$ we have (in view of $m_i \in \mathcal{Q}_L$ and (15)) that

$$(m_1 + m_2) + (p + Q) = m_1 + [m_2 + (p + Q)] = m_1 + [(m_2 + p) + Q] =$$
$$= [m_1 + (m_2 + p)] + Q ,$$

which by (15) is

$$[(m_1 + m_2) + p] + Q .$$

Hence $(m_1 + m_2) \in \mathcal{Q}_L$, and so (14) is a translation; that is $\tau_1 \circ \tau_2 \in T$. □

PROPOSITION 5 : *Let* T *be a group with respect to composition. Then* $\mathcal{Q}_L$ *is a group.*

PROOF : Let τ_1, τ_2 be translations as in (13) and let $\tau_1 \circ \tau_2 \in T$. Then there exists $m_3 \in \mathcal{Q}_L$ such that, $\forall x \in L$,

$$m_1 + (m_2 + x) = m_3 + x .$$

Hence we have $m_3 = m_1 + m_2$; that is

$$m_1 + (m_2 + x) = (m_1 + m_2) + x \qquad \forall x \in L .$$

Hence $\mathcal{Q}_L$ is closed under addition and associative. Moreover, if T is a group and $\tau_1 \in T$, then

$$\tau_1^{-1} \; : \; x' = - m_1 + x$$

is also a translation; hence

$$m_1 \in \mathcal{Q}_L \iff - m_1 \in \mathcal{Q}_L .$$

$\mathcal{Q}_L$ is then an associative subloop of L, that is a subgroup. □

Let us now suppose that the a.p. structure Σ is finite. If $\langle T \rangle$ contains proper dilatations, then $\langle T \rangle$ is a Frobenius group for which the kernel is the group T^* of all translations generated by T and a complement is the dilatation group $\mathcal{D}_0$ with fixed point $\underline{0} \equiv (\sigma)$. Then $\langle T \rangle = T^*\mathcal{D}_0$.

4.

We shall now conclude with a few remarks about right nucleus N_R and middle nucleus N_M of L. As in [4], N_R and N_M are defined in the following way:

$$N_R = \{ q \in L \mid \ell + (m + q) = (\ell + m) + q, \quad \forall \ell, m \in L \},$$

$$N_M = \{ p \in L \mid \ell + (p + q) = (\ell + p) + q, \quad \forall \ell, q \in L \}.$$

The proofs of the following propositions are straightforward.

PROPOSITION 6 : *Let* Q *be a subgroup of the partition* Π *such that* $Q \subset N_R$. *Then every line* $\underline{r} = (p + Q)$ $(\forall p \in L)$ *is mapped onto a parallel line by every mapping* $x' = \ell + x$ $(\forall \ell \in L)$. □

PROPOSITION 7 : *For all* p *in* N_M *all the lines* $\underline{r} = (p + Q)$ $(\forall Q \in \Pi)$ *are mapped onto parallel lines by every mapping* $x' = \ell + x$ $(\forall \ell \in L)$. □

If we consider the incidence structures $\Sigma_R = (P', R', \|', I)$ and $\Sigma_M = (P'', R'', \|'', I)$ such that $P' = N_R$, $P'' = N_M$ and R', R'', $\|'$, $\|''$ are induced by the lines and the parallelism of Σ, then Σ_R and Σ_M are translation a.p. structures, substructures of Σ. We note also that the cosets

$$C_R(p) = p + N_R, \quad C_M(p) = p + N_M \qquad (\forall p \in L)$$

represent again a.p. structures which are mutually parallel. Moreover, every map $x' = \ell + x$ $(\forall \ell \in L)$ acts as a translation on all $C_R(p)$, mapping them each on the other.

We shall now introduce, as for the left nucleus, the sets

$$\mathcal{Q}_R = \{ q \in L \mid \ell + (m+q) = (\ell + m) + q',\ q \in Q \in \Pi \Leftrightarrow q' \in Q,\ \forall \ell, m \in L \},$$

$$\mathcal{Q}_M = \{ p \in L \mid \ell + (p+Q) = (\ell + p) + Q,\ \forall \ell \in L,\ \forall Q \in \Pi \}.$$

PROPOSITION 8 : *If for all* $q \in Q \cap \mathcal{Q}_R$ $(\forall Q \in \Pi)$ *we have also that* $-q \in Q \cap \mathcal{Q}_R$, *then* $Q \cap \mathcal{Q}_R$ *is a group.*

PROOF : We have to prove that $Q \cap \mathcal{Q}_R$ is closed under addition. Let $q_1, q_2 \in Q \cap \mathcal{Q}_R$; then by condition (L 2) and by assumption,

$$\ell + [m + (q_1 + q_2)] = \ell + [(m+q_1) + q_2] = [\ell + (m+q_1)] + q_2' =$$
$$= [(\ell + m) + q_1'] + q_2' = (\ell + m) + (q_1' + q_2').$$

Since, by assumption $-q \in Q \cap \mathcal{Q}_R$, we have that $Q \cap \mathcal{Q}_R$ is a loop (subloop of Q), and so a group. □

BIBLIOGRAPHY

1. J. André, "Über Parallelstrukturen : Teil I, II, III", *Math. Z.*, 76 (1961), 85-102, 155-163, 240-256.
2. H. -J. Arnold, "Über eine Klasse von Spernerschen Quasimoduln", *Abh. Math. Sem. Univ. Hamburg*, 31 (1967), 206-217.
3. M. Biliotti, "Strutture di André e spazi di Sperner con traslazioni", to appear.

4. R.H. Bruck, *A Survey of Binary Systems* (Springer-Verlag, Berlin, 1966).
5. V. Havel, "Partitions in cartesian systems", *Časopis Pěst. Mat.*, 91 (1966), 246-252.
6. V. Havel, "Parallelisable partitions of groupoids", *Arch. Math. (Basel)*, 18 (1967), 118-121.
7. G. Korchmàros and M. Marchi, "Cappi con partizione e strutture di André", *Istit. Lombardo Accad. Sci. Lett. Rend. A*, to appear.
8. M. Marchi, "S-spazi e loro problematica", Seminario di Geometrie Combinatorie 1 (Università di Roma, 1977).
9. R. Permutti, "Sugli spazi affini generalizzati di E. Sperner", *Ricerche Mat.*, 11 (1962), 95-102.
10. R. Permutti, "Geometria affine su di un anello", *Atti Accad. Naz. Lincei Mem.*, 8 (1967), 259-287.
11. W. Seier, "Kollineationen von Translationsstrukturen", *J. Geom.*, 1 (1971), 183-195.
12. E. Sperner, "Zur Geometrie der Quasimoduln", Symposia Mathematica Vol.V, Rome 1969/70 (Academic Press, London, 1971), pp.421-438.
13. H. Wähling, "Kinematische Inzidenzloops", Kolloquium über Geometrie, 1975 (I), 37 (Technische Universität Hannover, 1976).
14. H. Wähling, "Projektive Inzidenzgruppoide und Fastalgebren", *J. Geom.*, 9 (1977), 109-126.

Istituto Matematico
Politecnico Milano
Piazza Leonardo da Vinci 32
20133 Milano
Italy

REGULAR CLIQUES IN GRAPHS AND SPECIAL $1\frac{1}{2}$ - DESIGNS

A. Neumaier

Many strongly regular graphs constructed from designs contain cliques with the property that every point not in the clique is adjacent to the same number of points of the clique. In the first section we give some examples and investigate various properties of such regular cliques. In particular, parameter relations and inequalities are discussed. Section two defines special $1\frac{1}{2}$-designs as a certain class of designs with connection numbers 0 and Λ, generalizing partial geometries. Notable examples of special $1\frac{1}{2}$-designs are transversal designs and classical polar spaces. It is shown that the point graph of a special $1\frac{1}{2}$-design is strongly regular, and the blocks are regular cliques in the point graph.

As applications we reprove a result of Higman on partial geometries with isomorphic point graphs, and improve an inequality of Cameron and Drake concerning the parameters of partial Λ - geometries.

1. REGULAR CLIQUES

In this paper, all graphs are finite, undirected, without loops or multiple edges. A *strongly regular graph* (SRG) is a graph with v *vertices (*or *points)* such that

(R1) every vertex is adjacent to k other vertices;

(R2) the number of vertices adjacent to two adjacent vertices is always λ ;

(R3) the number of vertices adjacent to two nonadjacent vertices is always μ .

A graph is called *regular*, respectively *edge-regular*, if only (R1), respectively (R1) and (R2) holds. A clique C

(i.e. a complete subgraph) is called *regular* if every point not in C is adjacent to the same number $e > 0$ of points in C; we call e the *nexus* of C. Of course, C is a maximal clique unless e equals the size of C.

If Γ is a graph with regular clique C, and we remove some edges disjoint from C, then in the new graph, C is still regular. In a complete graph, all cliques are regular. In the complement of the hexagon, there are regular cliques of sizes 2 and 3 (with nexus 1). In contrast to this, we have

THEOREM 1.1 : *Let Γ be an edge-regular graph which is not complete. If Γ has a regular clique, then*

(i) *all regular cliques have the same size K and nexus e;*

(ii) *every clique has size at most K;*

(iii) *the regular cliques are exactly the cliques of size K.*

PROOF : Let C be a regular clique of smallest size K, with nexus e. If we count in two ways the number of all edges xy with $x \notin C$, $y \in C$, and the number of all triangles xyz with $x \notin C$, $y,z \in C$, we find

$$K(k+1-K) = (v-K)e, \qquad (1)$$

$$K(K-1)(\lambda+2-K) = (v-K)e(e-1). \qquad (2)$$

Now let C be any clique of size K. For $x \notin C$, denote by e_x the number of vertices in C adjacent to x. Counting the same numbers as before, we find

$$K(k+1-K) = \Sigma e_x,$$

$$K(K-1)(\lambda+2-K) = \Sigma e_x(e_x-1),$$

where the sum extends over all $v-k$ vertices $x \notin C$. Hence by (1) and (2), $\Sigma(e_x-e)^2 = 0$; so $e_x = e$ for all $x \notin C$. Hence C is regular of nexus e.

If there is a clique of size $>K$ then any subclique C of size K is regular. But since C is not maximal, $e = K$, and since Γ is assumed to be regular, Γ is complete. □

Note: For similar results see Bose [1], Delsarte [6], Neumaier [9], Wilbrink and Brouwer [15].

Problem: Investigate the possibilities when Γ is regular but not edge-regular.

COROLLARY 1.2 : *Let Γ be an edge-regular graph containing a regular clique of size K and nexus e.*

(i) *There is a number $m > 1$ such that $(e-1, K-1)m$ is integral and*

$$v = K + \frac{(m-1)K(K-1)}{e}, \quad k = m(K-1), \ \lambda = K-2+(e-1)(m-1). \tag{3}$$

(ii) *If Γ is strongly regular, then*

$$\mu = em. \tag{4}$$

Moreover, m and

$$f = m(m-1)K(K-1)/e(K+m-1-e), \tag{5}$$

are integers.

(iii) *Among the strongly regular graphs which are neither complete nor complete multipartite, only finitely many contain a regular clique of given size K.*

PROOF : (i) is a consequence of (1) and (2) if we choose $m = k/(K-1)$. Note that $(e-1, K-1)m$ is the greatest common diviser of K and $\lambda - K + e + 1$. To prove (ii) we count in two ways the number of edges xy with x adjacent and y nonadjacent to a given point, and obtain $(v-1-k)\mu = k(k-1-\lambda)$. If we solve this for μ, and simplify with (i), we obtain $\mu = em$. Since m can be written as $m = K-1-e+\mu-\lambda$, it is an integer. From standard formulas for the eigenvalues and multiplicities of a SRG (see [8], [10], or [13]), we see that the eigenvalues of the adjacency matrix are $k = m(K-1)$, $K-1-e$, and $-m$, with respective (integral) multiplicities $1, f$, and $v-1-f$, where f is given by (5). In particular, $K = m-1-e$ divides $m(m-1)K(K-1)$, and hence

$$K+m-1-e \mid K(K-1)(K-e)(K-e-1) . \qquad (6)$$

If $K > e+1$, then we have only finitely many possibilities for e, and, by (6), for m. Hence v takes only finitely many values, and (iii) holds in this case. If $K \leq e+1$, then the following theorem completes the proof.

THEOREM 1.3 : *Let* Γ *be an edge-regular graph containing a regular clique* C *of size* K *and nexus* e.

(i) $e = K$ *if and only if* Γ *is complete.*

(ii) $e = K-1$ *if and only if* Γ *is complete multipartite with* K *classes of the same size.*

PROOF : (i) If $e = k$ then every point in C is adjacent to all other points, whence the valency is $v-1$. Hence Γ is complete.

(ii) If $e = K-1$, then each point not in C is nonadjacent to exactly one point of C. Hence no edge of C is in an induced subgraph whence by edge-regularity, there are no such induced subgraphs. For $a \in C$, denote by A_a the class consisting of a and all points which are not adjacent to a. Then the A_a partition the points of Γ. If $x \in A_a$, $y \in A_b$, $b \neq a$ then xy is an edge since otherwise xay is the forbidden subgraph. Now each point $a \in C$ is unjoined with just the points of $A_a - \{a\}$, whence by regularity, all A_a have same size s, and every point $x \in \Gamma$ is unjoined to exactly $s-1$ points. But the only candidates are the $s-1$ points $\neq x$ in the class of x. Hence no A_a contains an edge, and Γ is complete multipartite. Conversely, in a complete multipartite graph with K classes, the maximal cliques are just the sets containing one point from every class, and these are regular with size K and nexus $e = K-1$. □

Problem: (cf. Corollary 2.4) Is every edge-regular graph with a regular clique strongly regular?

LEMMA 1.4 : *Let* C *be a regular clique of size* K *with nexus* e *in an edge-regular graph with parameters (3). Then the points of* C*, together with the blocks* $B_a = \{x \in C \mid X \text{ adjacent to } a\}$ *for* $a \notin C$*, form a* $2-(K, e, (e-1)(m-1))$*-design.* □

This straightforward result is essentially due to Wilbrink and Brouwer [15]. Usually, the design will have repeated blocks.

LEMMA 1.5 : *Let* Γ *be an edge-regular graph which is not complete and which has regular cliques of size* K *and nexus* e*. Then*

(i) *two distinct regular cliques have at most* e *common points;*

(ii) *if two regular cliques have* $d \geq 2$ *common points,* $K \leq (m-1)(e-1) + d$;

(iii) *if* $K-1 > m(e-1)$, *every edge is in at most one regular clique.*

PROOF : (i) is obvious. If xy is an edge contained in two regular cliques which intersect in d points then x and y are adjacent to at least $2K-d-2 \leq \lambda$ points. This implies (ii), and (iii) if we observe that $d \leq e$. □

LEMMA 1.6 : *In an edge-regular graph which is not complete and which has regular cliques of size* K *and nexus* e*, suppose that any two distinct regular cliques have at most* d *common points, and two points are in at most* Λ *regular cliques. Then*

$$e - 1 \geq \frac{(\Lambda-1)(K-2)(K-d)}{(m-1)(K-2+(\Lambda-1)(d-2))} \tag{7}$$

and

$$\Lambda \leq \frac{(K-d)}{(K-2)^2 - \lambda(d-2)} \quad \text{if} \quad d-2 < \frac{(K-2)^2}{\lambda}. \tag{8}$$

PROOF : Choose an edge xy, and denote by a_z the number of distinct regular cliques containing x, y, z. If we count in two ways the number N_0 of points z adjacent to x and y, the number N_1 of pairs (z,A) consisting of a point $z \neq x,y$, and a regular clique A containing x,y,z, and the number N_2 of triples (z,A,B) consisting of a point $z \neq x,y$, and two distinct regular cliques A, B containing x,y,z, we find

$$\begin{aligned} N_0 &= \Sigma 1 = \lambda = K-2 + (m-1)(e-1), \\ N_1 &= \Sigma a_z = \Lambda(K-2), \\ N_2 &= \Sigma a_z(a_z - 1) \leq \Lambda(\Lambda - 1)(d-2), \end{aligned} \tag{9}$$

where the summation is over all points z adjacent to x and y. Now the average of the a_z is $a = N_1 / N_0$, and we have $0 \leq \Sigma (a_z - a)^2 = N_2 + (1 - 2a) N_1 + a^2 N_0 = N_1 + N_2 - N_1^2 / N_0$. Hence $N_1^2 \leq N_0(N_1 + N_2)$, and if we use (9) and solve for $e-1$ and Λ, we obtain (7) and (8). □

The proof also yields the following.

COROLLARY 1.7 : *Equality in (7) or (8) holds if and only if any two distinct regular cliques intersect in* $0, 1$, *or* d *points, and every triangle is in the same number of regular cliques.* □

Remarks : (1) By Lemma 1.5, we may always take $d = e$.

(2) The proof shows that the same result holds if we require that the regular cliques belong to a distinguished set of cliques.

(3) In some cases the condition $i(i+1) N_0 - 2i N_1 + N_2 = \Sigma (a_z - i)(a_z - 1 - i) \geq 0$ for all integers i will exclude some possibilities which are not excluded by (7) or (8).

EXAMPLE 1.8 : The two SRGs with $K = 4$, $m = 3$, $e = 2$ are obtained from the two groups X of order 4 as follows: the points are the triples $(a,b,c) \in X^3$ with $a + b + c = 0$, adjacent if they have the same entry in some place. Regular cliques are the 12 sets consisting of the points with fixed entry in some place (type 1), and the sets $\{(a,b,c,), (a, b+i, c+i), (a+i, b, c+i), (a+i, b+i, c)\}$ with $a + b + c = 0$, $2i = 0 \neq i$ (type 2).

Hence in the cyclic case, all edges are in a unique regular clique of type 1, and in at most one regular clique of type 2. In the noncyclic case, all edges are in a unique clique of each type. This agrees with Lemma 1.6, which states that $\Lambda \leq 2$.

EXAMPLE 1.9 : A SRG with parameters K, $m = K-1$, $e = K-2 \geq 1$ is obtained from a K-set X as follows: the points are the ordered pairs $(a,b) \in X^2$, adjacent if and only if they have distinct entries in both places. Regular cliques are the sets $\{(a, \pi a) \mid a \in X\}$, where π is a permutation of X. Hence two regular cliques may have up to $K-2$ common points, and every edge is in $\Lambda = (K-2)!$ regular cliques. Lemma 1.6 gives no bound for Λ unless $K = 3$ or $K = 4$, where the bound is exact.

EXAMPLE 1.10 : A SRG with parameters K, $m = 2K-3$, $e = K-2 \geq 1$ is obtained from a 2K-set X as follows: the points are the 2-subsets of X, adjacent if and only if they are disjoint. Regular cliques are the partitions of X into 2-subsets. Hence two regular cliques may contain up to $K-2$ common points, and every edge is in $\Lambda = 1 \cdot 3 \cdot 5 \ldots (2K-5)$ regular cliques. Lemma 1.6 gives no bound for Λ unless $K = 3$ or $K = 4$, where the bound is exact.

2. SPECIAL 1½ - DESIGNS

A *special 1½ - design* consists of a set of v points, a collection of b blocks, and an incidence relation ϵ between points and blocks such that

(S1) blocks contain K points, points are in R blocks,

(S2) two distinct points are in 0 or Λ blocks, $0 < \Lambda < R$, and both possibilities occur.

(S3) If B is a block then every point outside the block is adjacent to exactly e points on B.

Here two points are called *adjacent* if they are distinct and contained in some block. This defines a graph, the *point graph* of the design.

THEOREM 2.1 : *The point graph* Γ *of a special* $1\frac{1}{2}$*-design* $\mathcal{B}$ *is strongly regular. Every block of* $\mathcal{B}$ *is a regular clique in* Γ.

PROOF : For a point a, the number of pairs (A,x) with $a, x \in A$ is $k\Lambda = R(K-1)$, where k is the number of points adjacent to a. Hence k is independent of a. For an edge ab, the number of pairs (A,x) with $a, x \in A$, x adjacent to b, and $b \notin A$, is $(k-1-\lambda)\Lambda = (R-\Lambda)(K-e)$, where λ is the number of points adjacent to a and b. Hence λ is independent of ab. For a nonedge ab, the number of pairs (A,x) with $a, x \in A$ and b adjacent with x is

$$\mu\Lambda = Re, \tag{10}$$

where μ is the number of points adjacent to a and b. Hence μ is independent of ab. Therefore Γ is strongly regular. By (S3), the blocks are regular cliques in Γ. □

COROLLARY 2.2 : *There is an integer* $m \geq 2$ *such that the point graph has parameters given by Corollary 1.2, and*

$$v = K + (m-1)K(K-1)/e, \quad b = \Lambda m + \Lambda m(m-1)(K-1)/e, \quad R = \Lambda m.$$

PROOF : Apply Corollary 1.1. By (10), $R = \Lambda m$, and since the number of pairs (x,A) with $x \in A$ is $vR = bK$, the formula for b follows. □

THEOREM 2.3 : *Let* Γ *be a graph with a point- and edge-transitive automorphism group* G, *and let* C *be a regular clique in* Γ. *If we define the images of* C *under* G *as blocks we obtain a special* $1\frac{1}{2}$*-design with point graph* Γ.

PROOF : Point-transitivity implies (S1), edge-transitivity implies (S2), and by construction, all blocks are regular cliques whence (S3) holds. □

COROLLARY 2.4 : *A point- and edge-transitive graph which contains a regular clique is strongly regular.* □

A *transversal design* $T[K, \Lambda; m]$ consists of a set of points, a partition of the points into $K > 1$ classes of size m each, and a collection of blocks (point sets) such that every block contains exactly one point from every class, and any two points from different classes are in exactly Λ blocks. Neumaier [11] shows that every transversal design $T[K, \Lambda; m]$ is a special $1\frac{1}{2}$-design with $R = \Lambda m$, $e = K-1$.

THEOREM 2.5 : *For a special* $1\frac{1}{2}$*-design* $\mathcal{B}$, $e \le K-1$, *with equality if and only if* $\mathcal{B}$ *is a transversal design.*

PROOF : Since there are nonadjacent points, $e \le K-1$. If $e = K-1$ then the point graph is complete multipartite, and the regular cliques are just the sets which contain one point from every class. By Theorem 2.1, $\mathcal{B}$ is a transversal design. □

THEOREM 2.6 : *Points and maximal subspaces of a classical, finite polar space form a special* $1\frac{1}{2}$*-design.*

PROOF : For the axioms and basic properties of a polar space see e.g. Tits [14]. If the block set of $\mathcal{B}$ is the set of maximalsubspaces of a polar space of rank s over $GF(q)$ then the blocks are $(s-1)$-dimensional projective spaces whence $K = (q^s - 1)/(q - 1)$. The points of a block which are adjacent to a given point outside form an $(s-2)$-dimensional subspace of the block whence $e = (q^{s-1} - 1)/(q - 1)$. Every point is in the same number R of blocks since the automorphism group is transitivie on points. Every pair of points which is in a block is in a unique line, and since the automorphism group is transitive on lines, in the same number Λ of blocks. Hence $\mathcal{B}$ is a special $1\frac{1}{2}$-design. If $\mathcal{B}$ is not a polar space over $GF(q)$ then the polar space is a generalized quadrangle, i.e. a special $1\frac{1}{2}$-design with $\Lambda = 1$, $e = 1$. □

In some cases [3], [14], the blocks can be split into two sets $\mathcal{B}_1$, $\mathcal{B}_2$ such that every (s-2)-dimensional subspace is in exactly one block of $\mathcal{B}_1$, and one block of $\mathcal{B}_2$. In this case a similar proof shows that the *half polar spaces* $\mathcal{B}_1$ and $\mathcal{B}_2$ also form special $1\frac{1}{2}$-designs.

By a result of Tits [14], all polar spaces of rank > 2 can be embedded into projective spaces. Thas and De Clerck [13] classified all partial geometries consisting of some lines, and all points on these lines, of a projective space. It would be interesting to classify similarly the special $1\frac{1}{2}$-designs consisting of some subspaces, and all points on these subspaces, of a projective space, or a polar space.

Special $1\frac{1}{2}$-designs with $\Lambda = 1$ are just the partial geometries defined by Bose [1], and those with $\Lambda = 1$, $e = 1$ are the generalized quadrangles. If we take Λ identical copies of the blocks of a partial geometry $\mathcal{B}$ we obtain a special $1\frac{1}{2}$-design, called a *multiple* of $\mathcal{B}$. From Theorem 2.1 and Lemma 1.5, we now get immediately

THEOREM 2.7 :

(i) *A special* $1\frac{1}{2}$*-design* $\mathcal{B}$ *with* $K-1 > m(e-1)$ *is a multiple of a partial geometry.*

(ii) $e = 1$ *if and only if* $\mathcal{B}$ *is a multiple of a generalized quadrangle.* □

COROLLARY 2.8 : (Higman [8]) *If two partial geometries have the same point graph, and satisfy* $K - 1 > R(e-1)$, *then they are isomorphic.*

PROOF : The point graph of a partial geometry has $m = R$ whence the union of the blocks of two partial geometries with the same point graph is a special $1\frac{1}{2}$-design with $\Lambda' = 2$, $K' = K$, $m' = R$, $e' = e$ (count repeated blocks twice!). By Theorem 2.7 (i), every block is repeated twice whence the partial geometries are isomorphic. □

Remark: Corollary 2.8 is in some sense best possible. For, the points of the graph of Example 1.7 (noncyclic case) together with the regular cliques of type 1, respectively type 2, as blocks define two distinct partial geometries with $K = 4$, $R = 3$, $e = 2$, hence $K - 1 = R(e-1)$, with the same point graph.

We note another consequence of Theorem 2.1.

COROLLARY 2.9: *Every special* $1\frac{1}{2}$*-design is a* 2*-class partially balanced design with* $\lambda_1 = 0$. □

In general, the converse is not true, but Bridges and Shrikhande [2] proved a result (their Corollary 2.4) which easily implies

THEOREM 2.10: *A* 2*-class partially balanced design with* $\lambda_1 = 0$ *which has more points than blocks is a special* $1\frac{1}{2}$*-design.* □

This result again implies Theorem 2.6.

3. PARTIAL Λ-GEOMETRIES AND $1\frac{1}{2}$-DESIGNS

A $1\frac{1}{2}$*-design* consists of a set of points, a collection of blocks, and an incidence relation ϵ between points and blocks such that blocks have K points, points are in R blocks, and, for a given point a and block B, the number of flags (i.e. incident point-line pairs) (x,A) with $a \neq x \in B$, $a \in A \neq B$ is α if $a \notin B$, and β if $a \in B$. Neumaier [11] introduces also the parameter

$$N = R + K + \beta - \alpha - 1. \qquad (11)$$

LEMMA 3.1:

(i) *A special* $1\frac{1}{2}$*-design with parameters* Λ, K, m, e *is a* $1\frac{1}{2}$*-design with parameters*

$$K,\ R = \Lambda m,\ \alpha = \Lambda e,\ \beta = (\Lambda-1)(K-1),\ N = \Lambda(K+m-1-e). \qquad (12)$$

(ii) *Every* $1\frac{1}{2}$*-design with property* (S2) *is special.*

PROOF : (i) is straightforward. (ii) Let a be a point not on B, and suppose that there are e points on B adjacent with a. Through a and each such point there are exactly Λ blocks, whence $\alpha = \Lambda e$. Hence e is independent of a and B, i.e. (S3) holds. □

LEMMA 3.2 : *The parameters of a special* $1\frac{1}{2}$*-design satisfy*

$$\Lambda - 1 \geq \frac{(e-m)(m-1)}{m(K+m-1-e)} , \tag{13}$$

with equality if and only if any two blocks intersect in the same number of points, which is e/m .

PROOF : By Lemma 3.7 of Neumaier [11], $\alpha \geq R(K-N)$, with equality if and only if any two blocks intersect in the same number of points, which is $K - N = \alpha / R$. Now use the parameters (12). □

This lower bound is sharp, since there are many transversal designs which satisfy (13) with equality, namely the duals of affine 2-designs. An upper bound for Λ can often be obtained from Lemma 1.6, if there are no repeated blocks.

COROLLARY 3.3 : *For a special* $1\frac{1}{2}$*-design,* $e > m$ *implies* $\Lambda > 1$ *and* $K-1 \leq m(e-1)$.

PROOF : $\Lambda > 1$ by Lemma 3.2. If $K-1 > m(e-1)$, then by Theorem 2.7(i) we have a multiple of a partial geometry with $e > m$, contradicting 3.2 (with $\Lambda = 1$). □

The dual of a $1\frac{1}{2}$-design is obtained by interchanging the roles of points and blocks, and reversing incidence. It is again a $1\frac{1}{2}$-design with K, R interchanged, and some α, β, N. In particular, the dual of a special $1\frac{1}{2}$-design is special if and only if the axiom dual to (S2),

(S2*) two distinct blocks intersect in 0 or Λ^* points ($0 < \Lambda^* < K$), and both possibilities occur,

holds. In this case the dual parameters Λ^* and e^* can be calculated from (12) and its dual: By comparing the two expressions for α and β we obtain

$$(R-1)(\Lambda^*-1) = (K-1)(\Lambda-1), \quad \Lambda^* e^* = \Lambda e. \tag{14}$$

Cameron and Drake [3] define a *partial Λ-geometry* as a design with (S1), (S2), (S2*) and (S3), with $\Lambda^* = \Lambda$. For $\Lambda = 1$, this is just a partial geometry, and for $\Lambda > 1$, (14) implies $e^* = e$, $R = K$. Cameron and Drake [3] show that half polar spaces of rank 4 over $GF(q)$ are partial Λ-geometries with $\Lambda = q+1$, $K = R = q^3+q^2+q+1$, $e = q^2+q+1$. They also have examples of partial Λ-geometries with $\Lambda = 2$, $e = 3$, $K \in \{8, 24\}$. Drake [7] calls the number $i = K-e$ the *index* of the partial Λ-geometry, and constructs infinitely many partial Λ-geometries with $\Lambda > 1$ and index 1. On the other hand, we have

LEMMA 3.4: *For given $i > 1$, there are only finitely many partial Λ-geometries with $\Lambda > 1$ and index i.*

PROOF: From Theorems 1.1 and 2.1, we can see that, in the notation of Neumaier [10], the point graph has parameters m, $n = K+m-1-e = i+m-1$, $\mu = em$, whence its complement has parameters $\bar{m} = i$, $\bar{n} = i+m-1$, $\bar{\mu} = i(i-1)(m-1)/e > 0$. If $\bar{\mu} \in \{i(i-1), i^2\}$ then $e \le m-1$, $(\Lambda-1)m = K-m = e+i-m \le i-1$. Since $\Lambda > 1$, there are only finitely many possibilities for e and m, and hence for m, n and μ. Therefore, we have only finitely many such graphs. If $\mu \notin \{i(i-1), i^2\}$, then, by Theorem 5.1 of Neumaier [10], we also have only finitely many such graphs. Now (S2*), and Theorem 2.1 imply that the blocks are distinct regular cliques, whence there are only finitely many patial Λ-geometries with $\Lambda > 1$ and index i. □

It can be shown that for $\Lambda > 1$ there is no partial Λ-geometry with index 2, and the only possibilities with index 3 have parameters

$$v = 50, \quad \Lambda = 5, \quad K = 15, \quad e = 12, \quad \text{or}$$
$$v = 85, \quad \Lambda = 3, \quad K = 15, \quad e = 12.$$

On the other hand, De Clerck [5] gives a complete classification of the (infinitely many) partial Λ-geometries with $\Lambda = 1$ and index 2.

LEMMA 3.5 : *For a partial* Λ*-geometry with* $\Lambda > 1$ *we have*

$$K = R = \Lambda m, \tag{15}$$

$$e - 1 \geq A(m) = \frac{(\Lambda - 1)(m\Lambda - 2)}{m + \Lambda - 3}. \tag{16}$$

PROOF : (15) has already been proved. By (S2*) and Lemma 1.6, inequality (7) holds for $K = \Lambda m$, $d = \Lambda^* = \Lambda$, which implies (16).□

COROLLARY 3.6 : *For a partial* Λ*-geometry with* $\Lambda > 1$ *and index* $i > 1$, *we have*

$$i \leq m(m-2), \tag{17}$$

$$e \geq 3\Lambda - 3. \tag{18}$$

PROOF : By (16), $i = \Lambda m - e \leq \Lambda m - 1 - A(m) = (m-1)^2 - (m-1)(m-2)^2 / (\Lambda + m - 3)$. For $m = 2$, this implies $i \leq 1$, contrary to our assumptions. For $m \geq 3$, this implies $i < (m-1)^2$ whence $i \leq m(m-2)$. Since $A(m)$ increases as m increases, $e-1 \geq A(m) \geq A(3) = 3\Lambda - 5 + 2/\Lambda > 3\Lambda - 5$. Hence $e \geq 3\Lambda - 3$. □

Remark : Some instances with $e = 3\Lambda - 3$ can be excluded by using Remark (3) after Corollary 1.7 : The substitution $x = 1$ contradicts $\Lambda = 3$, $e = 6$, and $\Lambda = 4$, $e = 9$, and the substitution $x = 2$ contradicts $\Lambda = 7$, $e = 18$. With this method, Cameron and Drake [3] proved the result $e \geq 2\Lambda + 1$ if $\Lambda > 2$, which is better than (18) only for $\Lambda = 3$ and $\Lambda = 4$.

ACKNOWLEDGEMENT : I want to thank D.A. Drake for several stimulating discussions on the subject.

BIBLIOGRAPHY

1. R.C. Bose, "Strongly regular graphs, partial geometries, and partially balanced designs", *Pacific J. Math.*, 13 (1963), 389-419.
2. W.G. Bridges and M.S. Shrikhande, "Specially partially balanced incomplete block designs and associated graphs", *Discrete Math.*, 9 (1974), 1-18.
3. P.J. Cameron and D.A. Drake, "Partial λ - geometries of small nexus", (Combinatorial Mathematics, Optimal Design and their Applications, Fort Collins, 1978), *Ann. Discrete Math.*, 6 (1980), 19-29.
4. P.J. Cameron, J.M. Geothals and J.J. Seidel, "Strongly regular graphs having strongly regular subconstituents", *J. Algebra*, 55 (1978), 257-280.
5. F. De Clerck, "The pseudo-geometric and geometric (t,s,s-1)-graphs", to appear.
6. P. Delsarte, "An algebraic approach to the association schemes of coding theory", *Philips Res. Rep. Suppl.* No.10 (1973).
7. D.A. Drake, "Partial λ - geometries and generalized Hadamard matrices over groups", to appear.
8. D.G. Higman, "Partial geometries, generalized quadrangles and strongly regular graphs", *Geometria Combinatoria e sue Applicazioni,* Perugia 1970 (Università degli Studi di Perugia, 1971), pp.263-293.
9. A. Neumaier, "Quasi-residual 2-designs, 1½ - designs, and strongly regular multigraphs", *Geom. Dedicata,* to appear.
10. A. Neumaier, "Strongly regular graphs with smallest eigenvalue -m", *Arch. Math.*, 33 (1979), 392-400.
11. A. Neumaier, "t½ - designs", *J. Combin. Theory Ser. A.*, 28 (1980), 226-248.
12. J.J. Seidel, "Strongly regular graphs, an introduction", *Surveys in Combinatorics,* London Math. Soc. Lecture Note Series 38, (Cambridge 1979), pp.157-180.
13. J.A. Thas and F. De Clerck, "Partial geometries in finite projective spaces", *Arch. Math.* 30 (1978), 537-540.
14. J. Tits, *Buildings of Spherical Type and Finite BN-pairs,* Lecture Notes in Math. 386, (Springer-Verlag, Berlin, 1974).
15. H.A.Wilbrink and A.E. Brouwer, "A (57,14,1) strongly regular graph does not exist", Preprint ZW121 (Math. Centrum, Amsterdam, 1978).

Institut für Angewandte Mathematik
Albert-Ludwigs-Universität
Hermann-Herder-Strasse 10
7800 Freiburg im Breisgau
Federal Republic of Germany

BERICHT ÜBER HECKE ALGEBREN UND COXETER ALGEBREN EINDLICHER GEOMETRIEN

Udo Ott

1. HECKE ALGEBREN

Seien $\Omega_1, \Omega_2, \ldots, \Omega_n$ paarweise disjunkte Mengen, und sei I eine auf der *Grundmenge* $\Omega = \bigsqcup_{i=1}^{n} \Omega_i$ definierte symmetrische und reflexive Relation. Eine Teilmenge $F \subset \Omega$ heisst eine *Fahne*, wenn für je zwei Elemente $\omega, \tau \in F$ die Beziehung $\omega I \tau$ besteht. Wir wollen das Tupel $G = (\Omega_1, \ldots, \Omega_n, I)$ eine *Geometrie* vom Rang $rg(G) = n$ nennen, wenn die beiden folgenden Aussagen gelten:

(G1) *Aus* $\omega I \tau$ *und* $\omega, \tau \in \Omega_i$ *fur ein i folgt* $\omega = \tau$.
(G2) *Maximale Fahnen haben* n *Elemente.*

Im folgenden setzen wir voraus, dab das Tupel G eine Geometrie ist. Ist F eine Fahne von G, dann ist die Teilmenge

$$\Omega_F = \{ \omega \in \Omega \setminus F \mid F \cup \{\omega\} \text{ Fahne} \}$$

die Grundmenge einer Geometrie vom Rang $rg(G) - |F|$, die wir die in der Fahne F *abgeleitete Geometrie* G_F nennen wollen. Ähnlich ist für jede Teilmenge $J \subset \{1, 2, \ldots, n\}$ die Menge

$$\Omega_j = \bigsqcup_{j \in J} \Omega_j$$

die Grundmenge einer Geometrie vom Rang $|J|$, welche wir die *Teilgeometrie vom Typ* J nennen wollen. Zwei maximale Fahnen F und G der Geometrie G heissen *i-benachbart*, in Zeichen $F \underset{i}{\sim} G$, wenn sie sich um höchstens ein Element aus Ω_i unterscheiden, wenn also $|F \setminus G| = |G \setminus F| \le 1$ und $G \setminus F$, $F \setminus G \subset \Omega_i$ gilt. Offenbar ist die so definierte Relation der i-Nachbarschaft eine Äquivalenzrelation auf der Menge der maximalen Fahnen der Geometrie. Im folgenden sei R ein kommutativer Ring mit Einselement und ohne Nullteiler. Wir fassen die Menge $F_k(G)$ der k-elementigen Fahnen der Geometrie G als Basis eines freien R-Moduls V_G^k auf: $V_G^k = \langle F_k(G) \rangle$. Der Modul $V_G^n = V_G$ heisst der *Standardmodul* der Geometrie. Für jede Ziffer i mit $1 \le i \le n$ ist durch die Vorschrift

$$\sigma_i F = \sum_{\substack{G \underset{i}{\sim} F \\ G \neq F}} G, \quad F \in F_n(G)$$

eine lineare Abbildung $\sigma_i : V_G \to V_G$ des Standardmoduls in sich definiert.

Unter der über dem Ring R definierten *Hecke Algebra* der Geometrie G verstehen wir die von den linearen Abbildungen $\sigma_1, \sigma_2, \ldots, \sigma_n$ erzeugte Unteralgebra des Endomorphismenrings $\mathrm{Hom}_R(V_G, V_G)$, in Zeichen $H_R(G) = \langle \sigma_1, \sigma_2, \ldots, \sigma_n \rangle$.

Wenn wir im folgenden die

ANNAHME: Jede maximale Fahne ist zu genau $1 + q_i$ maximalen Fahnen i-benachbart

machen, dann gilt trivialerweise

LEMMA 1.1: *Für alle* i *gilt die Gleichung*

$$\sigma_i^2 = q_i 1 + (q_i - 1)\sigma_i$$

Ferner überlegt man leicht (etwa in Anlehnung an [4]):

LEMMA 1.2: *Sei* $m \ge 2$ *eine natürliche Zahl, und seien* i, j *zwei voneinander verschiedene Ziffern aus der Typenmenge* $\{1, 2, \ldots, n\}$. *Ist für alle Fahnen* H *mit* $|H| = n-2$ *und* $H \cap (\Omega_i \cup \Omega_j) = \emptyset$ *die abgeleitete Geometrie* G_H *ein verallgemeinertes* m-*Eck, dann gilt die Gleichung*

$$(\sigma_i \sigma_j)^k = (\sigma_j \sigma_i)^k, \quad \text{falls } m = 2k$$

oder die Gleichung

$$(\sigma_i \sigma_j)^k \sigma_i = (\sigma_j \sigma_i)^k \sigma_j, \quad \text{falls } m = 2k+1.$$

BEISPIEL : Sei $G = (\Omega_1, \Omega_2, I)$ ein vollständiger Graph mit $|\Omega_1| = 2+q$, $q \geq 2$, Ecken, dann ist $H_C(G)$ eine halbeinfache Algebra der Dimension $\dim_C H_C(G) = 7$.

2. COXETER ALGEBREN

Eine über der Typenmenge $T = \{1,2,\ldots,n\}$ definierte symmetrische Matrix $\alpha : T \times T \to N \cup \{\infty\}$ heisst ein *Coxeterdiagramm*, wenn $\alpha(i,i) = 1$ und $\alpha(i,j) \geq 2$ für alle $i \neq j$ gilt. Unter einer R-wertigen *Parameterfunktion* verstehen wir eine Abbildung

$$q : \begin{cases} T \to R \\ i \mapsto q_i \end{cases}.$$

Ein Coxeterdiagramm α veranschaulicht man häufig, indem man die Menge T als Eckenmenge eines Graphen betrachtet, in welchem zwei Ecken i und j durch eine mit der Ziffer $\alpha(i,j)$ bzw. dem Symbol ∞ belegte Kante oder durch eine Kante der Vielfachheit $\alpha(i,j) - 2$ verbunden sind. Ist zusätzlich eine Parameterfunktion $q : T \to R$ gegeben, dann belegen wir die Ecke i mit dem Parameter q_i.

BEISPIEL :

```
x      x      y                x      x      y
o------o======o     bzw.       o--3---o--4---o
1      2      3                1      2      3
```

beschreibt das Coxeterdiagramm

$$\begin{pmatrix} 1 & 3 & 2 \\ 3 & 1 & 4 \\ 2 & 4 & 1 \end{pmatrix}$$ und die Parameterfunktion $q : T \to R$, welche die Werte $q(1) = q(2) = x$ und $q(3) = y$ annimmt.

Gegeben sei nun ein Paar (α, q), bestehend aus einem Coxeterdiagramm α und einer auf der Typenmenge T definierten R-wertigen Parameterfunktion q. Unter der über dem Ring R definierten *Coxeter Algebra* vom Typ (α, q), in Zeichen Cox (α, q), wollen wir die von n Elementen $\Sigma_1, \Sigma_2, \ldots, \Sigma_n$ vermöge der Relationen

$$\Sigma_i^2 = q_i 1 + (q_i - 1)\Sigma_i \qquad \text{alle } i \in T,$$

$$(\Sigma_i \Sigma_j)^k = (\Sigma_j \Sigma_i)^k, \qquad \text{falls } \alpha(i, j) = 2k \qquad (*)$$

oder

$$(\Sigma_i \Sigma_j)^k \Sigma_i = (\Sigma_j \Sigma_i)^k \Sigma_j, \qquad \text{falls } \alpha(i, j) = 2k + 1,$$

definierte R - Algebra verstehen.

Nach [4] ist etwa die über dem Körper der komplexen Zahlen definierte Hecke Algebra eines verallgemeinerten m- Ecks mit den Parametern s und t zu der Coxeter Algebra vom Typ $\overset{s}{\circ}\!\!-\!\!-m-\!\!-\!\!\overset{t}{\circ}$ isomorph. Ein entsprechendes Theorem gilt nach [3] fur alle Geometrien vom Lie-Typ. Aus diesem Grunde liegt folgende Definition nahe. Eine von n Elementen $\sigma, \sigma, \ldots, \sigma_n$ erzeugte R - Algebra $A = R[\sigma, \ldots, \sigma_n]$ heisst vom *Coxeter Typ* α bzw. (α, q), wenn es einen Algebrahomomorphismus

$$\Phi : \operatorname{Cox}(\alpha, q) \to A$$

mit

$$\Phi : \Sigma_i \to a_i \sigma_i + b_i \quad \text{für gewisse } a_i, b_i \in R \text{ mit } a_i \neq 0$$

gibt. Gilt für ein i die Beziehung $(a_i, b_i) \neq (1, 0)$, so sagen wir, dass an der Stelle i eine Deformation vorliegt.

BEISPIELE : Der Einfachheit halber sei im folgenden $R = \mathbf{C}$.

(a) (Ohne Deformationen) Sei G eine Geometrie vom Lie-Typ $D_n(q)$, $n \geq 3$. Die Hecke Algebra der Geometrie ist vom Coxeter Typ

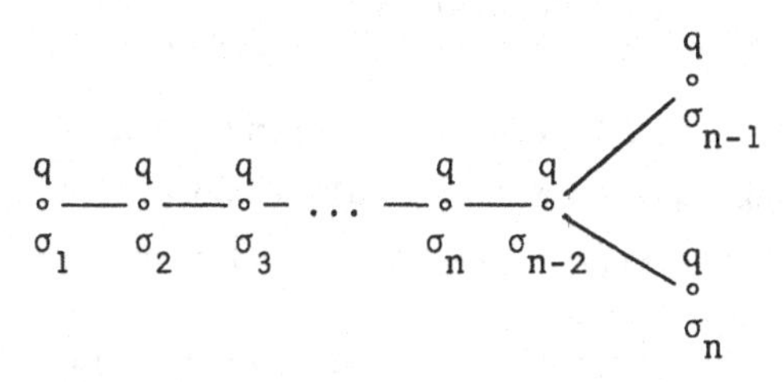

Nach [3] vermittelt die Zuordnung $\Sigma_i \to \sigma_i$ sogar einen Isomorphismus. Sei nun $\pi : G \to G$ eine Polaritat, welche die Mengen

Ω_{n-1} und Ω_n vertauscht. Bezeichen wir die von der Polaritat auf dem Standardmodul induzierte lineare Abbildung ebenfalls mit π, dann ist die Algebra $R[\sigma_1,\ldots,\sigma_{n-1},\pi]$ vom Coxeter Typ

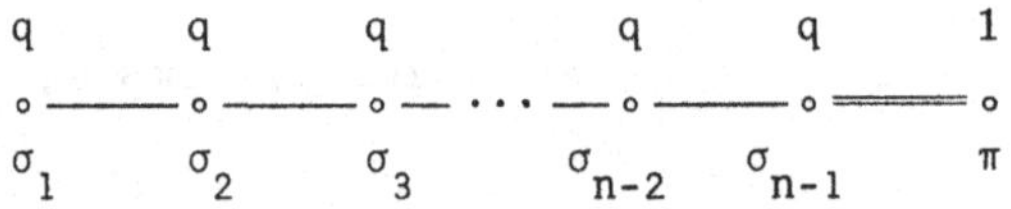

$$\overset{q}{\underset{\sigma_1}{\circ}} - \overset{q}{\underset{\sigma_2}{\circ}} - \overset{q}{\underset{\sigma_3}{\circ}} - \cdots - \overset{q}{\underset{\sigma_{n-2}}{\circ}} - \overset{q}{\underset{\sigma_{n-1}}{\circ}} = \overset{1}{\underset{\pi}{\circ}}$$

(b) (ohne Deformationen) Ist $\pi : G \to G$ eine Polarität eines verallgemeinerten m-Ecks der Ordnung q, dann ist in ähnlicher Weise die Algebra $R[\sigma_1, \pi]$ vom Coxeter Typ

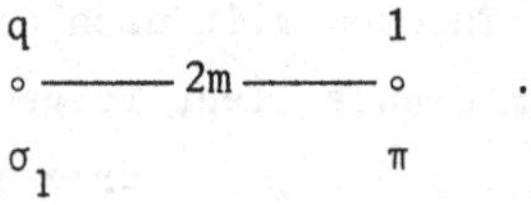

$$\overset{q}{\underset{\sigma_1}{\circ}} \overset{2m}{—\!—} \overset{1}{\underset{\pi}{\circ}} .$$

(c) (mit Deformation) Die oben schon erwähnte Hecke Algebra eines vollstandigen Graphen mit $2 + q$, $q \geq 3$ Ecken ist vom Coxeter Typ

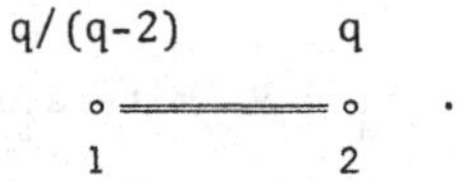

$$\overset{q/(q-2)}{\underset{1}{\circ}} = \overset{q}{\underset{2}{\circ}} .$$

3. GEOMETRIEN VOM COXETER TYP

Sei G eine Geometrie vom Rang n, und sei α ein mit einer Parameterfunktion q versehenes Coxeterdiagramm. Ist die Hecke Algebra der Geometrie vom Coxeter Typ (α, q), dann nennen wir die Geometrie vom entsprechenden Coxeter Typ.

BEISPIEL: Sei $G = (\Omega_1, \Omega_2, \Omega_3, I)$ eine Erweiterung einer projektiven Ebene der Ordnung $q \geq 4$ mit Ω_1 als Punkt-, Ω_2 als Geraden- und Ω_3 als Blockmenge (es gilt $|\Omega_1| = q^2 + q + 2$, $|\Omega_2| = \binom{|\Omega_1|}{2}$ und $|\Omega_3|(q+2) = |\Omega_1|(q^2+q+1)$). Diese Geometrie ist vom Coxeter Typ

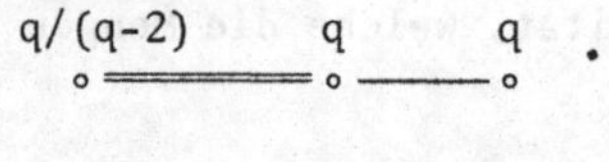

$$\overset{q/(q-2)}{\circ} = \overset{q}{\circ} - \overset{q}{\circ} .$$

Viele sporadische Gruppen operieren als fahnentransitive Gruppen auf Geometrien vom Coxeter Typ. Die folgenden Beispiele stützen sich im wesentlichen auf die in [1] angegebenen Geometrien

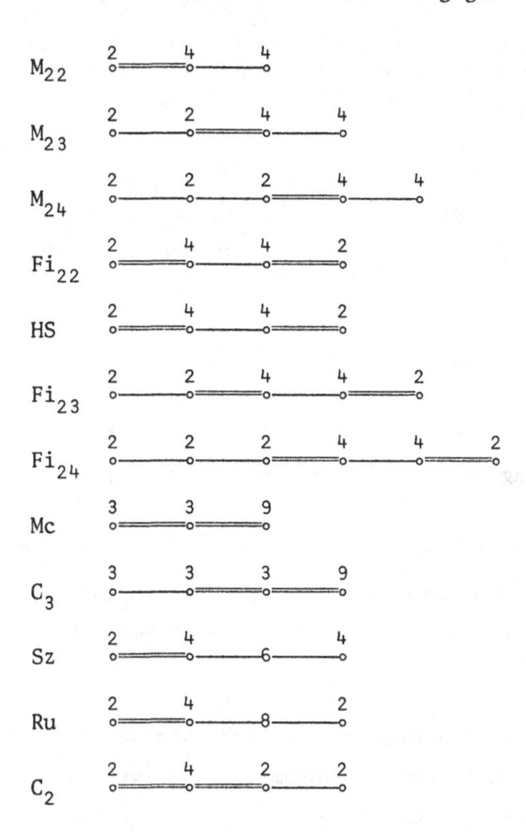

4. DARSTELLUNGEN

Im folgenden sei $G = (\Omega_1, \ldots, \Omega_n, I)$ eine endliche Geometrie vom Rang $n \geq 2$. Wir setzen ferner voraus, dass die Annahme mit $q_i \geq 1$ für alle $i = 1,2,\ldots,n$ gültig ist. Bezeichnen wir mit v das Element

$$v = \sum_{F \in F_n} F ,$$

dann gilt für jedes erzeugende Element σ_i der Hecke Algebra die Gleichung

$$\sigma_i v = q_i v .$$

Also gilt

LEMMA 4.1 : *Es gibt einen Algebrahomomorphismus ind :* $H_R(G) \to R$ *von der Hecke Algebra in den Ring* R, *welcher das Element* σ_i *auf das Element* $q_i \cdot 1$ *abbildet.*

Die im ersten Abschnitt definierten Moduln V_G^i sind homologischen Methoden zugänglich. Fur unsere Zwecke reicht ein Homomorphismus $\partial : V^n \to V^{n-1}$, der durch die Vorschrift

$$\partial F = \sum_{j=1}^{n} (-1)^{j+1} (F \setminus \Omega_j), \quad F \in F_n \tag{4.2}$$

definiert ist. Man überlegt leicht, dass für jedes Element $v \in \text{Kern}(\partial)$ die Gleichung $\sigma_i v = -v$ für alle $i = 1,2,\ldots,n$ gilt. Also haben wir (man beachte $q_i \geq 1$):

LEMMA 4.3 : *Es gibt einen Algebrahomomorphismus* $st : H_R(G) \to R$ *von der Hecke Algebra in den Ring* R, *welcher das Element* σ_i *auf das Element* -1 *abbildet.*

Für das Folgende sei $R = K$ ein algebraisch abgeschlossener Körper der Charakteristik Null. Wenn die Geometrie G vermöge eines Algebrahomomorphismus $\Phi : \text{Cox}(\alpha, t) \to H_K(G)$ vom Coxeter Typ (α, t) ist, dann nehmen aufgrund der definierenden Gleichungen die Funktionen $st\,\Phi$ und $ind\,\Phi$ an der Stelle Σ_i die Werte -1 und t_i an.

Im folgenden sei $A = \text{Cox}(\alpha, t)$ eine Coxeter Algebra mit $t_i \neq 0$, welche den Standardmodul V_G als Modul besitzt. Dann existiert der irreduzible Charakter $st : A \to K$ mit $st(\Sigma_i) = -1$ für alle $i = 1,2,\ldots,n$. Wir nehmen ferner an, dab auch ein irreduzibler Charakter $ind : A \to K$ mit $ind(\Sigma_i) = t_i$ existiert. Offenbar existiert dieser Charakter, falls im Falle $\alpha(i, j) \equiv 1$ (mod 2) die Beziehung $t_i = t_j$ besteht.

Ist L ein irreduzibler A-Modul, so bezeichnet $V_G(L)$ den von sämtlichen zu L isomorphen Untermoduln des Standardmoduls erzeugten Modul. Man nennt diesen Modul manchmal auch den durch L bestimmten *parabolischen* Untermodul. Die Dimensionen $\dim_k V_G(L)$ sind die *arithmetischen Invarianten* der Geometrie. In einigen Fallen kann die Dimension parabolischer Untermoduln leicht mit Hilfe von Orthogonalitätsrelationen berechnet werden. Die Herleitung dieser Orthogonalitätsrelationen stützt sich auf Eigenschaften der entsprechenden Coxetergruppe:

Sei (W, R) die durch das Coxeterdiagramm α definierte Coxetergruppe. Die Gruppe W hat das Erzeugendensystem $R = \{r_1, r_2, \ldots, r_n\}$ und ist durch die Relationen

$$\begin{aligned} r_i^2 &= 1 \qquad \text{alle } i \\ (r_i r_j)^{\alpha(i,j)} &= 1 \qquad \text{alle } i \neq j \end{aligned} \tag{4.4}$$

definiert. Jedes Element $w \in W$ hat eine Darstellung der Form $w = r_{i_1} r_{i_2} \ldots r_{i_k}$ mit $r_{i_j} \in R$. Ist bei einer solchen Darstellung die Anzahl k der Faktoren minimal, dann spricht man von einer reduzierten Darstellung des Elementes und nennt $k = l(w)$ die Länge von w.

Unter Werwendung wohlbekannter Eigenschaften der Coxetergruppe (man vgl. etwa [6], 3.4.3) leitet man folgenden Satz her:

SATZ 4.5: *Ordnet man jedem Element* $w = r_{i_1} r_{i_2} \ldots r_{i_k}$ *in reduzierter Darstellung das Element* $\wedge(w) = \Sigma_{i_1} \Sigma_{i_2} \ldots \Sigma_{i_k}$ *der Coxeter Algebra zu, dann ist* $\wedge : W \to A$ *eine wohldefinierte Abbildung auf ein Erzeugendensystem des* K*-Raumes* A.

Diesem Satz entnimmt man die Rechenregeln

$$\Lambda(w)\Sigma_i = \Lambda(w r_i) \qquad \text{falls} \quad l(w r_i) > l(w)$$

$$\Lambda(w)\Sigma_i = t_i \Lambda(w r_i) + (t_i - 1)\Lambda(w) \qquad \text{falls} \quad l(w r_i) < l(w)$$

$$\Sigma_i \Lambda(w) = t_i \Lambda(r_i w) + (t_i - 1)\Lambda(w) \qquad \text{falls} \quad l(r_i w) < l(w)$$

$$\Sigma_i \Lambda(w) = \Lambda(r_i w) \qquad \text{falls} \quad l(r_i w) > l(w) \tag{4.6}$$

Man überlegt weiter

SATZ 4.7 : *Sei* (W, R) *eine endliche Coxetergruppe. Ist* $\phi : V_1 \to V_2$ *eine* K *- lineare Abbildung von einem* A *- Modul* V_1 *in einen* A *- Modul* V_2*, dann ist die Abbildung*

$$S(\phi) = \sum_{w \in W} \frac{1}{\operatorname{ind}\Lambda(w)} \Lambda(w)\phi\, \Lambda(w^{-1})$$

A *- linear.*

BEWEIS : Man hat nur $S(\phi)\Sigma_i = \Sigma_i S(\phi)$ für alle $i = 1, 2, \ldots, n$ zu überlegen. Mit Hilfe der oben genannten Rechenregeln kann man das leicht nachrechnen.

Hieraus ergibt sich bekanntlich die Beziehung

$$0 = \sum_{w \in W} \frac{1}{\operatorname{ind}\Lambda(w)} \psi_1(\Lambda(w))\, \psi_2(\Lambda(w^{-1}))$$

fur Charaktere ψ_1 und ψ_2 nicht isomorpher irreduzibler A - Moduln. Wenn wir auf dem Raum der auf A definierten K - wertigen Funktionen die symmetrische Bilinearform

$$(\psi_1, \psi_2) = \sum_{w \in W} \frac{1}{\operatorname{ind}\Lambda(w)} \psi_1(\Lambda(w))\, \psi_2(\Lambda(w^{-1}))$$

einführen, so gilt also die

FOLGERUNG 4.8 : *Die Coxetergruppe* (W, R) *sei endlich. Sind* ψ_1, ψ_2 *Charaktere nicht isomorpher irreduzibler* A *- Moduln,* *so gilt* $(\psi_1, \psi_2) = 0$.

Ist V irgendein A-Modul und ist Φ der zugehörige Charakter, so gilt unter derselben Voraussetzung

$$(\Phi, \psi) = m(\psi)\,(\psi, \psi), \tag{4.9}$$

wobei $m(\psi)$ die Vielfachheit des durch den irreduziblen Charakter ψ bestimmten Moduls in V ist. Auf diese Weise lässt sich die Dimension parabolischer Moduln bestimmen.

BEISPIEL: Ist G ein (irreduzibles) endliches Gebäude, so gilt für den Charakter Φ des Standardmoduls die Gleichung $\Phi(\Lambda(w)) = 0$ für alle $w \neq 1$, also

$$|F_n|\,\psi(1) = m(\psi)\,(\psi, \psi).$$

Wendet man diese Formel etwa auf die linearen Darstellungen der Coxeter Algebra an, so ergibt sich leicht die in [2] angegebene Liste für die Dimensionen der entsprechenden parabolischen Moduln.

5. ANWENDUNGEN

Hier berichten wir kurz über Anwendungen, die irgendwo anders erscheinen werden.

Zunächst jedoch noch eine allgemeine Bemerkung:

Die Elemente $e_i = (1+\sigma_i)/(1+q_i)$ der Hecke Algebra einer Geometrie sind idempotent. Die Algebra $e_i H_K(G) e_i$ nennt man die i-te Verbindungsalgebra.

(a) Sei $G = (\Omega_1, \Omega_2, I)$ eine Geometrie mit den Parametern q_1, q_2, für welche $q_1 q_2 \neq 1$ gelten möge. Die Elemente aus Ω_1 sollen Punkte, die aus Ω_2 Geraden heissen. Unter einem Weg ω der Länge k von einem Punkt A zu einem Punkt B versteht man eine Folge verschiedener Punkte

$$A = A_0 A_1 A_2 \ldots A_{k-1} A_k = B$$

mit der Eigenschaft, dab aufeinanderfolgende Punkte auf einer Geraden liegen.

Sei $n = 2m+1$, $m \geq 1$. Die Geometrie G heibt ein *Quasi-n-Eck*, wenn je zwei Punkte durch genau einen Weg von höchstens der Lange m verbunden sind, und wenn es zwei Punkte gibt, welche nur durch einen Weg der Lange m verbunden sind.

Trivialerweise ist eine solche Geometrie vom Coxeter Typ $\overset{q_1}{\circ}\!\!-\!\!-\!\!\infty\!\!-\!\!-\!\!\overset{q_2}{\circ}$. In Anlehnung an [4] studiert man hier (über $\mathbb{C}$) Darstellungen der Form

$$\Sigma_1 \mapsto \begin{pmatrix} -1 & a \\ 0 & q_1 \end{pmatrix} \quad \Sigma_2 \mapsto \begin{pmatrix} q_2 & 0 \\ b & -1 \end{pmatrix} ,$$

wobei $a\,b = q_1 + q_2 + 2\sqrt{q_1\,q_2}\,\cos\alpha$ und α eine Lösung der Gleichung

$$0 \quad \sqrt{\frac{q_2}{q_1}}\,\sin\,(m+1)\,\alpha \;+\; \sin\,m\,\alpha$$

ist. Eine Analyse dieser auf die erste Verbindungsalgebra eigenschränkten Darstellungen liefert dann das Ergebnis, dab es nur für $n = 3$ oder $n = 5$ Quasi-n-Ecke gibt.

(b) Ist $G = (\Omega_1, \Omega_2, I)$ ein verallgemeinertes Hexagon der Ordnung $q \neq 1$ und $\pi : G \to G$ eine Polarität des Hexagons, dann ist die Algebra $\mathbb{C}\langle \sigma_1 , \pi \rangle$, wie schon weiter oben erwähnt, vom Coxeter Typ $\overset{q}{\circ}\!\!-\!\!-\!\!12\!\!-\!\!-\!\!\overset{1}{\circ}$. Die Coxeter Algebra von diesem Typ besitzt einen irreduziblen zweidimensionalen Modul. Die Berechnung seiner Vielfachheit in dem Standardmodul (gemass 4.9) ergibt dann das Ergebnis, dass ein verallgemeinertes Hexagon der Ordnung $q \neq 1$ höchstens dann eine Polarität gestattet, wenn die Zahl $3q$ ein Quadrat ist.

(c) (Zusammen mit M. Ronan) Schliesslich kann man mit den angegebenen Methoden beweisen, dab die 2-Überlagerung einer endlichen Titsgeometrie vom Typ $\circ\!=\!=\!\circ\!-\!-\!\circ$ oder $\circ\!-\!-\!5\!-\!-\!\circ\!-\!-\!\circ$ ein Gebäude ist. Das bedeutet, dass auf die in [6], Theorem 1 angegebenen Voraussetzungen im Falle der Endlichkeit verzichtet werden kann.

LITERATUR

1. F. Buekenhout, "Diagrams for geometries and groups", *J. Combin. Theory Ser. A.*, 27 (1979), 121-151.
2. C.W. Curtis, N. Iwahori and R. Kilmoyer, "Hecke algebras and characters of parabolic type of finite groups with (B,N)-pairs", *Inst. Hautes Études Sci. Publ. Math.*, 40 (1971), 81-116.
3. N. Iwahori, "On the structure of a Heckering of a Chevalley group over a finite field", *J. Fac. Sci. Univ. Tokyo Sect. IA Math.*, 10 (1964), 215-236.
4. R. Kilmoyer and L. Solomon, "On the theorem of Feit-Higman", *J. Combin. Theory Ser. A.*, 15 (1973), 310-322.
5. J. Tits, *Buildings of Spherical Type and Finite BN-pairs*, Lecture Notes in Math. 386 (Springer-Verlag, Berlin, 1974).
6. J. Tits, "A local approach to buildings", manuscript.

Institut D für Mathematik
Technische Universität Braunschweig
Pockelsstr. 14
3300 Braunschweig
Federal Republic of Germany

ON BUILDINGS AND LOCALLY FINITE TITS GEOMETRIES

U. Ott and M.A. Ronan

In a recent paper [3] on buildings, Tits introduced the concept of a *chamber system of type* M, where M is a diagram whose rank 2 constituents are generalized polygons, and in this note we will call such chamber systems *Tits geometries*. They will be called *locally finite* if the number of chambers containing a cell of codimension 1 is finite. The following theorem is proved in [3].

THEOREM (Tits) : *If* G *is a Tits geometry for which the universal cover of every subgeometry corresponding to a* C_3 *or* H_3 *subdiagram is a building, then the universal* 2-*cover of* G *is a building.*

C_3 o———o═══o H_3 o———o—5—o .

Moreover in [3] Tits shows how to construct free geometries of type C_3 and H_3 whose universal covers are not buildings, owing to the fact that there are no thick H_3 buildings, and that a free geometry of type C_3 (or H_3) may have planes which are not Moufang. However, if one restricts to the locally finite case, then an H_3 Tits geometry must be thin, because by the theorem of Feit and Higman [1], there are no finite generalized 5-gons except those having parameters (1, 1), and therefore the universal cover of a finite H_3 Tits geometry is the icosahedron, which is a (weak) building. In this note we will prove the following:

THEOREM 1 : *The universal cover of a locally finite* C_3 *Tits geometry is a building.*

COROLLARY 1 : *If* G *is a locally finite Tits geometry, then its universal* 2-*cover is a building.*

NOTATION : Let C denote a connected C_3 Tits geometry which is its own universal cover, and with finite parameters x and y as indicated by the diagram

$$\begin{array}{ccc} x & x & y \\ \circ\!\!-\!\!-\!\!-\!\!- & \circ\!=\!=\!=\!= & \circ \\ r_1 & r_2 & r_3 \end{array} .$$

Let $W = \langle r_1, r_2, r_3 \rangle$ denote the appropriate Coxeter group whose reflections correspond to the nodes of the diagram as shown above. Let $w = r_{i_1} \dots r_{i_k}$ be a reduced word in W, and let

$$\ell_1(w) = |\{ i_j \mid i_j = 1 \text{ or } 2 \}|$$

$$\ell_2(w) = |\{ i_j \mid i_j = 3 \}| .$$

Let A be the appropriate Coxeter algebra, and $\Lambda : W \to A$ the map defined in [2]. Let V be the standard A-module with basis the chambers of C, and let ϕ be the character of A on V.

LEMMA : *Using the above notation* $\phi(\Lambda(w)) = 0$ *if* $w \neq 1$.

PROOF : Let γ be a gallery of type w in C from some chamber c to d. If $c = d$, then since C is its own universal cover it follows that γ is homotopic to the constant gallery at c, and this implies $w = 1$ (see [3] Section 3). Therefore $w \neq 1$ implies $c \neq d$, so the matrix of the action of $\Lambda(w)$ on V has only zeros on the diagonal, and the result follows. □

Now from [2] Section 4, using the Lemma above, we see that the number of chambers $|C| = \phi(1) = (\phi, \text{ind}) =$ $= m(\text{ind}, \text{ind})$, where m is a positive integer. But

$$(\text{ind}, \text{ind}) = \sum_{w \in W} x^{\ell_1(w)} y^{\ell_2(w)} =: p(x, y),$$

and since C is connected it is clear that $|C| \leq p(x, y)$. It therefore follows that $|C| = p(x, y)$.

PROOF OF THEOREM 1 : As we have just established that $|C| = p(x, y)$, it follows that any two distinct reduced galleries of the same type starting at some common chamber $c \in C$ necessarily have distinct end chambers. Thus condition (R_c) of Tits [3] Section 5.2 is satisfied, and by Theorem 3 of [3] it follows that C is a building. □

PROOF OF COROLLARY 1 : This follows immediately from Theorem 1 and Tits' Theorem above, together with the discussion following it. □

BIBLIOGRAPHY

1. W. Feit and G. Higman, "The non-existence of certain generalized polygons", *J. Algebra,* 1 (1964), 114-131.
2. U. Ott, "Bericht über Hecke Algebren und Coxeter Algebren endlicher Geometrien", these proceedings.
3. J. Tits, "Local characterizations of buildings", to appear.

Institut D. für Mathematik
Technische Universität Braunschweig
Pockelsstr. 14
3300 Braunschweig
Federal Republic of Germany

Mathematics Department
University of Illinois
at Chicago Circle
Chicago
Illinois 60680
U.S.A.

MOUFANG CONDITIONS FOR FINITE GENERALIZED QUADRANGLES

S.E. Payne and J.A. Thas*

A generalized quadrangle $S = (P, B, I)$ of order (s, t) is said to satisfy the Moufang condition $(M)_p$ for some point p provided the following holds : for any two lines A, B of S incident with p, the group of collineations of S fixing A and B pointwise and p linewise is transitive on the lines $(\neq A)$ incident with a given point x on A $(x \neq p)$. Those S satisfying $(M)_p$ for a prescribed set of points p, and in some cases certain other hypotheses, are studied. Sample results : If s is prime and S of order (s, s^2) satisfies $(M)_p$ for some point p, then S is isomorphic to the elliptic quadric $Q(5, s)$. If S satisfies the dual condition $(\hat{M})_L$ for every line L through some coregular point p, and if $s > 1$ and $t > 1$, then $s = q^{h'}$ and $t = q^{h''}$ with q a prime power and $h' = h''$ or $h''a = h'(a+1)$ with a odd. If in particular s is prime, then S is isomorphic to $Q(4, s)$ or $Q(5, s)$; if in particular every line is regular, then $t = s^2$ or S is isomorphic to $Q(4, s)$.

1. INTRODUCTION

A general reference for this work is [17], although a review of several definitions and pertinent results will be given in 2.

Let $S = (P, B, I)$ be a generalized quadrangle (GQ) of order (s, t). For a fixed point p define the following condition:

$(M)_p$ *For any two lines* A, B *of* S *incident with* p, *the group of collineations of* S *fixing* A *and* B *pointwise and* p *linewise is transitive on the lines* $(\neq A)$ *incident with a given point* x *on* A $(x \neq p)$.

Then S is said to satisfy condition (M) provided it satisfies $(M)_p$ for all points $p \in P$. For a fixed line $L \in B$, let $(\hat{M})_L$ be the condition that is the dual of $(M)_p$, and let $(\hat{M})$ be the dual of (M). If S satisfies $(\hat{M})_L$ for each line L incident with a point p, it is said to satisfy the condition $(\hat{M})_p$. If S satisfies both (M) and $(\hat{M})$, it is called a *Moufang* GQ, and by a celebrated theorem of Tits [18], [20] must be one of the classical examples. The proof uses deep results on algebras, and there is some interest in finding ways both to weaken the hypotheses and to avoid the heavy results on algebras. These two goals have provided the motivation for this work.

Section 2 presents a variety of combinatorial results, some of which are new. Section 3 provides some basic information concerning collineations. If there is a group G_p fixing each line incident with p and acting regularly on the set of points not collinear with p, then, with respect to p, S is an *elation* GQ (EGQ) *with elation group* G_p. Briefly we say that $S^{(p)}$ is an EGQ. A double theme of this work is the determination of both what conditions force a GQ S to be an EGQ, and the determination of conditions forcing an EGQ to be classical.

Section 4 reviews earlier results on Moufang conditions and presents some "easy" conclusions. In Section 5 some theorems on EGQ's are obtained, and in the important Section 6 we consider generalized quadrangles satisfying $(\hat{M})_p$ for some point p. It appears that there is a strong analogy between the theory of the quadrangles satisfying $(\hat{M})_p$ and the theory of translation planes. In the final Section 7 it is shown that if S of order (s, t), $s > 1$ and $t > 1$, satisfies $(M)_p$ for a point p that belongs to no unicentric triad, then p is coregular and $S^{(p)}$ is an EGQ. It follows as a corollary that if $t = s^2$ and s is prime, and S satisfies $(M)_p$ for some point p, then S must be isomorphic to the GQ $Q(5, s)$ arising from an elliptic quadric in $PG(5, s)$.

2. BASIC CONCEPTS AND RESULTS

A *generalized quadrangle* (GQ) of order (s,t) $(s \geq 1,\ t \geq 1)$ is an incidence structure $S = (P, B, I)$ with pointset P, lineset B, and symmetric point-line incidence relation I satisfying the following axioms.

A1. *Each point (resp. line) is incident with* $1+t$ *lines (resp.* $1+s$ *points).*

A2. *Two points are incident with at most one line.*

A3. *If* x *is a point not incident with a line* L, *there is a unique point-line pair* $(y,M) \in P \times B$ *with* $xIMIyIL$.

When the points x, y (resp. lines L, M) of S are collinear (resp. concurrent), we write $x \sim y$ (resp. $L \sim M$); otherwise $x \not\sim y$ (resp. $L \not\sim M$). For $(x,L) \in P \times B$ put $x^{\perp} = \{y \in P \mid\mid y \sim x\}$ and $L^{\perp} = \{M \in B \mid\mid L \sim M\}$. Note: $x \in x^{\perp}$ and $L \in L^{\perp}$. The *trace* of a pair of distinct points (x,y) is defined to be the set $x^{\perp} \cap y^{\perp}$ and is denoted $tr(x,y)$ or $\{x,y\}^{\perp}$. More generally, if $A \subset P$, A "perp." is defined by $A^{\perp} = \cap\{x^{\perp} \mid\mid x \in A\}$. For $x \neq y$, the *span* of the pair (x,y) is

$$sp(x,y) = \{x,y\}^{\perp\perp} = \{u \in P \mid\mid u \in z^{\perp} \text{ for all } z \in x^{\perp} \cap y^{\perp}\}.$$

And the *closure* of the pair (x,y) is

$$cl(x,y) = \{z \in P \mid\mid z^{\perp} \cap sp(x,y) \neq \emptyset\}.$$

If $x \sim y$ or if $x \not\sim y$ and $|sp(x,y)| = t+1$, we say the pair (x,y) is *regular*. The point x is *regular* provided (x,y) is regular for all $y \in P$, $y \neq x$. The pair (x,y), $x \not\sim y$, is *antiregular* provided $|z^{\perp} \cap \{x,y\}^{\perp}| \leq 2$ for all $z \in P - \{x,y\}$. Finally, a point x is *antiregular* provided (x,y) is antiregular for all $y \in P$ with $x \not\sim y$.

There is a point-line duality for GQ (of order (s,t)) which interchanges "point" and "line", interchanges s and t, in any definition or theorem. Normally, we assume without further notice that the dual of a given theorem or definition is also given. A point x is *coregular* provided each line incident with x is regular.

Let (L_0, L_1) be a regular pair of nonconcurrent lines, and write $tr(L_0, L_1) = \{M_0, \ldots, M_s\}$, $sp(L_0, L_1) = \{L_0, \ldots, L_s\}$. The configuration consisting of the $2(1+s)$ lines of $tr(L_0, L_1) \cup sp(L_0, L_1)$ together with the $(1+s)^2$ points incident with these lines is called a *grid*. A dual configuration consisting of $2(1+t)$ points and $(1+t)^2$ lines is a *dual grid*.

A *triad (of points)* is a triple of pairwise noncollinear points. Given a triad $T = (x, y, z)$, a *centre* of T is just a point of $T^{\perp}$. We say T is *acentric*, *centric*, or *unicentric* according as $|T^{\perp}|$ is zero, positive, or equal to 1.

PROPOSITION 2.1 : *Let* S *be a* GQ *of order* (s,t) *with a fixed pair* (x,y) *of noncollinear points.*

(i) *If* $1 < s < t$, *then* (x,y) *is neither regular nor antiregular.*

(ii) *The pair* (x,y) *is regular with* $s = 1$ *or* $s \geq t$ *iff each centric triad* (x,y,z) *has exactly* 1 *or* $t+1$ *centres. When* $s = t$ *this is the case iff each triad* (x,y,z) *is centric.*

(iii) *If* $s \geq t$, *then* (x,y) *belongs to no unicentric triad iff either* $t = 1$ *or* $s = t$ *and* (x,y) *is antiregular. Hence for* $s = t$ *the pair* (x,y) *is antiregular iff each triad* (x,y,z) *has* 0 *or* 2 *centres.*

(iv) *If* $s > 1$ *and* $|sp(x,y)| > 2$ *then* (x,y) *does belong to some unicentric triad.*

(v) *If* $1 < t = s^2$, *then each triad has exactly* $1+s$ *centres.*

(vi) *If* (x,y) *is antiregular with* $s = t$, *then* s *is odd.*

PROOF : For all but (iv) and (vi) Section III of [17] is a convenient reference. For (iv), let $z \in sp(x,y) - \{x,y\}$. If w is any point for which $z \sim w$, $z \neq w$, $x \not\sim w \not\sim y$, then (x,y,w) is a unicentric triad. Part (vi) was mentioned without proof in [10], so we include a proof here for the sake of completeness.

Let $tr(x,y) = \{u_0, \dots, u_s\}$ with (x,y) antiregular. For $i = 0,1$, let $x \, I \, L_i \, I \, u_i \, I \, M_i \, I \, y$, and let $K \in tr(L_0, M_1)$, $L_1 \neq K \neq M_0$. Let $v_2, \dots, v_s$ be the points of K not collinear with x or y. Let $u_i \sim v_j$, for some $i \geq 2$. Then (x,y,v_j) is a triad with centre u_i, and hence with exactly one other centre $u_{i'}$. It follows that $u_2, \dots, u_s$ occur in pairs of centres of triads of the form (x,y,v_j), each pair being uniquely determined by either of its members. Hence $s-1$ is even. □

PROPOSITION 2.2 : *Let* S *be a* GQ *of order* (s,t), $1 < s$, $1 < t$, *having a coregular point* x, *implying* $s \leq t$.

(i) *If* (x,y,z) *is a triad with* m *centres, then* $m \equiv 1 + t \pmod 2$.

(ii) *If* t *is odd, then* $|sp(x,y)| = 2$ *for all* $y \not\sim x$.

(iii) *If* $s = t$, *then* x *is regular iff* s *is even.*

(iv) *If* $s = t$, *then* x *is antiregular iff* s *is odd.*

PROOF : Let $u_1, \dots, u_m$ be all the centres of a triad (x,y,z) with $tr(x,y) = \{u_1, \dots, u_m, u_{m+1}, \dots, u_{t+1}\}$. We assume $m < 1 + t$. For $i > m$, let L_i be the line through x and u_i, and M_i the line through y and u_i. Let K be the line through z meeting L_i, and N the line through z meeting M_i. Let M be the line through y meeting K and L the line through x meeting N. Since the line L_i through x is regular, M must meet L in some point $u_{i'} \in tr(x,y)$, $m+1 \leq i' \leq t+1$, $i' \neq i$. In this way with each point $u_i \in \{u_{m+1}, \dots, u_{t+1}\}$ there corresponds a point $u_{i'} \in \{u_{m+1}, \dots, u_{t+1}\}$, $i \neq i'$, and clearly this correspondence is involutory. Hence the number of points of $tr(x,y)$ that are not centres of (x,y,z) is even. This completes the proof of (i), from which (ii) follows by part (iv) of 2.1 . Now assume $s = t$. If x is regular, then any triad (x,y,z) has 1 or $1+s$ centres, and by part (i) s is even. Conversely, if s is even then by part (i) any triad (x,y,z) is centric, hence x is regular. This proves part (iii). To prove (iv), first note that if $s=t$ and x is antiregular, then by part (vi) of 2.1, s is odd. Now suppose x is coregular and $s=t$ is odd. Let (x,y,z) be any triad containing x. By (i) the number of centres of (x,y,z) must be even. Hence by part (iii) of 2.1, it follows that x is antiregular. □

In [10] it was shown that if $s=t$ the antiregularity of any point x gives rise to a projective plane of order s which is Desarguesian iff S is isomorphic to the GQ $Q(4, s)$ arising from the nonsingular quadric in $PG(4, s)$. As each projective plane of order less than 9 is Desarguesian, we have the following corollary of 2.2.

COROLLARY 2.3 : *If* S *is a* GQ *of order* (s,s), s *odd,* $3 \leq s \leq 7$, *with a coregular point, then* S *is isomorphic to* $Q(4, s)$. □

(Note : in 2.3, the hypotheses s odd and $3 \leq s \leq 7$ may be replaced by the hypothesis $2 \leq s \leq 7$ since the GQ of orders (2,2) and (4,4) are known to be unique, and no GQ of order (6,6) can have a regular line. The latter follows because a regular line would force the existence of a projective plane [17] of order 6, which is precluded by the Bruck-Ryser theorem.)

3. COLLINEATIONS

Let $S = (P, B, I)$ be a GQ of order (s,t), and let θ be a collineation of S. Put $P_\theta = \{x \in P \mid\mid x^\theta = x\}$, $B_\theta = \{L \in B \mid\mid L^\theta = L\}$, and let I_θ be the restriction of I to $P_\theta \times B_\theta \cup B_\theta \times P_\theta$. The following result is well-known (c.f. [7]).

PROPOSITION 3.1 : *The substructure* $S_\theta = (P_\theta, B_\theta, I_\theta)$ *of the fixed elements of* θ *must be given by at least one of the following:*

(i) P_θ *is a set of pairwise noncollinear points, and* $B_\theta = \emptyset$.

(i)' B_θ *is a set of pairwise nonconcurrent lines, and* $P_\theta = \emptyset$.

(ii) P_θ *contains a point* x *such that* $x \sim y$ *for every point* $y \in P_\theta$ *and each line of* B_θ *is incident with* x.

(ii)' B_θ *contains a line* L *such that* $L \sim M$ *for every line* $M \in B_\theta$, *and each point of* P_θ *is incident with* L.

(iii) $P_\theta = P_\theta^1 \cup P_\theta^2$, *where* P_θ^i *is a set of pairwise noncollinear points,* $i = 1,2$, *and each point of* P_θ^1 *is collinear with each point of* P_θ^2. *The set* B_θ *consists of the lines joining points of* P_θ^1 *with points of* P_θ^2.

(iii)' $B_\theta = B_\theta^1 \cup B_\theta^2$, *where* B_θ^i *is a set of pairwise nonconcurrent lines,* $i = 1,2$, *and each line of* B_θ^1 *is concurrent with each line of* B_θ^2. *The set* P_θ *consists of the points at which lines of* B_θ^1 *meet lines of* B_θ^2.

(iv) S_θ *is a subquadrangle of order* (s', t'), $2 \le s' \le s$, $2 \le t' \le t$. □

REMARK : In case (iv) we have $s = s'$ or $s \ge s't'$, and dually $t = t'$ or $t \ge s't'$ [4], [13].

A collineation θ of S is said to be a *whorl about a point* p of S provided θ fixes each line of S incident with p. Then Proposition 3.1 has the following corollary.

COROLLARY 3.2 : *Let* θ *be a nonidentity whorl about* p, *and assume* $t > 1$. *Then one of the following must occur :*

(i) $y^\theta \ne y$ *for each point* $y \in P - p^\perp$.

(ii) *There is a point* y, $y \not\sim p$, *for which* $y^\theta = y$. *If* $T = tr(p,y)$, $U = sp(p,y)$, *then* $T \cup \{p,y\} \subset P_\theta \subset T \cup U$ *and* $L \in B_\theta$ *iff* L *joins a point of* T *with a point of* $U \cap P_\theta$.

(iii) *The substructure* $S_\theta = (P_\theta, B_\theta, I_\theta)$ *of fixed elements forms a subquadrangle of* S *with order* (s', t), *where* $2 \le s' \le s/t \le t$, *so* $t < s$. □

Let θ be a whorl about the point p. If $\theta = id.$, or if θ fixes no point outside $p^{\perp}$, then θ is called an *elation about* p. If θ fixes each point of $p^{\perp}$, then θ is a *symmetry about* p. It follows from Corollary 3.2 that for $s > 1$ any symmetry about p is an elation about p. For $s > 1$ the symmetries about p form a group of order dividing t, and p is called a *centre of symmetry* provided its symmetry group has order t. If p is a centre of symmetry, then p must be regular. Dually, for $t > 1$ a line whose symmetry group has order s must be regular and is called an *axis of symmetry*.

In general, it seems to be an open question as to whether or not the set of elations about a point must be a group. For $s = t$, Thas [14] shows that the set of elations about a coregular point p is indeed a group, but a corresponding result for $s < t$ is still lacking. If there is a group G of elations about p acting regularly on $P - p^{\perp}$, we say S is an *elation generalized quadrangle* (EGQ) *with elation group* G *and base point* p. Briefly we say that $S^{(p)}$ is an EGQ. Most of the present article deals with EGQ in some form or other.

Let $S^{(p)}$ be an EGQ of order (s,t) with elation group G, $1 < s$, $1 < t$. If G contains a full group of t symmetries about p, we say $S^{(p)}$ is a *skew-translation generalized quadrangle* (STGQ) with *skew-translation group* G. Here p is a centre of symmetry and $s \geq t$. The new examples of Kantor [2] are of this type (c.f. [9]). In the remainder of this article we shall be more interested in the following alternate kind of EGQ.

If p is coregular and $s = t$, Thas [14] defines the EGQ $S^{(p)}$ to be a *translation generalized quadrangle* (TGQ) *with base point* p. The elation group G, called a *translation group* in this case, contains a full group of s symmetries about each line through p and is elementary abelian. In [10] it was shown that if $s = t$ for a GQ S, and at least three lines through p are axes of symmetry, then each line through p is an axis of symmetry and $S^{(p)}$ is a TGQ whose translation group is generated by the symmetries about the lines through p. Hence the following more general definition of TGQ agrees with that of [14] when $s=t$. If $S^{(p)}$ is an EGQ whose elation group G contains a full group of s symmetries about each line through p, we say $S^{(p)}$ *is a* TGQ *(with base point* p *and) with translation group* G. Then necessarily p is coregular.

4. GLOBAL MOUFANG CONDITIONS

Recall the Moufang conditions $(M)_p$, (M), $(\hat{M})_L$, $(\hat{M})$, $(\hat{M})_p$ (and its dual $(M)_L$) given in the introduction. Note that a line L is an axis of symmetry iff L is a regular line for which S satisfies $(\hat{M})_L$.

Thas [16] has proved the following two results.

THEOREM 4.1 : *Let* S *be a* GQ *of order* (s,s), $s>1$, *having an antiregular point* p *for which* S *satisfies* $(M)_p$. *Then* $S^{(p)}$ *is a* TGQ *(and* s *is odd).* □

THEOREM 4.2 : *Let* S *be a* GQ *of order* (s,t) *satisfying* (M). *Then one of the following must occur:*

(i) *All points of* S *are regular (so* $s=1$ *or* $s \geq t$*).*

(ii) $|sp(x,y)| = 2$ *for all points* x,y *with* $x \not\sim y$.

(iii) $S \cong H(4,s)$ *(the* GQ *arising from a nonsingular Hermitian variety in* $PG(4,s)$*).* □

We offer a companion result.

THEOREM 4.3 : *Let* S *be a* GQ *of order* (s,t) *satisfying* $(\hat{M})$. *Then one of the following must occur:*

(i) *All points of* S *are regular (so* $s=1$ *or* $s \geq t$*).*

(ii) $|sp(x,y)| = 2$ *for all points* x,y *with* $x \not\sim y$.

(iii) $S \cong H(4,s)$.

PROOF : The idea of the proof is to show that S satisfies the following property (H) : If (x,y,z) is a triad with $x \in cl(y,z)$, then $y \in cl(x,z)$. Then by theorems IV.7 and VIII.3 of [17] the desired conclusion follows. So let (x,y,z) be a triad with $x \sim w$, $w \in sp(z,y)$. Let u be the centre of (y,w,z) collinear with x. If $uz = L$, then $(\hat{M})_L$ guarantees that there is a collineation θ fixing all lines through u, all points on L, and mapping w to x. It follows that $y^\theta \in sp(x,z)$, so that $y \in cl(x,z)$. □

Using Theorems 4.2 and 4.3, we obtain an easy corollary. (For further remarks concerning the characterization of the symplectic geometry $W(s)$ in $PG(3,s)$ as the unique GQ of order (s,s) with all points regular, and as isomorphic to the dual of the geometry $Q(4, s)$ in $PG(4, s)$, see Section V of [17]).

COROLLARY 4.4 : *Let* S *be a* GQ *of order* (s,t), $s > 1$ *and* $t > 1$, *satisfying* (M), *so that the dual* S^* *satisfies* $(\hat{M})$. *Then one of the following must occur, where* q *is an arbitrary prime power:*

(i) $S \cong H(4, s)$ *and* $(s,t) = (q^2, q^3)$.

(ii) $S \cong H^*(4,t)$ *(the dual of* $H(4, t)$*), and* $(s,t) = (q^3, q^2)$.

(iii) $S \cong W(s)$ *and* $(s,t) = (q,q)$.

(iv) $S \cong Q(4, s)$ *and* $(s,t) = (q,q)$.

(v) *All points are regular,* $s > t$, *and* $|sp(L,M)| = 2$ *for all lines* L,M, $L \not\sim M$.

(vi) *All lines are regular,* $s < t$, *and* $|sp(x,y)| = 2$ *for all points* x,y, $x \not\sim y$.

(vii) $|sp(x,y)| = |sp(L,M)| = 2$ *for all points* x,y, $x \not\sim y$, *and all lines* L,M, $L \not\sim M$. □

The following immediate corollary indicates just how close we are at this point to having an independent proof of the theorem of Tits that a finite Moufang GQ must be classical.

COROLLARY 4.5 : *Let* S *be a* GQ *of order* (s,t), $1 < s \le t$, *satisfying both* (M) *and* $(\hat{M})$. *Then one of the following holds:*

(i) $S \cong H(4, s)$ *and* $(s,t) = (q^2, q^3)$.

(ii) *Either* S *or its dual is isomorphic to* $W(s)$ *and* $(s,t) = (q,q)$.

(iii) *All lines are regular,* $s < t$, *and* $|sp(x,y)| = 2$ *for all* $x, y \in P$, $x \neq y$.

(iv) $|sp(x,y)| = |sp(L,M)| = 2$ *for all* $x, y \in P$, $L, M \in B$, $x \neq y$, $L \neq M$. □

In Section 6 we shall show in case (iii) that $t = s^2$, with s a prime power, that (for each point p) $S^{(p)}$ is a TGQ with elementary abelian translation group, and that if s is prime then $S \cong Q(5, s)$.

5. ELATION GENERALIZED QUADRANGLES

Let $S^{(p)}$ be an EGQ of order (s,t) with elation group G, and let y be an arbitrary but fixed point of $P - p^{\perp}$. Let $L_0, \ldots, L_t$ be the lines incident with p, and define z_i and M_i by $L_i \, I z_i \, I M_i \, I y$, $0 \le i \le t$. Put $S_i = \{ \theta \in G \mid\mid M_i^{\theta} = M_i \}$, $S_i^* = \{ \theta \in G \mid\mid z_i^{\theta} = z_i \}$, and $J = \{ S_i \mid\mid 0 \le i \le t \}$. Then $|G| = s^2 t$; J is a collection of $1 + t$ subgroups of G, each of order s; for each i, $0 \le i \le t$, S_i^* is a subgroup of order st containing S_i as a subgroup. Moreover, the following two conditions are satisfied:

F1. $S_i S_j \cap S_k = 1$, for distinct i, j, k.

F2. $S_i^* \cap S_j = 1$, for distinct i, j.

Hence J is a 4-*gonal family* for G in the sense of [9], and S is canonically isomorphic to the GQ $S(G,J)$ described below. Points of $S(G,J)$ are of three kinds:

(i) elements of G;

(ii) right cosets $S_i^* g$, $g \in G$, $i \in \{0, \ldots, t\}$;

(iii) a symbol (∞).

Lines of $S(G,J)$ are of two kinds:

(a) right cosets $S_i g$, $g \in G$, $i \in \{0, \ldots, t\}$;

(b) symbols $[S_i]$, $i \in \{0, \ldots, t\}$.

A point g of type (i) is incident with each line $S_i g$, $i = 0, \ldots, t$. A point $S_i^* g$ of type (ii) is incident with $[S_i]$ and with each line $S_i h$ contained in $S_i^* g$. The point (∞) is incident with each line $[S_i]$ of type (b), and with no line of type (a).

Of course y^g corresponds to g, z_i^g corresponds to $S_i^* g$, p corresponds to (∞), M_i^g corresponds to $S_i g$, and L_i corresponds to $[S_i]$. By II.5 of [9] we have:

PROPOSITION 5.1 :

(i) S_i *is a group of symmetries about* L_i *iff* $S_i \lhd G$.

(ii) L_i *is regular iff* $S_i S_j = S_j S_i$ *for all* j, $0 \le i, \ j \le t$. □

As an immediate corollary we obtain the following result.

COROLLARY 5.2 : *An* EGQ *with an abelian elation group is a* TGQ. □

In what follows, an EGQ $S^{(p)}$ will be viewed as a coset geometry $S(G,J)$ without further comment when it seems convenient to do so.

THEOREM 5.3 : *Let* $S = (P,B,I)$ *be a* GQ *of order* (s,t) *with* $1 < s \le t$, *and let* $|sp(x,p)| = 2$ *for all* $x \not\sim p$ *with* p *a fixed point. If* G *is a group of whorls about* p *acting transitively on* $P - p^\perp$, *then the following holds:*

(i) *If* $|G| > s^2 t$, *then* G *is a Frobenius group on* $P - p^\perp$, *so that the set of elations about* p *in* G *is a normal subgroup of order* $s^2 t$ *acting regularly on* $P - p^\perp$, *i.e.* $S^{(p)}$ *is an* EGQ *with a normal subgroup of* G *as elation group.*

(ii) *If* G *is generated by elations about* p, *then* G *acts regularly on* $P - p^\perp$, *i.e.* $S^{(p)}$ *is an* EGQ.

PROOF : By Corollary 3.2, each nonidentity element of G fixes at most one point of $P - p^\perp$, since $|sp(x,p)| = 2$ for all $x \not\sim p$ and $s \le t$. If $x \in P - p^\perp$ and $|G_x| > 1$, then G acts on $P - p^\perp$ as a Frobenius group (c.f. [3]). Hence the subset of elations in G is the kernel of G as a Frobenius group, and (i) follows.

Suppose G is generated by elations and assume that the set E of all elations in G is not a group. Then in G there are elations θ_1 and θ_2 for which $\theta_1\theta_2$ is a nonidentity whorl with one fixed point. Hence G is a Frobenius group on $P - p^\perp$ and E is a group, a contradiction. It follows that $E = G$ acts regularly on $P - p^\perp$. □

Suppose that G_1 is a group of elations about p ($1 < s \le t$ and $|sp(x,p)| = 2$ for all $x \not\sim p$) acting regularly on $P - p^\perp$, and let θ be an elation about p. Then $G = \langle G_1, \theta \rangle$ must equal G_1 by part (ii) of Theorem 5.3. Hence we have

COROLLARY 5.4 : *If* $S^{(p)}$ *is an* EGQ *of order* (s,t) *with elation group* G, $1 < s \le t$ *and* $|sp(x,p)| = 2$ *for all* $x \not\sim p$, *then* G *is the set of all elations about* p. □

6. TRANSLATION GENERALIZED QUADRANGLES

THEOREM 6.1 : *Let* $S = (P,B,I)$ *be a* GQ *of order* (s,t), $1 < s$, $1 < t$. *Suppose* S *satisfies* $(\hat{M})_p$ *for a coregular point* p. *Then* $S^{(p)}$ *is a* TGQ *with abelian translation group* G.

PROOF : For $s = t = 2$, S is isomorphic to $W(2)$, and so we may assume $t > 2$. Let $L_0, \dots, L_t$ be the lines through p and S_i the group of symmetries about L_i. Our hypothesis is that $|S_i| = s$ for each i, i.e. L_i is an axis of symmetry and $s \le t$. Each element of S_i is easily seen to commute with each element of S_j, for $i \ne j$. For each i, $0 \le i \le t$, put $G_i = \langle S_j \;\|\; 0 \le j \le t,\; i \ne j \rangle$, $G = \langle S_i \;\|\; 0 \le i \le t \rangle$. So $[S_i, G_i] = 1$ and $G = S_i G_i$. One goal is to show that $G = G_i$, from which it will follow that S_i is abelian and hence G is abelian.

Step 1 : G_i *is transitive on the points of* $\Gamma(p) = P - p^\perp$.
For notational convenience we show that G_0 is transitive on $\Gamma(p)$. Let $x_1, \dots, x_s$ be the points of L_0 other than p. If a point

y of $\Gamma(p)$ is collinear with x_j, there is a symmetry about L_1 moving y to a point collinear with x_1. Hence we need only to show that G_0 is transitive on $x_1^{\perp} \cap \Gamma(p)$. Let M_1, M_2 be two distinct lines through x_1, $L_0 \neq M_i$, and let $y_i \mathrm{I} M_i$, $y_i \neq x_1$, $i = 1,2$. It suffices to show that y_1 and y_2 are in the same G_0-orbit. First suppose that some point u in $tr(y_1, y_2)$ is collinear with x_j, $2 \leq j \leq s$. Let L_{j_i} be the line through p meeting the line $y_i u$, $i = 1,2$ (remark that $j_i \neq 0$). As y_i and u are in the same S_{j_i}-orbit, $i = 1,2$, it follows that y_1 and y_2 are in the same G_0-orbit. On the other hand, if each point in $tr(y_1, y_2)$ is in $p^{\perp}$, let y_3 be a point with $y_3 \sim x_1$, $y_3 \not\sim p$, $y_3 \not\sim y_1$, $y_3 \not\sim y_2$, $y_3 \notin sp(y_1, y_2)$ (such a point exists since $t > 2$). Hence by the previous case y_3 and y_2 are in the same G_0-orbit, as are y_3 and y_1. It follows that G_0 is transitive on $\Gamma(p)$. Of course G must also be transitive on $\Gamma(p)$.

Step 2: $G = G_i$, *so that* G *is abelian.*
As $|\Gamma(p)| = s^2 t$, if $y \in \Gamma(p)$ we have $|G| = s^2 t k$ where $k = |G_y|$ and $|G_i| = s^2 t m$ where $m = |(G_i)_y|$. Clearly $m \mid k$, say $mr = k$. Then $s^2 t k = |G| = |S_i G_i| = |S_i|\,|G_i| / |S_i \cap G_i| = s^3 t m / |S_i \cap G_i|$, implying $r \mid S_i \cap G_i| = s$. Hence $r \mid s$ and $r \mid k$. Let q be a prime dividing r. Then there must be a collineation $\theta \in G_y$ having order q. Let M be the line through y meeting L_0 at x_i. Clearly θ fixes L_0 and M. The orbits of θ on M consist of cycles of length q and fixed points including y and x_i. As $q \mid s$, there are at least $q+1$ points of M fixed by θ. Moreover, each point of $tr(x,p)$ is fixed by θ. Considering the possible substructures of fixed elements allowed by Corollary 3.2, if $\theta \neq \mathrm{id.}$, we have a contradiction. Hence no prime q divides r, i.e. $r = 1$ and $G = G_i$.

Step 3: $k = 1$, *i.e.* G *is a group of elations about* p *and* $S^{(p)}$ *is a* TGQ.

By elementary permutation group theory an abelian transitive group is regular, hence $|G| = s^2 t$ and $k = 1$. □

Note : In this section we shall assume always $s > 1$ and $t > 1$.

COROLLARY 6.2 : *The translation group of a* TGQ $S^{(p)}$ *is uniquely defined. The elements of the translation group are called the translations about* p.

PROOF : Let $S^{(p)}$ be a TGQ with translation group G. Then with the notations of Corollary 6.1, $\langle S_i \parallel 0 \le i \le t \rangle$ is a subgroup of G. Since $|G| = |\langle S_i \parallel 0 \le i \le t \rangle| = s^2 t$, we have $G = \langle S_i \parallel 0 \le i \le t \rangle$. □

As an immediate corollary of Corollary 5.4, we have

COROLLARY 6.3 : *If* $|sp(x,p)| = 2$ *for all* $x \not\sim p$, *then the translation group* G *of the* TGQ $S^{(p)}$ *is the set of all elations about* p. □

Now let $S^{(p)}$ be a TGQ with (abelian) translation group G. As above, $L_0, \ldots, L_t$ are the lines through p and S_i is the group of symmetries about L_i. In the coset geometry notation of 6, $\theta \in G$ fixes a point $S_i^* g$ of $[S_i]$ iff $\theta \in g^{-1} S_i^* g = S_i^*$ iff θ fixes all points of $[S_i]$. In other words, $\theta \in G$ fixes one point of L_i iff θ fixes all points of L_i, and S_i^* is the *point-stabilizer* of L_i.

THEOREM 6.4 : *The substructure* $S_\theta = (P_\theta, B_\theta, I_\theta)$ *of the fixed elements of the translation* $\theta \neq$ id., *must be given by one of the following*

(i) P_θ *is the set of all points on* r *lines,* $1 \le r \le t+1$, *through* p *and* B_θ *is the set of all lines through* p.

(ii) $P_\theta = \{p\}$ *and* B_θ *is the set of all lines through* p.

(iii) P_θ *is the set of all points on one line* L_i *through* p, *and* B_θ *is the set of all lines concurrent with* L_i.

PROOF : By the remark preceding Theorem 6.4 and by Proposition 3.1 we have possibilities (i), (ii) or P_θ is the set of all points on one line L_i through p, and B_θ consists of at least $t+2$ lines concurrent with L_i. In the last case, let $L^\theta = L$, $p \not\mathrm{I}\, L$, and assume $x^\theta = y$, with $x \,\mathrm{I}\, L$, $x \not\sim p$. Since the translation group acts regularly on $P - p^\perp$, θ is the symmetry about L_i defined by $x^\theta = y$. Hence we have case (iii). □

As a corollary we have

COROLLARY 6.5 : *If* $S^{(p)}$ *is a* TGQ, *then* S *satisfies* $(M)_p$.

PROOF : Let $L_i \,\mathrm{I}\, p \,\mathrm{I}\, L_j$, $i \neq j$, $x \,\mathrm{I}\, L_i$, $x \neq p$, $A \,\mathrm{I}\, x \,\mathrm{I}\, B$, $A \neq B \neq L_i \neq A$. On L_j we choose a point y, $y \neq p$, and we define the points z and u to be the points of A and B resp. which are collinear with y. If θ is the translation defined by $z^\theta = u$, then $x^\theta = x$, $y^\theta = y$, $A^\theta = B$, and Theorem 6.4 tells us that θ fixes L_i and L_j pointwise and p linewise. It follows that S satisfies $(M)_p$. □

THEOREM 6.6 : *Let* $S^{(p)}$ *be a* TGQ *with* S_i, S_i^*, G, *etc., as above. Then the following hold*

(i) $|S_i^* \cap S_j^*| = t$, *if* $0 \le i < j \le t$.

(ii) $|S_i^* \cap S_j^* \cap S_k^*| \ge t/s$, *if* $0 \le i < j < k \le t$.

PROOF : We have $|S_i^* \cap S_j^*| = |\mathrm{tr}(z_i, z_j)| - 1 = t$, $z_i \,\mathrm{I}\, L_i$, $z_i \neq p$, $z_j \,\mathrm{I}\, L_j$, $z_j \neq p$ (c.f. 5).

For i, j, k distinct, $|(S_i^* \cap S_j^*)S_k^*| = |S_i^* \cap S_j^*|\,|S_k^*| / |S_i^* \cap S_j^* \cap S_k^*| \le |G|$ implies (ii). □

Theorem 6.6 implies that if $S^{(p)}$ is a TGQ, then any triad having at least two centres and having p as centre, must have at least $1+t/s$ centres. For $t > s$ it is tempting to conjecture that $1+t/s$ is the exact number of centres, partly because this is the case in the only known examples. Here $t = s^2$ and S is the GQ arising from an ovoid in three-dimensional space. However, when $s=t$ and $S \cong Q(4, s)$, then each triad of points has 0 or 2 centres if s is odd and 1 or $1+s$ centres if s is even.

Let $S^{(p)}$ be a TGQ with S_i, S_i^*, G, etc., as above. We define *the kernel* K of $S^{(p)}$ or of the 4-gonal family $J = \{ S_i \mid\mid 0 \le i \le t \}$ to be the set of all endomorphisms α of G satisfying $S_i^\alpha \subset S_i$, $i = 0,\dots,t$. For the usual addition and multiplication of endomorphisms K *is a ring.*

The only GQ with $s=2$ are $W(2)$ and $Q(5, 2)$, and so in what follows we shall assume $s > 2$.

THEOREM 6.7 : *With the usual addition and multiplication of endomorphisms the kernel* K *of the* TGQ $S^{(p)}$ *is a field. It follows that* $S_i^\alpha = S_i$ *and* $S_i^{*\alpha} = S_i^*$ *for all* i *and for all* $\alpha \in K - \{0\}$. *Moreover, if* $|sp(x,p)| = 2$ *for all* $x \not\sim p$, *then the multiplicative group of* K *is isomorphic to the group of all whorls about* p *with fixed point* y, $y \not\sim p$.

PROOF : If each element $\alpha \in K - \{0\}$ is an automorphism of G, then clearly K is a field. So let us assume that $\alpha \in K - \{0\}$ is not an automorphism of G. Then $\langle S_0,\dots, S_t \rangle = G \supsetneqq G^\alpha = \langle S_0^\alpha, \dots, S_t^\alpha \rangle$, and so $S_i^\alpha \neq S_i$ for some i. Let $g^\alpha = 1$, $g \in S_i - \{1\}$. If i, j, k are mutually distinct and $g' \in S_j$ with $\{g'\} \neq S_j \cap S_k^* g^{-1}$, then $gg' = hh'$, with $h \in S_k$ and $h' \in S_\ell$ for a uniquely defined $\ell \neq k, j$. Hence $h^\alpha h'^\alpha (g'^{-1})^\alpha = 1$. It follows that $h^\alpha = h'^\alpha = (g'^{-1})^\alpha = 1$, and so $g'^\alpha = 1$. Since $|\mathrm{Ker}(\alpha) \cap S_j| \ge s-1 > s/2$, we have $S_j \subset \mathrm{Ker}(\alpha)$. There easily follows that $S_0 \cup \dots \cup S_t \subset \mathrm{Ker}(\alpha)$, and so $\mathrm{Ker}(\alpha) = G$, i.e. $\alpha = 0$, a contradiction. Hence $\alpha \in K - \{0\}$ is an automorphism of G, and the first part of the theorem is proved. Note that for each $\alpha \in K - \{0\}$ there holds $S_i^\alpha = S_i$ for all i. Now let $g \in S_i^* - S_i$. Then $S_i g \cap S_j = \emptyset$ for all $j \neq i$, hence $S_i g^\alpha \cap S_j = \emptyset$ for all $j \neq i$ and for all $\alpha \in K - \{0\}$, i.e. $g^\alpha \in S_i^*$ for all $\alpha \in K - \{0\}$. Consequently $S_i^{*\alpha} = S_i^*$ for all i and for all $\alpha \in K - \{0\}$.

Now let $|sp(x,p)| = 2$ for all $x \not\sim p$. If $\alpha \in K - \{0\} = K^*$, then $(S_i g)^\alpha = S_i g^\alpha$ and $(S_i^* g)^\alpha = S_i^* g^\alpha$. There easily follows that α defines a whorl about (∞) and with fixed point 1 of the GQ $S(G,J)$, with $J = \{S_i \,\|\, 0 \le i \le t\}$ and G the translation group of $S^{(p)}$. Conversely, let us consider a whorl of $S(G,J)$ about (∞) and with fixed point 1. This whorl defines a permutation α of G, and we will prove that $\alpha \in K^*$, i.e. that $(gh)^\alpha = g^\alpha h^\alpha$ for all $g, h \in G$.

Let us firstly remark that with every $u \in G$ there corresponds an elation θ_u of $S(G,J)$ about (∞), defined by $g^{\theta_u} = gu$, $(S_i g)^{\theta_u} = S_i(gu)$, $(S_i^* g)^{\theta_u} = S_i^*(gu)$, $(\infty)^{\theta_u} = (\infty)$, $[S_i]^{\theta_u} = [S_i]$. Now we have $(gh)^\alpha = (1^{\theta_{gh}})^\alpha = (1^{\theta_g\theta_h})^\alpha = 1^{\alpha\alpha^{-1}\theta_g\alpha\alpha^{-1}\theta_h\alpha} = 1^{(\alpha^{-1}\theta_g\alpha)(\alpha^{-1}\theta_h\alpha)}$. Clearly $\alpha^{-1}\theta_g\alpha$ and $\alpha^{-1}\theta_h\alpha$ are elations about (∞). We have $1^{\alpha^{-1}\theta_g\alpha} = g^\alpha$ and $1^{\alpha^{-1}\theta_h\alpha} = h^\alpha$. In $S(G,J)$ we have $|sp((\infty), u)| = 2$ for all $u \in G$, and so, from Corollary 6.3, every elation about (∞) is a translation. Hence $\alpha^{-1}\theta_g\alpha = \theta_{g^\alpha}$ and $\alpha^{-1}\theta_h\alpha = \theta_{h^\alpha}$. Consequently $(gh)^\alpha = 1^{\theta_{g^\alpha}\theta_{h^\alpha}} = 1^{\theta_{g^\alpha h^\alpha}} = g^\alpha h^\alpha$. Now the last part of the theorem easily follows. □

Now we introduce the *vector space* (G,F) *with* F *a subfield of the kernel*. The vectors of (G,F) are the elements of G and the field of scalars is the subfield F of K. Addition in (G,F) is the multiplication of the group G, and scalar multiplication is defined by $g\alpha = g^\alpha$, $g \in G$ and $\alpha \in F$. It is easy to verify that we have indeed a vector space over F.

THEOREM 6.8 : *If* $S^{(p)}$ *is a* TGQ *of order* (s,t) *with translation group* G*, then* $s = q^n$, $t = q^m$, *with* q *a prime power, and* G *is elementary abelian.*

PROOF : Let us consider the vector space (G,F), and suppose that $|F| = q$. Since the translation group G is the additive group of a vector space, it is an elementary abelian group.

Evidently S_i and S_i^* can be considered as subspaces of (G,F). Hence $|S_i| = s = q^n$ and $|S_i^*| = st = q^{n+m}$. □

Next we introduce the *generalized quadrangle* $T(n,m,q)$. In $PG(2n+m-1, q)$ we consider a set $O(n,m,q)$ of $q^m + 1$ $(n-1)$-dimensional subspaces $PG^{(o)}(n-1, q), \ldots, PG^{(q^m)}(n-1, q)$, every three of which generate a $PG(3n-1, q)$, and such that each element $PG^{(i)}(n-1, q)$ of $O(n,m,q)$ is contained in a $PG^{(i)}(n+m-1, q)$ having no point in common with $(PG^{(o)}(n-1, q) \cup \ldots \cup PG^{(q^m)}(n-1, q)) - PG^{(i)}(n-1, q)$. It is easy to see that $PG^{(i)}(n+m-1, q)$ is uniquely defined, $i = 0, \ldots, q^m$. Now we embed $PG(2n+m-1, q)$ in a $PG(2n+m, q)$. Points of the generalized quadrangle $T(n,m,q)$ are of three types:

(i) the points of $PG(2n+m, q) - PG(2n+m-1, q)$;

(ii) the $(n+m)$-dimensional subspaces of $PG(2n+m, q)$ which intersect $PG(2n+m-1, q)$ in a $PG^{(i)}(n+m-1, q)$;

and (iii) the symbol (∞).

Lines of the generalized quadrangle $T(n,m,q)$ are of two types:

(a) the n-dimensional subspaces of $PG(2n+m, q)$ which intersect $PG(2n+m-1, q)$ in a $PG^{(i)}(n-1, q)$;

and (b) the elements of $O(n,m,q)$.

Now we define the incidence. A point of type (i) is incident only with lines of type (a); here the incidence is that of $PG(2n+m, q)$. A point of type (ii) is incident with all lines of type (a) contained in it and with the unique element of $O(n,m,q)$ in it. The point (∞) is incident with no line of type (a) and all lines of type (b). It is an easy exercise to show that $T(n,m,q)$ is a GQ of order (q^n, q^m).

THEOREM 6.9 : *The* GQ $T(n,m,q)$ *is a* TGQ *with base point* (∞) *and for which* $GF(q)$ *is a subfield of the kernel. Moreover the translations of* $T(n,m,q)$ *induce the translations of the affine space* $AG(2n+m, q) = PG(2n+m, q) - PG(2n+m-1, q)$. *Conversely, every* TGQ *for which* $GF(q)$ *is a subfield of the kernel is isomorphic to a* $T(n,m,q)$. *It follows that the theory of* TGQ *is equivalent to the theory of the sets* $O(n,m,q)$.

PROOF : A translation of $AG(2n+m, q)$ defines in a natural way an elation about (∞) of $T(n,m,q)$. It follows that $T(n,m,q)$ is an EGQ with elation group G, where G is isomorphic to the translation group of $AG(2n+m, q)$. With the q^n translations of $AG(2n+m, q)$ with centre in $PG^{(i)}(n-1, q)$ there correspond q^n symmetries of $T(n,m,q)$ about the line $PG^{(i)}(n-1, q)$ of type (b). Hence $T(n,m,q)$ is a TGQ with base point (∞), and the translations of $T(n,m,q)$ induce the translations of $AG(2n+m, q)$. Also, it is not difficult to prove that $GF(q)$ is a subfield of the kernel of $T(n,m,q)$ (with the group of all homologies of $PG(2n+m, q)$ with centre $y \notin PG(2n+m, q)$ and axis $PG(2n+m-1, q)$ there corresponds in a natural way a subgroup of the multiplicative group of the kernel).

Conversely, let us consider a TGQ $S^{(p)}$ with translation group G, for which $GF(q) = F$ is a subfield of the kernel. If $s = q^n$ and $t = q^m$, then $[(G,F) : F] = 2n+m$. Hence with $S^{(p)}$ there corresponds an affine space $AG(2n+m, q)$. The cosets $S_i g$ of a fixed S_i are the elements of a parallel class of n-dimensional subspaces of $AG(2n+m, q)$, and the cosets $S_i^* g$ of a fixed S_i^* are the elements of a parallel class of (n+m)-dimensional subspaces of $AG(2n+m, q)$. The interpretation in $PG(2n+m, q)$ proves the last part of the theorem. □

THEOREM 6.10 : *Let us consider a set* $O(n,m,q)$, *as defined above. Then the following hold:*

(i) $ma = n(a+1)$ *with* a *odd, or* $n = m$;

(ii) *if the point* x *does not belong to an element of* $O(n,m,q)$, *then the number* π_x *of* $PG^{(i)}(n+m-1, q)$ *containing* x *is independent of* x *iff* $m = 2n$. *When* $m = 2n$ *this number* π_x *equals* $q^n + 1$.

PROOF : Let H be a hyperplane of $PG(2n+m-1, q)$ containing $PG^{(i)}(n-1, q)$, but not containing $PG^{(i)}(n+m-1, q)$. Suppose there are ℓ_1 spaces $PG^{(j)}(n-1, q)$, $j \neq i$, for which $H \cap PG^{(j)}(n-1, q)$ is (n-1)-dimensional, and ℓ_2 spaces $PG^{(j)}(n-1, q)$, $j \neq i$, for which $H \cap PG^{(j)}(n-1, q)$ is (n-2)-dimensional. Since any $PG(n,q)$ through $PG^{(i)}(n-1, q)$, but not contained in $PG^{(i)}(n+m-1, q)$, has exactly one point in common with $(PG^{(o)}(n-1, q) \cup \ldots \cup PG^{(q^m)}(n-1, q)) - PG^{(i)}(n-1, q)$, we have $q^{n+m-2} + q^{n+m-3} + \ldots + 1 = q^{m-2} + q^{m-3} + \ldots + 1 + \ell_1(q^{n-1} + q^{n-2} + \ldots + 1) + \ell_2(q^{n-2} + q^{n-3} + \ldots + 1)$ with $\ell_1 + \ell_2 = q^m$. Hence $\ell_1 = q^{m-n}$.

Now we consider a hyperplane $PG(n-2, q)$ of $PG^{(o)}(n-1, q)$, and we count the hyperplanes H of $PG(2n+m-1, q)$ containing $PG(n-2, q)$ and $q^{m-n} + 1$ of the spaces $PG^{(j)}(n-1, q)$, $j \neq 0$. There arises $q^{m-n}(q^{m-n} + 1) \mid q^m(q^m - 1)q^{m-n}$, and so $q^{m-n} + 1 \mid q^m(q^m - 1)$. Now we assume $n \neq m$. Then $q^{m-n} + 1 \mid q^m - 1$, i.e. $q^{m-n} + 1 \mid q^n + 1$. It easily follows that $n = (m-n)a$ with a odd, and (i) is proved.

Consider a point x that does not belong to an element of $O(n,m,q)$, and call π_x the number of $PG^{(i)}(n+m-1, q)$ containing x. Then we count the number of ordered pairs $(x, PG^{(i)}(n+m-1, q)$ through $x)$, and we obtain

$$\sum_x \pi_x = (q^m + 1)(q^{n+m-1} + q^{n+m-2} + \ldots + q^n) .$$

Next we count the number of ordered triples $(x, PG^{(i)}(n+m-1, q)$ through x, $PG^{(j)}(n+m-1, q)$ through $x)$, $i \neq j$, and we obtain

$$\sum_x \pi_x (\pi_x - 1) = (q^m + 1)q^m(q^{m-1} + q^{m-2} + \ldots + 1) .$$

If d is the number of points x, then

$$d \sum_x \pi_x^2 - (\sum_x \pi_x)^2 = (q^{m-1} + \ldots + 1)(q^{n-1} + \ldots + 1)q^m(q^m + 1)(q^{2n} - q^m) .$$

Hence $d \sum_x \pi_x^2 - (\sum_x \pi_x)^2 = 0$ iff $m = 2n$. If $m = 2n$, then $\pi_x = \sum_x \pi_x / d = q^n + 1$ for all x. □

The preceding theorem has some important corollaries.

COROLLARY 6.11 : *If* $S^{(p)}$ *is a* TGQ *of order* (s,t) *then we have the following*

(i) $s = q^n$ *and* $t = q^m$ *with* q *a prime power, and* $m = n$ *or* $ma = n(a+1)$ *with* a *odd.*

(ii) *If* s *is prime, then* $S \cong Q(4,s)$ *or* $S \cong Q(5,s)$.

(iii) *If all lines are regular, then* $S \cong Q(4,s)$ *or* $t = s^2$.

PROOF : Part (i) is an immediate corollary of Theorems 6.8 - 6.10. If s is a prime, then $n = 1$, $a = 1$ and $m = 2$, or $n = m = 1$. The GQ T(1, 1, q) (resp. T(1, 2, q)) arises by Tits' construction [1] from an oval (resp. ovoid) in PG(2,q) (resp. PG(3,q)). Since q is prime the oval (resp. ovoid) is a conic (resp. an elliptic quadric). Hence $S \cong Q(4,s)$ or $S \cong Q(5,s)$.

Now we assume that all lines are regular. If $s = t$, then $S \cong Q(4,s)$. So we suppose that $s \neq t$. By counting the number of sp(L,M), $L \not\sim M$, in S we obtain $(s+1) \mid (t+1)(st+1)t^2$, i.e. $(s+1) \mid (t^2-1)t^2$. Since $(s+1, t) = 1$, there holds $s+1 \mid t^2-1$. So $q^n+1 \mid (q^{n+(n/a)}-1)(q^{n+(n/a)}+1)$ or $q^n+1 \mid q^{2n/a}-1$. Hence $n < 2n/a$, i.e. $a = 1$ and $t = s^2$. □

As an immediate corollary of Corollaries 4.5 and 6.11, we have

COROLLARY 6.12 : *Let* S *be a* GQ *of order* (s,t), $1 < s \leq t$, *satisfying both* (M) *and* $(\hat{M})$. *Then one of the following holds:*

(i) $S \cong H(4,s)$ *and* $(s,t) = (q^2, q^3)$.

(ii) *Either* S *or its dual is isomorphic to* W(s) *and* $(s,t) = (q,q)$.

(iii) $t = s^2$ *with* s *a prime power.*

(iv) $|sp(x,y)| = |sp(L,M)| = 2$ *for all* $x, y \in P$, $L, M \in B$, $x \not\sim y$, $L \not\sim M$.

THEOREM 6.13 : *If* $S^{(p)}$ *is the* TGQ *corresponding to the set* O(n, 2n, q), *then it is the Tits quadrangle arising from an ovoid in threedimensional space iff*

(i) *each* $PG(3n-1, q)$ *containing at least three elements of* $O(n, 2n, q)$, *contains exactly* $q^n + 1$ *elements of* $O(n, 2n, q)$; *or*

(ii) *for each point* x *not contained in an element of* $O(n, 2n, q)$ *the* $q^n + 1$ *spaces* $PG^{(i)}(3n-1, q)$ *containing* x *have exactly* $(q^n - 1)/(q - 1)$ *points in common.*

PROOF : First suppose that each $PG(3n-1, q)$ containing at least three elements of $O(n, 2n, q)$ contains exactly $q^n + 1$ elements of $O(n, 2n, q)$. Then $O(n, 2n, q)$ is a $[n-1]$-ovoid [11], [12] of $PG(4n-1, q)$, and then from [12] follows that $S^{(p)}$ arises from an ovoid of $PG(3, q^n)$. Conversely, if $S^{(p)}$ arises from an ovoid O' of $PG(3, q^n)$ then the kernel of $S^{(p)}$ is $GF(q^n)$. Interpretation of $S^{(p)}$ in $PG(3n-1, q)$, the $(3n-1)$-dimensional space over the subfield $GF(q)$ of the kernel, shows that $O(n, 2n, q)$ has the desired property.

The GQ $S^{(p)}$ is isomorphic to a GQ arising from an ovoid O' of $PG(3, q^n)$ iff the point p is 3-regular [15], i.e. iff for each point x not contained in an element of $O(n, 2n, q)$ the $q^n + 1$ spaces $PG^{(i)}(3n-1, q)$ containing x have exactly $(q^n - 1)/(q - 1)$ points in common. This proves part (ii). □

We *conjecture* that a TGQ $S^{(p)}$ with $t > s$ and all lines regular (hence $t = s^2$) arises from an $O(n, 2n, q)$ satisfying (i) of Theorem 6.13. We have already a proof for $n = 2,3$ and we hope to prove it soon in the general case. *If this conjecture is true, then we get as an immediate corollary: If* S *is a* GQ *of order* (s,t), $1 < s \leq t$, *satisfying* (M) *and* $(\hat{M})$, *then* S *is "classical" or* $|sp(x,y)| = |sp(L,M)| = 2$ *for all* $x, y \in P$, $L, M \in B$, $x \not\sim y$, $L \not\sim M$.

As a final theorem in this section, we propose the following

THEOREM 6.14 : *A* TGQ $S^{(p)}$ *with* $|sp(x,p)| = 2$ *for all* $x \not\sim p$ *is a Tits quadrangle, i.e. arises from an oval in a plane or an*

ovoid in three-dimensional space, iff the group of all whorls about p *with fixed point* y, $y \neq p$, *has order* $s-1$.

PROOF: If $S^{(p)}$ is a Tits quadrangle, then $S^{(p)}$ is a $T(1,1,q)$ or a $T(1,2,q)$. Since the $q-1$ homologies of $PG(3,q)$ (resp. $PG(4,q)$) with centre u, $u \neq (\infty)$, and axis $PG(2,q)$ (resp. $PG(3,q)$) define $q-1$ whorls of $T(1,1,q)$ (resp. $T(1,2,q)$) about (∞) with fixed point u, the group of all whorls of $S^{(p)}$ about p with fixed point y, $y \neq p$, has order $s-1$. Conversely, assume that the group of all whorls of $S^{(p)}$ about p with fixed point y, $y \neq p$, has order $s-1$. Since $|sp(x,p)| = 2$ for all $x \neq p$ this means that the kernel of $S^{(p)}$ has order s. Hence we may take $n = 1$, and then $m \in \{1,2\}$. It follows that $S^{(p)}$ is a $T(1,1,s)$ or a $T(1,2,s)$. □

Notice that there is a point x, $x \neq p$, for which $|sp(x,p)| > 2$ iff the $q^m + 1$ "tangent spaces" $PG^{(i)}(n+m-1,q)$ of $O(n,m,q)$ have at least one point in common.

7. THE MOUFANG CONDITION $(M)_p$

Until further notice, we suppose that $S = (P,B,I)$ *is a* GQ *of order* (s,t), $1 < s$, $1 < t$, *satisfying* $(M)_p$ *for a point* p *that belongs to no unicentric triad.* Then $|sp(x,p)| = 2$ for all $x \neq p$. Let G_p denote the group of collineations generated by the whorls about p guaranteed to exist by $(M)_p$.

THEOREM 7.1: *The point* p *is coregular, so that* $s \leq t$.

PROOF: Let $L, M \in S$, $L \neq M$, $p\,I\,L$. Let N_1, N_2, N_3 be distinct lines in $tr(L,M)$ with $p\,I\,N_1$. We must show that any line concurrent with N_1 and N_2 is also concurrent with N_3. If this were not the case there would be points y_i on N_i, $i = 1,2,3$, with $y_1 \sim y_2$, $y_1 \sim y_3$, $y_2 \not\sim y_3$. Then (p, y_2, y_3) would be a triad with centre y_1, and hence by hypotheses with an additional centre u. Since S satisfies $(M)_p$ there must be a $\theta \in G_p$

fixing p linewise, py_1 and pu pointwise, and mapping y_1y_2 to y_1y_3. Define $v_i \in P$ by $N_i \mathrm{I} v_i \mathrm{I} M$, $i = 1,2,3$. It follows that $y_2^\theta = y_3$, $N_2^\theta = N_3$, $M^\theta = M$, and $v_2^\theta = v_3$. Let $w \in P$ be defined by $v_2 \sim w \mathrm{I}\, pu$. Then $(wv_2)^\theta = wv_3$, giving a triangle (w, v_2, v_3). This impossibility shows that the pair (L,M) must be regular and p must be coregular. □

THEOREM 7.2 : *The* GQ $S^{(p)}$ *is an* EGQ *with elation group* G_p, *and* G_p *is the set of all elations about* p.

PROOF : The first step is to show that G_p is transitive on $\Gamma(p) = P - p^\perp$. If (p,x,y) is a centric triad, it must have at least two centres u and v. By $(M)_p$ there is a collineation θ fixing p linewise, pu and pv pointwise, and moving ux to uy. Clearly $x^\theta = y$. Suppose $x, y \in \Gamma(p)$, $x \sim y$. Put $M = xy$, and let $p \mathrm{I} L$, $L \not\sim M$. Define u_i, $i = 1,2,3$, by $p \sim u_1 \mathrm{I} M$, $y \sim u_2 \mathrm{I} L$, and u_3 is any element of $\mathrm{tr}(u_1, u_2) - \{p,y\}$ (use $t > 1$). As (p, x, u_3) and (p, y, u_3) are centric triads, x and u_3, resp. y and u_3, are in the same G_p-orbit. Hence x and y are in the same G_p-orbit. Finally, suppose (p,x,y) is an acentric triad. Let $u \in \mathrm{tr}(x,y)$, so $u \in \Gamma(p)$. Then x, u, y are all in the same G_p-orbit by the preceding case. Hence G_p is transitive on $\Gamma(p)$.

Since $1 < s \le t$ and $|\mathrm{sp}(x,p)| = 2$ for all $x \not\sim p$, and since, by the remarks following Corollary 3.2, G_p is generated by elations about p, the GQ $S^{(p)}$ is an EGQ with elation group G_p by Theorem 5.3. By Corollary 5.4, the group G_p is the set of all elations about p. □

Note that if $s = t$ then, by the original definition and results of Thas [14], $S^{(p)}$ is a TGQ, and Theorem 7.2 is a generalization of Theorem 4.1. However, if $s < t$ we still lack a proof that $S^{(p)}$ is a TGQ.

LEMMA 7.3 : *Let* (p, x_1, x_2) *be a centric triad, and let* θ *be the unique elation about* p *mapping* x_1 *to* x_2. *Then the fixed* *points of* θ *are precisely those points lying on lines joining* p *with points of* $\{p, x_1, x_2\}^{\perp}$. *The fixed lines of* θ *are just those lines through* p .

PROOF : Since the triad (p, x_1, x_2) is centric it has at least two centres u and v . By $(M)_p$ and Theorem 7.2, the elation θ fixes p linewise and pu and pv pointwise, i.e. θ is one of the elations guaranteed to exist by $(M)_p$. Then by [16] p.159, the fixed points of θ are the points on ℓ (≥ 2) lines through p and the fixed lines of θ are just the lines through p . It is evident that the ℓ lines of fixed points are the lines joining p with the points of $\{p, x_1, x_2\}^{\perp}$. □

LEMMA 7.4 : *Let* (p, x_1, x_2) *be an acentric triad (implying* t *is odd by* 2.2*), and let* $\theta \in G_p$ *map* x_1 *to* x_2 . *Then the following hold:*

(i) *If* θ *has a fixed point* z, $z \neq p$, *then* pz *is incident with all fixed points, and a line is fixed iff it is incident with a fixed point.*

(ii) *If* s *is prime, then* p *is the unique fixed point.*

PROOF : First suppose $z^{\theta} = z$, $z \neq p$. Clearly $z \sim p$. If there is a line M through z , $M \neq pz$, with $M^{\theta} \neq M$, let $y \, \mathrm{I} \, M$, $y \neq z$. Then (p, y, y^{θ}) is a centric triad, so that θ fixes exactly those points lying on lines joining p with points of $\{p, y, y^{\theta}\}^{\perp}$, including all points of pz . But if $x_i \sim z_i \, \mathrm{I} \, pz$, $i = 1,2$, then $z_1^{\theta} = z_2$, forcing (p, x_1, x_2) to have a centre. Hence $M^{\theta} = M$. If a point z' not on pz were fixed by θ , the same argument would show that all lines through z' must be fixed, so each point of tr(z, z') must be fixed, an impossibility. If a line K, $K \not\sim pz$, were fixed by θ , the point on K collinear with p would be a fixed point not on pz . This completes the proof of (i).

If p is not the only fixed point, let $p \neq z = z^{\theta}$, so pz is the unique line through p with fixed points other than p .

Now we observe that $\langle\theta\rangle$ acts semiregularly on the s points different from z on any line M not pz through z. Hence if $|\langle\theta\rangle| = n$ then $n \mid s$. If s is prime, then all $\langle\theta\rangle$-orbits have length 1 or s. But if $x_1 \sim z_1 \, I \, pz$, clearly $z_1^\theta \neq z_1$, and the $\langle\theta\rangle$-orbit containing z_1 would have to contain all s points of pz different from p. This impossibility implies p is the only point fixed by θ. □

LEMMA 7.5 : *Let* $\theta \in G_p$ *be such that* $x \sim x^\theta$ *for all* $x \in \Gamma(p)$, *and hence for all* $x \in P$. *Then* θ *is a symmetry about some line through* p.

PROOF : Suppose $1 \neq \theta \in G_p$ with $x \sim x^\theta$ for all $x \in P$. Let $x \in \Gamma(p)$, so $x \sim x^\theta \neq x$, and put $M = xx^\theta$. Define $L \in B$ by $p \, I \, L \sim M$. Since $L^\theta = L$, clearly $M^\theta = M$. Let z be any point of L, $z \neq p$, $z \not{I} M$, and let $K \, I \, z$, $K \neq L$. Let $x \, I \, N \, I \, y \, I \, K$, and let $p \, I \, R \sim N$. Since M and R are fixed by θ (and p is coregular), both $tr(M,R)$ and $sp(M,R)$ are invariant under θ. But $y \sim y^\theta$ forces $K = yz$ to be fixed. This says any line K meeting L at a point other than z or p is fixed. Interchanging roles of K and M then shows that θ fixes every line meeting L. So θ is a symmetry about L, and hence fixes only those lines meeting L and only those points on L. □

The preceding three results have the following corollary.

THEOREM 7.6 : *There are precisely the following four kinds of collineations* $\theta \in G_p$.

(i) $\theta = id$.

(ii) (p, x, x^θ) *is a centric triad for all* $x \in \Gamma(p)$.

(iii) (p, x, x^θ) *is an acentric triad for some* $x \in \Gamma(p)$ *and, for all* $y \in \Gamma(p)$, (p, y, y^θ) *is an acentric triad or* $y \sim y^\theta$.

(iv) θ *is a nonidentity symmetry about some line through* p. □

We close the section with some corollaries.

COROLLARY 7.7 : *Let* $S = (P, B, I)$ *be a* GQ *of order* (s,t), $s > 1$, $t > 1$, *and suppose that* S *satisfies* $(M)_p$ *for some point* p, *with* G_p *defined to be the group of whorls about* p *generated by the elations about* p *guaranteed to exist by* $(M)_p$.

(i) *If* p *is coregular and* t *is odd, then* $S^{(p)}$ *is an* EGQ *with elation group* G_p.

(ii) *If each triad containing* p *has at least two centres, then* $S^{(p)}$ *is a* TGQ.

(iii) *If* $t = s^2$, *then* $S^{(p)}$ *is a* TGQ. *Hence, for* $t = s^2$, *also* $(\hat{M})_p$ *is satisfied.*

(iv) *If* $t = s^2$ *and* s *is prime, then* $S \cong Q(5,s)$.

PROOF : If p is coregular and t is odd, then each triad containing p has an even number of centres. Hence the general hypotheses of this section hold and the proof of (i) is completed by Theorem 7.2. Now suppose each triad containing p has at least two centres, so that $S^{(p)}$ is again an EGQ by Theorem 7.2. Since each triad containing p is centric, each nonidentity element of G_p must be either type (ii) or type (iv) as described in Theorem 7.6. If $x \sim y$, $x \neq y$ and $x, y \in \Gamma(p)$, then the unique elation θ for which $x^\theta = y$ is a nonidentity symmetry about some line through p. It follows that each line through p is an axis of symmetry, hence $S^{(p)}$ is a TGQ. This proves (ii), which together with Corollary 6.5 has (iii) as a corollary. By (iii) and Corollary 6.11 (ii), also (iv) follows. □

BIBLIOGRAPHY

1. P. Dembowski, *Finite Geometries* (Springer-Verlag, Berlin, 1968).
2. W.M. Kantor, "Generalized quadrangles associated with $G_2(q)$", *J. Combin. Theory Ser. A*, 29(1980), 212-219.
3. D.S. Passman, *Permutation Groups* (Benjamin, New York, 1968).
4. S.E. Payne, "A restriction on the parameters of a subquadrangle", *Bull. Amer. Math. Soc.*, 79 (1973), 747-748.

5. S.E. Payne, "Finite generalized quadrangles : a survey", *Proc. Internat. Conf. Projective Planes*, Washington State Univ., 1973 (Washington State Univ. Press, Pullman, 1973), pp. 219-261.
6. S.E. Payne, "Generalized quadrangles of even order", *J. Algebra*, 31 (1974), 367-391.
7. S.E. Payne, "Skew translation generalized quadrangles", *Proc. Sixth Southeastern Conf. on Combinatorics, Graph Theory and Computing*, Florida Atlantic Univ., 1975 (Utilitas Math., Winnipeg, 1975), pp. 485-504.
8. S.E. Payne, "An inequality for generalized quadrangles", *Proc. Amer. Math. Soc.*, 71 (1978), 147-152.
9. S.E. Payne, "Generalized quadrangles as group coset geometries", to appear.
10. S.E. Payne and J.A. Thas, "Generalized quadrangles with symmetry", *Simon Stevin*, 49 (1975-1976), 3-32; 81-103.
11. J.A. Thas, "The m-dimensional projective space $S_m(M_n(GF(q)))$ over the total matrix algebra $M_n(GF(q))$ of the $n \times n$-matrices with elements in the Galois field GF(q)", *Rend. Mat.*, 4 (1971), 459-532.
12. J.A. Thas, "Geometric characterization of the [n-1]-ovaloids of the projective space PG(4n-1, q)", *Simon Stevin*, 47 (1974), 97-106.
13. J.A. Thas, "A remark concerning the restriction on the parameters of a 4-gonal subconfiguration", *Simon Stevin*, 48 (1974-1975), 65-68.
14. J.A. Thas, "Translation 4-gonal configurations", *Atti Accad. Naz. Lincei Rend.*, 56 (1974), 303-314.
15. J.A. Thas, "Combinatorial characterizations of generalized quadrangles with parameters $s = q$ and $t = q^2$", *Geom. Dedicata*, 7 (1978), 223-232.
16. J.A. Thas, "Generalized quadrangles satisfying at least one of the Moufang conditions", *Simon Stevin*, 53 (1979), 151-162.
17. J.A. Thas and S.E. Payne, "Classical finite generalized quadrangles : a combinatorial study", *Ars Combin.*, 2 (1976), 57-110.
18. J. Tits, "Classification of buildings of spherical type and Moufang polygons : a survey", *Teorie Combinatorie*, Tomo I, Rome, 1973. Atti dei Convegni Lincei 17 (Accad. Naz. Lincei, Rome, 1976), pp. 229-246.
19. J. Tits, *Buildings of Spherical Type and Finite BN-pairs*, Lecture Notes in Math. 386, (Springer-Verlag, Berlin, 1974).
20. J. Tits, "Quadrangles de Moufang", to appear.

Department of Mathematics
Miami University
Oxford
Ohio 45056
U.S.A.

Seminar of Geometry and Combinatorics
University of Ghent
Krijgslaan 271
9000 Gent
Belgium

EMBEDDING GEOMETRIC LATTICES IN A PROJECTIVE SPACE

Nicolas Percsy

1. INTRODUCTION

We have obtained a necessary and sufficient condition for a geometry to be embeddable in a (generalized) projective space. This condition essentially requires that all intervals above a point are embeddable in such a space and that these embeddings are "compatible" along the lines of the geometry.

All geometries considered here are geometric lattices; embeddings are isometric in Kantor's sense (see [14]).

This result provides a unique general proof of the following known theorems on locally embeddable geometries of dimension at least four : Mäurer's result on locally affine geometries (the so-called Möbius spaces), Kantor's strong embedding theorem and his other embedding theorems involving universal properties (see [16] and [13], [14]). The result can be extended to the dimension three case, generalizing Kahn's work on Möbius, Laguerre and Minkowski planes (see [8] to [11]).

Let us mention one of its new applications : Kantor's concept of strong embedding can be extended to a larger class of geometries; for instance, it is possible to prove the embeddability of infinite locally affino-projective geometries of dimension at least four.

The notions of geometry and embedding are defined in §2 and §3 respectively. The main result can be found in §4, where more historical details and an idea of the proof are given. This proof is based on a general embedding lemma stated in §5. Finally, §6 presents some applications; it also gives a precise connection between the main result in §4 and the known theorems mentioned above.

2. DEFINITION OF A GEOMETRY

The terminology used here for lattices can be found in Birkhoff [3] (see also Welsh [20]; notice that the geometries introduced below need not be finite).

By a *geometry*, we mean an (upper) semimodular lattice (of finite length) with least element 0 and greatest element 1. If G is such a geometry, its elements are often called *subspaces*. For any $S \in G$, all maximal chains from 0 to S have the same number $d(S)+2$ of elements; $d(S)$ is the *dimension* of S. (This is one less than the dimension (or height) in the sense of Birkhoff [3].) If $d(S)=i$, we say that S is an *i-space;* 0-, 1- and 2-spaces are also called *points, lines* and *planes* respectively. The *dimension* or *rank* $d(G)$ of G is defined to be $d(1)$.

If G is an atomic lattice (i.e. if every element of G is a join of points), the geometry G is said to be *atomic*. Such a geometry is thus a geometric lattice. The latter concept is equivalent to other well-known ones: matroids (also called incidence spaces, see Welsh [20]), dimensional closure spaces (whose points are closed, see Buekenhout [4]), combinatorial geometries (Crapo and Rota [6]), incidence structures (Buekenhout [5]) belonging to the following diagram

L L ... L .

Our main results are concerned with atomic geometries; but for technical reasons (see §5), some geometries, appearing in the proofs, are not atomic.

Let G be a geometry and $S \in G$ a subspace. The interval $\{X \in G \mid X \geq S\}$ is obviously a geometry, sometimes called *local geometry* or *residual geometry at S*; it is denoted by G_S and has dimension $d(G) - d(S) - 1$. Let P be a property (of geometries); we say that *G has property P locally* if G_p has P, for all points $p \in G$. Let k be an integer such that $k \leq d(G)$; the lattice of all subspaces of G whose dimension is less than k, together with 1, clearly forms a geometry; this geometry is usually called the *truncation of G to dimension k*.

The following concept is a well-known generalization of classical projective spaces. A *generalized projective space* is a set P of *points*, together with distinguished subsets called *lines*, such that the following axioms hold :

(P1) any two distinct points are on a unique line;

(P2) each line contains at least two points;

(P3) if a line intersects two sides of a triangle (not at their intersection), it also intersects the third one.

A generalized projective space satisfying the following strengthening of (P2) is a *projective space* (in the usual sense) :

(P2') each line contains at least three points.

A *subspace* of P is a subset which contains every line intersecting it in at least two points. The set of all subspaces of a generalized projective space constitute, under inclusion, an atomic modular lattice (which is an atomic geometry, provided the dimension of the lattice is finite). Conversely, every atomic geometry, which is a modular lattice, (respectively a simple modular lattice), is the subspace lattice of a finite dimensional generalized projective space (respectively a finite dimensional projective space) (see Birkhoff [3, IV, §7]).

3 . EMBEDDINGS AND EMBEDDABILITY

An *isometric embedding* or *isometry* (Kantor [14]) from a geometry G into a geometry H is a mapping $\theta : G \to H$ satisfying the following conditions :

(E1) θ is injective;

(E2) $1^\theta = 1$;

(E3) $X \le Y$ (in G) if and only if $X^\theta \le Y^\theta$ (in H), for any two distinct $X, Y \in G$;

(E4) for all $Z \in G \setminus \{1\} : d(Z^\theta) = d(Z)$.

Actually, Kantor's definition contains a further axiom, demanding that θ be join-preserving; this is a consequence of the above conditions :

PROPOSITION 1 : *Any isometry* $\theta : G \to H$ *has the following property :*

(E5) *for any family* X_i *of elements of* G, *if the join* $\bigvee_i X_i$ *is different from 1, then* $(\bigvee_i X_i)^\theta = \bigvee_i X_i^\theta$.

PROOF: See Percsy [17] .

The condition (E5), together with (E1), (E2), (E4) provides an equivalent definition of an isometry; actually (E4) is redundant (for more details, see Percsy [17]). The restriction $Z \neq 1$ in (E4) permits us to consider the canonical injection of a truncation of a geometry into that geometry as an isometry. An isometry of atomic geometries satisfying $d(1^\theta) = d(1)$ is an injective strong mapping in the sense of Welsh [20] .

The following obvious properties are useful :

(1) The composition of isometries is an isometry.

(2) A one-to-one isometry is an isomorphism (defined to be an order-preserving bijection).

(3) If $\theta : G \to H$ is an isometry, then for every $S \in G$, the restriction of θ to G_S is an isometry from G_S to $H_{S\theta}$.

We are interested in "embedding theorems", i.e. theorems providing conditions for the embeddability of a geometry into some generalized projective space. Actually, those theorems hold for "infinite dimensional geometries". But in verifying the embeddability of such geometries, we need only check the hypotheses for finite dimensional subspaces. This allows us to restrict our study to the finite dimensional case (as is done in §1) and motivates the following definitions.

Let P be a generalized projective space (respectively a projective space) whose subspace lattice has (possibly infinite) dimension $d \geq k$, where k is an integer. The lattice of all subspaces of P, whose dimension is less than k, form an atomic geometry (non modular, unless $d = k$) of rank k. Such a geometry is called a *generalized projective geometry* (respectively a *projective geometry)* of rank k; its diagram (in the sense of Buekenhout [5]) is clearly

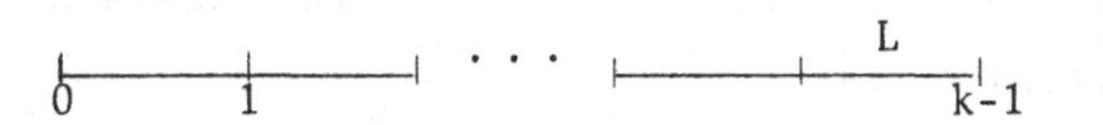

4 . THE MAIN EMBEDDING THEOREM

The most famous embedding is certainly the extension of an affine space into a projective space. This embedding is carried out by adding ideal subspaces to the affine space; an ideal i-space is defined to be a partition of the space by $(i+1)$-spaces, any two of which are in a common $(i+2)$-space (i.e. they are parallel). Batten [1], [2] has determined all atomic geometries of rank at least 4 which can be embedded by a similar method (i.e. by using the same definition for ideal subspaces) : these geometries turn out to be locally projective, as affine spaces are (see [1]). Actually, the embeddability of such geometries has been proved (by a similar method) implicitly by Mäurer [16], explicitly by Wille [21] and, for not necessarily atomic geometries, by Kantor [13] .

Mäurer's paper characterizes Möbius spaces; it implicitly proves that all locally affine atomic geometries of rank at least 4 are embeddable. There are two main steps in the proof. First, each local geometry G_p , at a point p , which is affine, is naturally extended into a projective space $\overline{G}_p$; these projective spaces can be glued together in order to get a larger geometry E , in which G is embedded, and whose local geometries are the projective $\overline{G}_p$. The second step is the previous result : such a locally projective E is embeddable.

A natural idea in generalizing this is to consider a locally embeddable geometry G (of rank at least 4) : more precisely, for each point p , the local geometry G_p is assumed to be embeddable in some projective geometry $P(p)$ via an embedding $\hat{p} : G_p \to P(p)$ (here, G_p need not be an affine space deduced from $P(p)$, nor the whole space $P(p)$ as previously). Kantor [13], [14] (see also [12] and [15]) obtained deep results in that direction by assuming that the locally embeddings $\hat{p}$ satisfy some universal properties ("categorical embedding" [14]) or that the G_p are "large chunks" of the $P(p)$ ("strong embedding" [13]); for more details, see §6 . The proofs contain two parts, similar to those of Mäurer's proof. Actually, Kantor was not inspired by Mäurer's work, but by other results, as explained in [12], [13] ; his theorems contain Mäurer's one as a very particular case.

The above results, and this similarity in their proof, led us to the following question : given a locally embeddable geometry G , when can the projective P(p) spaces be glued together in order to embed G in a locally projective geometry E (and hence, in a projective geometry by [13] or [16], [21], [2]) ? It turned out that the right answer is not to assume some property for each embedding $\hat{p}$, as in Kantor's or Mäurer's results, but to require that these local embeddings "behave well" towards one another, that they are "compatible". We obtained in this way an embedding lemma, which holds actually for more general geometries than the semi-modular lattices and the projective spaces considered here (see §5).

As a consequence, we deduced the following necessary and sufficient condition of embeddability for a geometry of rank at least 4 . As shown in §6, the above-mentioned known results can be deduced from it.

THEOREM 1 : *An atomic geometry* G *of rank* $n \geq 4$ *is embeddable in some generalized projective geometry if and only if it satisfies the following conditions :*

(L) G *is locally embeddable :*
for each point p, *there exists an embedding* $\hat{p} : G_p \to P(p)$ *of the local geometry* G_p *at* p *into some rank* $n-1$ *generalized projective geometry* P(p) *(depending on* p *).*

(C) *For any two distinct points* $p, q \in G$, *the embeddings of* $G_{p \vee q}$, *via* $\hat{p}$ *and* $\hat{q}$, *are projectively equivalent: more precisely, it exists an isomorphism* σ_{pq} *of projective geometries such that the following diagram commutes*

$$G_{p\vee q} \begin{array}{c} \xrightarrow{\hat{p}} \\ \xrightarrow[\hat{q}]{} \end{array} \begin{array}{l} P(p)_{\hat{p}(p\vee q)} \\ P(q)_{\hat{q}(p\vee q)} \end{array} \Big\} \; \sigma_{pq}$$

and these σ_{pq} *must be chosen in such a way that the following conditions hold:*

(C1) *for any two distinct points* $a, b \in G : \sigma_{ab} = \sigma_{ba}^{-1}$.

(C2) *if* a, b, c *are three distinct points on a line* L,
$\sigma_{bc} \circ \sigma_{ab}$ *and* σ_{ac} *coincide on* $P(a)_{\hat{a}(L)}$.

(C3) *if* a, b, c *are three distinct points on a plane* C *,*
$\sigma_{bc} \circ \sigma_{ab}$ *and* σ_{ac} *coincide on* $P(a)_{\hat{a}(C)}$.

Moreover, if G *satisfies the above assumptions, there exists an embedding* θ *from* G *into a generalized projective geometry* P *of rank* n *, such that for all points* $p \in G$ *,* P(p) *is isomorphic to the local geometry* $P_{\theta(p)}$ *of* P .

PROOF : Suppose G satisfies the hypothesis above. By the embedding lemma of §5, G is embeddable in a locally projective geometry of rank at least 4 (which is not atomic); the latter geometry is embeddable following Kantor [13, §3] (see also Mäurer [16] or, for atomic geometries, Wille [21] and Batten [2]) .

It is straightforward to check that the above conditions are necessary; also the last part of the theorem is obvious.

The rank three case will be discussed in the next section. Let us remark that the interest of the last part of Theorem 1 is the following: if we are able to prove that all the P(p) are (non-generalized) projective spaces, or all finite, all pappian, all with the same ground field,..., then so is P .

5 . A FUNDAMENTAL EMBEDDING LEMMA

As mentioned above, Theorem 1 is based on the following embedding lemma. It is quite remarkable that this lemma holds if G is locally embeddable in arbitrary geometries P(p), *not necessarily projective*. Moreover, it holds for geometries G which are more general than those introduced in §2 (for instance polar spaces). However, in order to simplify the lemma's statement, we shall restrict it to the particular geometries defined above; for more details, see Percsy [19].

EMBEDDING LEMMA : *Let* G *be an atomic geometry of rank* $n \geq 3$. *Assume that for each point* p *, there is an embedding* $\hat{p}$ *of* G_p *into some geometry* P(p) *of rank* n - 1 *(not necessarily projective). If these embeddings satisfy the "compatibility" conditions* (C), (C1), (C2) *of Theorem* 1, *and also condition* (C3) *provided that* $n \geq 4$, *then there is a rank* n *geometry* E *and an embedding* $\theta : G \to E$ *such that*

(i) *for each point* $e \in E$, *there is a point* $p \in G$ *such that* $\theta(p) = e$;

(ii) *for each point* $p \in G$, *there is an isomorphism* σ_p *between* $P(p)$ *and* $E_{\theta(p)}$ *such that* $\sigma_p \circ \hat{p}$ *coincides with* θ *on* G_p.

PROOF : See Percsy [19]. The main idea of the proof is an idea of Mäurer [16] and Kantor [13] already mentioned in §4 : the $P(p)$ are glued together in order to obtain the right geometry E. This "gluing" is made possible here thanks to the compatibility conditions, which are new. Also the proof is quite different and more general than those of the above-mentioned authors.

What about the rank 3 case? The preceeding lemma holds if the rank of G is 3, but Theorem 1 does not. The reason for this is that locally projective geometries of rank 3 are not necessarily embeddable. Counter-examples are known; see for instance Kantor [13]. Recently, Kahn [10], [11] obtained a necessary and sufficient condition for the embeddability of a rank 3 geometry which is locally a projective plane : it is essentially the "bundle condition", which is well known in Möbius geometries.

By using Kahn's result, our embedding lemma provides an embedding theorem for rank 3 geometries : one has just to assume, besides the compatibility conditions on G, Kahn's extra condition on the geometry E obtained by applying the lemma. Of course, this is really interesting only if the latter condition can be "translated" to properties of the original geometry G. This can be done at least for those geometries whose local embeddings $\hat{p} : G_p \to P(p)$ satisfy the following axiom : for any point $x \in G_p$, there is at most one line L of $P(p)$ such that $L \cap \hat{p}(G_p) = \{x\}$; see Percsy [19]. We obtain in this way a generalization of similar results obtained by Kahn in the particular case of inversive planes, Laguerre planes and Minkowski planes ([8] to [11], especially the latter).

6. APPLICATIONS

This section is devoted to some applications of Theorem 1, especially to extensions of Kantor [13] and [14]. They are based on a simplified version of Theorem 1, which holds in an important particular case, and whose proof is not too hard.

We need a definition. An embedding $\theta : G \to P$ is said to be *rigid* if no nontrivial automorphism of P fixes all the elements of G^{θ}. Kantor [14] also defines a similar notion of rigidity, but his definition involves a universal property in a certain class of embeddings; here, such a technical development is not needed.

COROLLARY 1 : *Let* G *be an atomic geometry of rank* $n \geq 4$. *If* G *satisfies conditions* (L) *and* (C) *of Theorem* 1, *and the further axiom* (R) *below, then it is embeddable.*

(R) *For any plane* $C \in G$, *and for each point* p *on* C, *the embedding* $\hat{p} : G_C \to P(p)_{\hat{p}(C)}$ *is rigid.*

Now, Kantor's "categorical embedding" theorems [14] are an easy consequence of Corollary 1. Actually, they can be extended in the following way. Let us denote by C_n an arbitrary class of "rank n" embeddings (G, P, θ), where G is a rank n geometry, P is a rank n projective geometry and $\theta : G \to P$ is an embedding. For instance, such a class can be obtained as follows.

EXAMPLES : (1) Let K be a given field; $C_n(K)$ is defined as the class of all embeddings of an arbitrary rank n atomic geometry G into the rank n truncation of a projective space over K.

(2) We may consider, in finite geometry, the class of all (G, P, θ), where G and P are finite.

(3) Consider the class of all rank n (G, P, θ), with G and P arbitrary, such that θ satisfies one of Kantor's strong embedding conditions : each point and each line of P is the intersection (in P) of all members of G^{θ} containing it.

(4) Consider the class of all (G, P, θ) such that P is an n-dimensional projective space (not degenerate) and that there is a hyperplane of P containing all points (of P) not in G^{θ}. Geometries G admitting such an embedding are usually called

affino-projective spaces.

We say that a geometry G is *uniquely embeddable with respect to a class* C_n, if all embeddings of G, belonging to that class, are projectively equivalent (in the sense of condition (C) in Theorem 1). For instance, all affino-projective spaces are uniquely embeddable with respect to the class (4) above. We can state now the announced result, which follows easily from Corollary 1.

COROLLARY 2 : *Let* G *be an atomic geometry of rank* $n \geq 4$ *and let* C_{n-2} *be a class of embeddings. If* G *satisfies axiom* (L) *of Theorem* 1, *axiom* (R) *above and condition* (U) *below, it is embeddable.*

(U) *For each line* L *and each point* p *on* L, *the embedding* $(G_L, P(p)_{\hat{p}(L)}, \hat{p})$ *is a member of* C_{n-2} *and* G_L *is uniquely embeddable with respect to* C_{n-2}.

It is easily seen that this corollary is a generalization of Kantor [14]. Notice that we are free to choose the class C_{n-2} arbitrarily; it is thus possible to obtain a great number of applications. We may also drop here both restricting hypotheses appearing in Kantor [14] : i.e. all embeddings correspond to projective spaces over a given field, and that field may not be isomorphic to one of its proper subfields.

Kantor's strong embedding theorem [13] is also a consequence of Corollary 2 : it is obtained by considering embedding classes like Example 3. It can also be generalized to a wider class of geometries; for the technical details, see Percsy [18].

Let us mention finally an application of Corollary 2 which is not actually contained in Kantor's results.

COROLLARY 3 : *Any atomic geometry of rank at least* 4, *which is locally an affino-projective space (in the sense of Example* (4)*), is embeddable.*

PROOF : Use Corollary 3 by taking, as C_{n-2}, the class of example (4).

This corollary contains, as a particular case, Mäurer's result on locally affine geometries [16], used by him to characterize Möbius geometries. More generally, Corollary 3 can be used to simplify Dienst's axioms for the geometry of a semi-quadratic set [7]: this provides a characterization of Hermitian, orthogonal and symplectic quadrics in terms of the associated Möbius, Laguerre or Minkowski geometry (which is locally affino-projective).

BIBLIOGRAPHY

1. L.M. Batten, "d-partition geometries", *Geom. Dedicata* 7 (1978), 63-69.
2. L.M. Batten, "Embedding d-partition geometries in generalized projective space", *Geom. Dedicata* 7 (1978), 163-174.
3. G. Birkhoff, *Lattice Theory*, Third edition (Amer. Math. Soc., Providence, 1967).
4. F. Buekenhout, "Espaces à fermeture", *Bull. Soc. Math. Belg.* 19 (1967), 147-178.
5. F. Buekenhout, "Diagrams for geometries and groups", *J. Comb. Theory Ser. A*, 27 (1979), 121-151.
6. H.H. Crapo and G. -C. Rota, *On the Foundations of Combinatorial Theory: Combinatorial Geometries*, Preliminary edition, (M.I.T. press, Cambridge, Mass., 1970).
7. K.J. Dienst, "Schnitt und Zykelgeometrien", *Geom. Dedicata* 6 (1977), 23-53.
8. J. Kahn, "Finite inversive planes satisfying the bundle theorem", to appear.
9. J. Kahn, "Inversive planes satisfying the bundle theorem", *J. Combin. Theory Ser. A*, 29 (1980), 1-19.
10. J. Kahn, *"Locally projective planar lattices which satisfy the bundle theorem"*, Ph.D. thesis, Ohio State University, 1979.
11. J. Kahn, "Locally projective planar lattices which satisfy the bundle theorem", to appear.
12. W.M. Kantor, "Some highly geometric lattices", *Teorie Combinatorie*, Tomo I, Rome 1973. Atti dei Convegni Lincei 17 (Accad. Naz. Lincei, Rome, 1976), pp.183-191.
13. W.M. Kantor, "Dimension and embedding theorems for geometric lattices", *J. Combin. Theory Ser. A*, 17 (1974), 173-195.
14. W.M. Kantor, "Envelopes of geometric lattices", *J. Combin. Theory Ser. A*, 18 (1975), 12-26.
15. W.M. Kantor, "Relationships with finite geometries and finite groups", to appear in a new edition of [6].
16. H. Mäurer, "Ein axiomatischer Aufbau der mindestens 3-dimensionalen Möbius-Geometrie", *Math. Z.*, 103 (1968), 282-305.
17. N. Percsy, *"Plongement de géométries"*, Thèse de doctorat, Université Libre de Bruxelles, 1980.

18. N. Percsy, "An extension of Kantor's strong embedding", in preparation.

19. N. Percsy, "Locally embeddable geometries", submitted.

20. D.J.A. Welsh, *Matroid Theory* (Academic Press, London 1976).

21. R. Wille, "On incidence geometries of grade n", *Geometria Combinatoria e sue Applicazioni*, Perugia 1970 (Università degli Studi di Perugia, 1971), pp.421-426.

Département de Mathématique
Université de l'Etat à Mons
Av. Maistriau 15
7000 Mons
Belgium

COVERINGS OF CERTAIN FINITE GEOMETRIES

Mark A. Ronan

1. INTRODUCTION

The purpose of this article is to investigate the Tits universal 2-covers (see Section 2 for a definition) of the sporadic group geometries appearing in the work of Buekenhout, mainly in [1]. The main results are listed in Table 1, Section 3 deals with topological preliminaries, and Sections 4-7 go into the details of individual cases. Sometimes we allow more generality than is necessary for our goal, and in doing so we hope to provide methods which may be used in cases other than those considered here.

2. DEFINITIONS, NOTATION AND SUMMARY OF RESULTS

The concept of a universal 2-cover is due to Tits [13] and we repeat the definition here in the more specialized framework of diagram geometries, see [1]. If G is a diagram geometry, then a *2-cover* is a geometry G together with a type preserving map $p : \overline{G} \to G$ such that p is an isomorphism when restricted to residues of rank 2; in particular $\overline{G}$ belongs to the same diagram as G. The *universal* 2-cover $p : \tilde{G} \to G$ has the property that if $f : \overline{G} \to G$ is any 2-cover, then there exists a 2-cover $g : \tilde{G} \to \overline{G}$ making the following diagram commute:

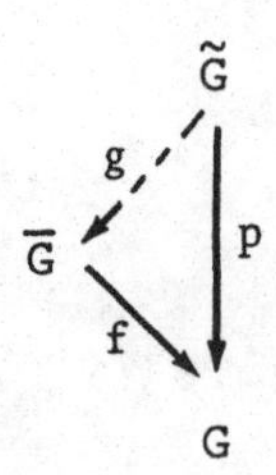

The existence of a universal 2-cover is proved in [6], and in this paper we will always use the symbol $\tilde{G}$ to stand for the universal 2-cover of G.

In [6] it is shown that there is a 1-1 correspondence between 2-covers of G and conjugacy classes of subgroups of a certain *2-fundamental group* $\pi^2(G)$. In particular, the universal 2-cover $\tilde{G}$ corresponds to the trivial subgroup of $\pi^2(G)$, and $G = \tilde{G}$ if and only if $\pi^2(G) = 1$. If $|G|$ denotes the number of maximal flags (or *chambers*) of G, then

$$|\tilde{G}| = |G| \cdot |\pi^2(G)| .$$

Moreover, if Aut G is transitive on chambers, then Aut $\tilde{G}$ is also.

In Table 1 we give the order of $\pi^2(G)$ and also the Euler characteristic $\chi(G)$ considering G as a simplicial complex. The parameters are also given - the number q_i beneath the i th-node means that each flag of cotype i lies in $q_i + 1$ chambers. By the diagram $\underset{s}{\circ}\!\!-\!\!\!-\!\!\!-\!\!\underset{t}{\circ}$, for $s \le t$, we understand a rank 2 geometry of points and lines with $s+1$ points per line, $t+1$ lines per point, and such that any two points lie on a unique line. If $s = t$ one has a projective plane, and if $s = 1$ one has a complete graph.

In all the geometries of Table 1, the vertices corresponding to the left most node of the diagram will be called *points*.

3. SOME TOPOLOGICAL ASPECTS

Given a diagram geometry G one forms a simplicial complex, which by abuse of notation we also denote by G, by taking as simplexes all flags of G with the obvious inclusion relation, and we will talk of topological properties of G without further explanation. If the *rank* r(G) is defined to be the number of nodes of the diagram, then $r(G) = 1 + \dim G$, considering G as a simplicial complex.

If $r(G) = 2$, then plainly $G = \tilde{G}$ by definition. If $r(G) = 3$, then $\tilde{G}$ is clearly the universal topological cover

TABLE 1

Group	Diagram Geometry G	$\lvert \pi^2(G) \rvert$	$\chi(G)$
M_{11}	1 — 9	1	- 44
M_{12}	1 — 1 — 9	1	166
M_{22}	1 — 4 — 4	1	561
M_{23}	1 — 1 — 4 — 4	1	- 3,519
M_{24}	1 — 1 — 1 — 4 — 4	1	17,711
Suz	1 — 9 = 3	1	7,302,880
·1	1 — 1 — 9 = 3		- 3,318,257,398,400
F_{22}	1 — 4 — 4 = 2	1	- 63,097,515
F_{23}	1 — 1 — 4 — 4 = 2	1	3.7×10^{11}
F_{24}	1 — 1 — 1 — 4 — 4 = 2	3	6.3×10^{17}
HS	1 — 4 — 4 — 1	1	- 21,250
Mc	1 — 3 = 9	1	77,275
·3	1 — 1 — 3 = 9	2	- 5,704,874
·2	1 — 2 = 4 — 4	2	- 263,387,375
J_2	1 — 2 —(6)— 2	∞	1,675
Suz	1 — 4 —(6)— 4	∞	2,839,617
Ru	1 — 2 —(8)— 4	∞	5,348,035
He	triangle 1, 4, 4 (4 = 4)	∞	147,833
Ly	5 — 5 —(6)— 5	∞	110,404,109,376
Suz	4 = 2 = 2	∞	5,859
M_{24}	2 = 2 — 6	1	21,505

of G, but if $r(G) > 3$, then in general $\tilde{G}$ will be a branched (or ramified) cover with branch set of codimension at least 2 in G. However, apart from the $\cdot 2$ example in Section 7, all the covers considered here turn out to be unramified; in this connection the following lemma will be extremely useful.

LEMMA 1 : *If* $r(G) \geq 4$ *and all residues of rank at least* 3 *are simply-connected, then* $\tilde{G}$ *is the universal topological cover of* G. *If, moreover,* G *is simply-connected, then* $\tilde{G} = G$.

PROOF : This is simply lemma (1.8) of [6] applied to diagram geometries. □

Now let I be the indexing set of G; that is to say I is in bijective correspondence with the set of nodes of the diagram, and each vertex of G has type i for some $i \in I$. Following Tits [12], for any flag f we let $Sh_i(f)$ (Sh for shadow) denote the set of vertices of type i incident with f. Now given any set of vertices $\{v_0, \ldots, v_k\}$ of type i we let

$$L(v_0, \ldots, v_k) = \left\{ \text{flags } f \mid v_0, \ldots, v_k \in Sh_i(f) \right\} .$$

Clearly $L(v_0, \ldots, v_k)$ is a subcomplex of G and one checks it is equal to $\text{link } v_0 \cap \ldots \cap \text{link } v_k$.

DEFINITION 1 : We say G satisfies the *i-contractibility condition* if given any set $\{v_0, \ldots, v_k\}$ of vertices of type i, then $L(v_0, \ldots, v_k)$ is contractible or is the null set.

Suppose G satisfies the following condition:

(*) For any set $\{v_0, \ldots, v_k\}$ of vertices of type i for which $L(v_0, \ldots, v_k) \neq \emptyset$ there is some vertex $x \in L(v_0, \ldots, v_k)$ such that every flag which is maximal in $L(v_0, \ldots, v_k)$ contains x.

Then it easily follows that the i-contractibility condition is satisfied because one conically contracts $L(v_0, \ldots, v_k)$ to x.

In particular (*) is satisfied in all irreducible buildings for every $i \in I$, though it is never satisfied for diagram geometries whose diagram is disconnected.

DEFINITION 2 : Define a new simplicial complex $N_i(G)$ (called the *i-nerve* of G) by taking as vertices of $N_i(G)$ all vertices of G of type i, and as simplexes all subsets $\{v_0, \ldots, v_k\}$ of type i such that $L(v_0, \ldots, v_k) \neq \emptyset$. The 1-skeleton of $N_i(G)$ will be called the *i-graph* of G. If vertices of type i are called points we will talk of the *point-nerve*, and *point-graph* of G.

LEMMA 2 : *If* G *satisfies the i-contractibility condition, then* $N_i(G)$ *is homotopically equivalent to* G.

PROOF : The simplexes of a simplicial complex X may be viewed as a small category whose morphisms are inclusion maps, and in this way the barycentric subdivision X' becomes its classifying space, see [5]. Now regarding G' with reverse inclusion, the map $Sh_i : G' \to N_i(G)'$ can be thought of as a functor, and one can apply Theorem A of Quillen [5] since for any vertex $Y = (v_0, \ldots, v_k)$ of $N_i(G)'$, $Y \backslash Sh_i = L(v_0, \ldots, v_k)$ is contractible (using the notation $Y \backslash f$ of [5]). The result follows. □

As mentioned before, if the diagram of G is disconnected, then G does not satisfy the i-contractibility condition, so Lemma 2 is of no help in this case. However:

LEMMA 3 : *If the diagram of* G *is disconnected and* $r(G) \geq 3$, *then* G *is simply-connected.*

PROOF : Let $i, j, k \in I$ correspond to nodes of the diagram such that the connected component containing k does not contain i or j. By using the residual connectivity of a diagram geometry (see [1]) one sees that any circuit in G based at a vertex of type i is homotopic to a circuit γ whose vertices are alternately of types i and j. But clearly γ lies in link v for any vertex v of type k, and is therefore null-homotopic. Thus any circuit is null-homotopic and B is simply-connected. □

4. GEOMETRIC LATTICES AND THE MATHIEU GROUPS

The first five geometries of Table 1, for the Mathieu groups, are geometric lattices where the points are the atoms. If Λ is a geometric lattice we let $B\Lambda$ denote the simplicial complex whose simplexes are the chains of Λ (i.e. $B\Lambda$ is the classifying space of Λ treated as a small category). The following lemma was essentially first proved by Folkman [4], whose proof only lacks the observation that $B\Lambda$ is simply-connected if Λ has rank at least 3.

LEMMA 4: *If Λ is a geometric lattice of rank n, then $B\Lambda$ is homotopy equivalent to a bouquet of $(n-1)$-spheres.*

PROOF: Following Folkman, consider $N = N_A(\Lambda)$ where A denotes the atoms. Let $\sum$ denote a standard $(v-1)$-simplex where v is the number of atoms. Then for $i < n-1$ one has

$$\pi_i(N) = \pi_i(N^{i+1}) = \pi_i(\textstyle\sum^{i+1}) = \pi_i(\textstyle\sum) = 0 ,$$

where X^k is the k-skeleton of X; the first and third equalities are elementary topological facts, the second is due to the fact that any $i+1$ atoms have a join in Λ, and the fourth follows from the fact that $\sum$ is contractible. Since $B\Lambda \simeq N$ (for example, use Lemma 2) we see that $B\Lambda$ is $(n-1)$-dimensional and $(n-2)$-connected, hence is homotopically a bouquet of $(n-1)$-spheres. □

APPLICATION: Consider now the first five geometries in Table 1. Except for the rank 2 geometry for M_{11} for which $G = \tilde{G}$ trivially, the others are simply-connected by Lemma 4. Moreover, by Lemmas 3 and 4 every residual geometry of rank ≥ 3 is simply-connected, and hence by Lemma 1, $G = \tilde{G}$ in each case.

5. SOME RANK 3 PERMUTATION GROUP GEOMETRIES (F_{22}, F_{23}, HS, Mc)

Let N denote the point-nerve and Γ the point-graph of G as in Definition 2 of Section 3. If $x \in \Gamma$, let Γ_x denote the set of vertices of Γ adjacent to x.

LEMMA 5: *Suppose Γ has diameter 2 and satisfies the following conditions:*

(a) Every set of three mutually adjacent points is a simplex of N.

(b) Given $y \notin \Gamma$, then $\Gamma_x \cap \Gamma_y$ is connected.

(c) Given $y, z \notin \Gamma_x$, $y \in \Gamma_z$, then $\Gamma_x \cap \Gamma_y \cap \Gamma_z \neq \emptyset$.

Then N is simply-connected.

PROOF: By condition (a) every circuit of length 3 is null-homotopic in N. By condition (b) every circuit of length 4 may be split into triangles, and is therefore null-homotopic by the preceding sentence. By condition (c) every circuit of length 5 may be split into circuits of length 3 and 4, and is therefore null-homotopic also. Since Γ has diameter 2, it follows that every circuit may be split into circuits of length at most 5, so every circuit is null-homotopic, and the result follows. □

APPLICATION: Lemma 5 may be applied, using Lemma 2, to the geometries G given in Table 1 for F_{22}, F_{23}, Mc and HS, showing them all to be simply-connected. The preceding results on M_{22} and M_{23}, and Lemma 3, show that all residues of rank ≥ 3 are simply-connected, and so by Lemma 1, $G = \tilde{G}$. For the F_{24} geometry condition (b) of Lemma 5 is not satisfied and we will see in Section 7 that it is not simply-connected but admits a 3-fold universal cover, corresponding to the Schur multiplier of F'_{24}.

6. COVERINGS OF 2-GRAPHS

This discussion of 2-graphs is relevant for the $\cdot 3$ and $\cdot 2$ geometries in Table 1, the former because its 276 points form a 2-graph and the latter because it contains the $O_6^-(2)$ 2-graph on 28 vertices. Background material on 2-graphs is contained in the excellent survey article [9] by Seidel. We briefly recall that a *2-graph* is a mod 2 3-cocycle on a standard n-simplex (i.e. a set T of 3-subsets such that every 4-subset contains 0, 2 or 4 of them). A *regular 2-graph* T is a 2-graph for which every 2-subset lies in a constant number of 3-subsets of T. We call T *non-trivial* if it is neither the complete set nor the null set of 3-subsets.

Given a non-trivial 2-graph T, let ΔT denote the 2-dimensional simplicial complex obtained by taking the elements of T together with their faces; clearly the 1-skeleton of ΔT is a complete graph. Let us define the *fundamental group* $\pi(T)$ of T to be $\pi_1(\Delta T)$.

PROPOSITION 6 : *If* T *is a non-trivial regular* 2*-graph, then* $\pi(T) = Z_2$.

PROOF : As in [9] we can regard T as a non-trivial 2-fold cover K of the complete graph K_n on n vertices. Now form a simplicial complex ΔK by filling in all triangles of K. Then one sees that ΔK is a 2-fold cover of ΔT, and we will show that ΔK is simply-connected. Any circuit γ of ΔK when projected to ΔT can be split into triangles, and hence γ may be split (in ΔK) into triangles and hexagons. The triangles are null-homotopic because by definition they are boundaries of 2-simplexes, so consider a hexagon $o'p'q'o''p''q''$ which projects to the triangle opq of ΔT.

In the switching class of graphs corresponding to T, see [9], take the one with o as an isolated vertex and call it Γ; since opq is not in T, p is not adjacent to q in Γ. As T is regular it follows that Γ is a strong graph (see [9] section 7), and as T is non-trivial one easily checks that p and q may be placed on a minimal circuit $psqtr$. Clearly the triples ptq, rsq, tps, qpr, srt all lie in T since they each contain only one edge of Γ. Now let r' (resp. s', resp. t') denote the unique vertex of ΔK above r (resp. s, resp. t) and adjacent to p' (resp. o'', resp. q''). Then one has triangles $o'p'r'$, $p'r'q'$, $r'q's'$, $q's'o''$, $s'o''p''$, $p''s't'$, $s't'r'$, $t'r'o'$, $o't'q''$, $t'q''p''$ showing, as in Figure 1, that $o'p'q'o''p''q''$ is null-homotopic in ΔK. This completes the proof. □

REMARK: Without the assumption of regularity in Proposition 6 there are counterexamples; take, for instance, the non-trivial 2-graph on four vertices.

APPLICATION: If G is the $\cdot 3$ geometry in Table 1, then the 2-skeleton of $N_P(G)$ (P denotes the points) is ΔT where T is the 2-graph on 276 vertices. Using the result of the previous section on Mc, together with Lemmas 1 and 3, we see that $\tilde{G}$ is a topological cover of G, and using Lemma 2, $\pi^2(G) = \pi_1(G) = \pi_1(\Delta T) = Z_2$. Moreover, the 2-skeleton of $\tilde{G}$ is the ΔK of Proposition 6, and we easily see that Aut $\tilde{G}$ preserves the fibers above each point of G (i.e. of ΔT), and so Aut $\tilde{G}$ normalizes the group $\pi^2(G) = Z_2$ which acts on these fibers. Since all automorphisms of G lift to $\tilde{G}$ we have Aut $\tilde{G}/Z_2 \cong \cdot 3$, see [6], and since the Schur multiplier of $\cdot 3$ is trivial, we have Aut $\tilde{G} = \cdot 3 \times Z_2$.

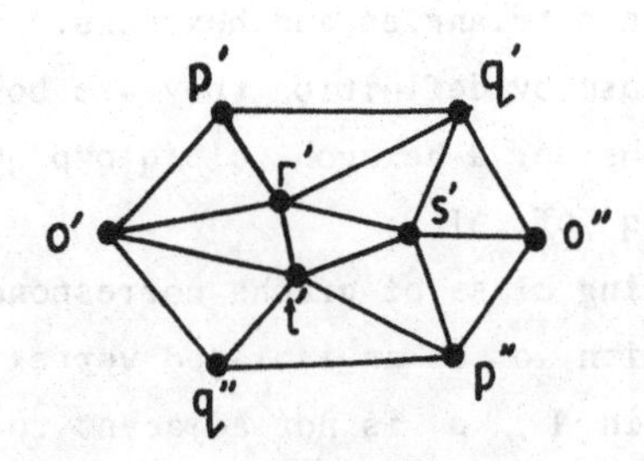

FIGURE 1

7. SOME SPECIAL CASES

7.1 Ramification and the ·2 geometry

For the ·2 geometry G in Table 1 we construct a 2-fold universal 2-cover $\tilde{G}$ which is ramified at all vertices corresponding to nodes at the right-hand end of the diagram, which we call *blocks*. We briefly recall the construction of G given by Buekenhout in [1]. In the Leech lattice Λ fix an interval of type 2 (namely 0 and $P = 4e_0 + 4e_\infty$) and let A denote the set of 4,600 points of Λ forming a 222 triangle with 0 and P

Given $x, y \in A$, set $(x,y) = x \cdot y / 16$ where $x \cdot y$ is the usual Euclidean dot product, and one checks that $(x,y) = 4$ if and only if $(P - x, y) = 2$. As points of G, Buekenhout takes *diameters* $\bar{x} = \{x, P - x\}$ of A; lines are pairs $\bar{x}\bar{y}$ such that $(x,y) = 2$ or 4; planes are cliques of four points, and blocks are cliques of 28 points forming a 2-graph.

It is easy to check that the point graph of G satisfies the conditions of Lemma 5, and so G is simply-connected. However, by Proposition 6, we see that the residue of a block is not simply-connected but has a 2-fold universal cover. In fact this appears as the residue of a block in $\tilde{G}$ which we proceed to describe. Points of $\tilde{G}$ are points of A; lines are pairs xy such that $(x,y) = 4$; planes are cliques of four points, and blocks are the sets of 56 points used to define the blocks of G in [1]; the covering map $p : \tilde{G} \to G$ sends x to $\bar{x}$. Clearly the residue of a block B of $\tilde{G}$ is a non-trivial 2-fold cover of the block pB of G, and by Proposition 6 is therefore simply-connected; the residue of a point is the $U_6(2)$ building, which is also simply-connected, and the other two rank 3 residues are simply-connected by Lemma 3. Since G is simply-connected and the map $p : \tilde{G} \to G$ induces a monomorphism of fundamental groups π_1 (see [6]), it follows that $\tilde{G}$ is simply-connected, and one now applies Lemma 1 to see that $\tilde{G}$ is indeed the universal 2-cover of G. As in the case of ·3 above, the Schur multiplier of ·2 is trivial, and one sees that $\mathrm{Aut}\,\tilde{G} = \cdot 2 \times Z_2$; this is the subgroup of $\cdot\infty$ (the group of the Leech lattice) fixing the set $\{0, P\}$. It should be observed that $\tilde{G}$ does not satisfy the intersection condition in [1] because two blocks can intersect in a set of eight points comprising two disjoint planes (this set of points is called a *cube* in [1], using the adjacency relation $(x,y) = 2$), or in a set of two non-collinear points x and $P - x$.

7.2 The Suzuki group geometry on $3\,U_4(3)$

The geometry G with diagram o——o══o for Suz in Table 1 has as its points the cosets of $3\,U_4(3)$ in Suz. These points may be thought of as a set of Z_3 subgroups of Suz as in the work of Stellmacher [11], and the blocks are the maximal cliques. The existence of G as a diagram geometry was recently observed by Buekenhout (private communication) following a suggestion of B. Fischer. We will show that G is simply-connected, hence $G = \tilde{G}$, and to do this we examine the point-graph Γ of G, and use N to denote the point-nerve of G.

The point-graph Γ of G is distance regular of diameter 4, and fixing some point p we let a, b, c, d denote points at distances 1, 2, 3, 4 from p in suborbits A, B, C, D respectively. Denote the points of X collinear with y by $X(y)$, and the graph has the following structure:

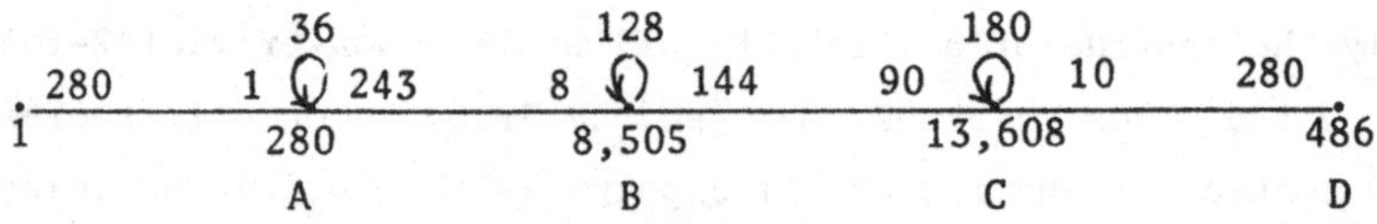

36 128 180

280 1 243 8 144 90 10 280

1 280 8,505 13,608 486

A B C D

$A(b) = K_{4,4}$ (a subquadrangle of the $U_4(3)$ quadrangle, with two points per line), $D(c)$ is a set of 10 mutually opposite points of the $U_4(3)$ quadrangle, with automorphism group $PSL_2(9)$.
It is therefore easily seen that the subgraph $B(c)$ on 90 points is connected, as are obviously $A(b)$ and $C(d)$; thus it is clear that any minimal 4, 6 or 8-gon may be split into 3-gons and

5 or 7-gons. Moreover, if γ is any minimal 5 (resp. 7)-gon containing p, then the two points of γ distant 2 (resp. 3) from p are collinear with a common point distance 1 (resp. 2) from p (to see this note that all edges of B (resp. C) are equivalent under the group fixing p, and every point of A (resp. B) is adjacent to an edge of B (resp. C)), and therefore any 5 or 7-gon can be split into smaller m-gons. Now since Γ has diameter 4 and there are no minimal 9-gons we see that any circuit can be split into 3-gons, and since by definition of G any 3-gon lies in a block and is therefore null-homotopic in N, we see using Lemma 2 that G is simply-connected.

7.3 The F_{24} Geometry

If G is the F_{24} geometry in Table 1, then we will show that its universal 2-cover $\tilde{G}$ is a 3-fold cover of G with automorphism group the non-split extension $3F_{24}$ of F_{24} by Z_3. We recall that the points of G are a class D of involutions (3-transpositions) in F_{24}, see [3], two being collinear if they commute, and the subspaces of G are sets of mutually collinear points (cliques), maximal ones having 24 points, see [1]. The group $3F_{24}$ clearly contains a class $\tilde{D}$ of involutions such that the natural map from $3F_{24}$ to F_{24} restricts to a map $p : \tilde{D} \to D$ where for $d \in D$, $p^{-1}(d)$ is the set of three involutions in the group $Z_3 \langle d \rangle \cong S_3$. If C is a subspace of G, then C is a clique of points in D, and $\langle C \rangle$ is an elementary abelian 2-group, so the extension $Z_3 \langle C \rangle$ splits, and one sees that $p^{-1}(C)$ is a set of three mutually disjoint cliques each isomorphic to C. We let these cliques be our subspaces of $\tilde{G}$, where the points of $\tilde{G}$ are the elements of $\tilde{D}$. Clearly we have an induced map $p : \tilde{G} \to G$ which is a topological covering, and since $3F_{24}$ is a non-split extension, the covering is non-trivial. We now show that $\tilde{G}$ is simply-connected.

Consider the point-graph D of G, for which conditions (a) and (c) of Lemma 5 are satisfied. Given two non-collinear points $d, e \in D$, then $D_d \cap D_e$ has three connected components each with 1,080 points (here D_d means the points collinear with d), and there are two types of minimal circuits dxey of length four; those for which x and y lie in the same component of $D_d \cap D_e$, which are null-homotopic, and those for which x and y lie in different components, which are evidently not null-homotopic, otherwise G would be simply-connected, by Lemma 5, contradicting the existence of $\tilde{G}$. As pointed out to me by B. Fischer, the latter type of circuits, here called *essential*, are permuted transitively by F_{24}, and each may be embedded in an S_7 subgroup. This enables one to see that such circuits when lifted to circuits of length 12 in $\tilde{G}$ become null-homotopic; indeed let $\gamma = ((12), (34), (15), (36))$ be an essential circuit in S_7. Then Figure 2 shows that 3γ is null-homotopic, and hence the 3-fold lift of γ to $\tilde{G}$ is null-homotopic. Since any circuit of D may be split into triangles and essential circuits, it follows that any circuit of $\tilde{D}$ may be split into triangles and 12-circuits both of which are null-homotopic. Therefore $\tilde{G}$ is simply-connected, and by the preceding results on F_{23} and M_{24} and by Lemma 3 it follows that all residues of rank ≥ 3 are simply-connected, so by Lemma 1 $\tilde{G}$ is the universal 2-cover as required.

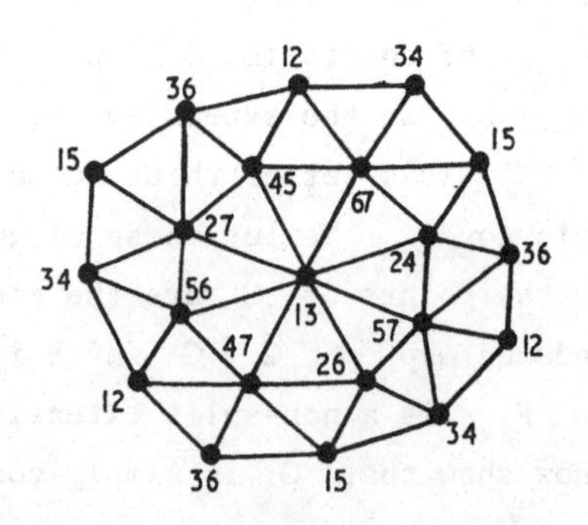

FIGURE 2

7.4 The 2-local Geometry for M_{24}

The geometry G for M_{24} which is last in the list in Table 1 is called *2-local* because the stabilizers of the vertices are precisely the maximal 2-local subgroups of M_{24}, see [8]. The set of points and lines forms a near hexagon, described in [10], and so if Γ is the point-graph, then any circuit of length 3 in Γ lies in a line, any circuit of length 4 lies in a quad, there are no minimal circuits of length 5, and any circuit may be split into circuits of length at most 6. Now if $\gamma = (a, b, c, d, e, f)$ is a minimal circuit of length 6, then one checks (using the Steiner system for which the points are octads) that there exist collinear points x and y such that x is collinear with b and f, and y with c and e (possibly $x = a$ or $y = d$). Thus one sees that γ splits into circuits of length 4 and is thus null-homotopic. This shows that G is simply-connected and so $G = \tilde{G}$.

7.5 Non-spherical examples

For those rank 3 geometries in Table 1 for which $\pi^2(G)$ is infinite one applies the results of [7] to see that $\tilde{G}$ must be contractible and infinite, owing to the fact that its minimal circuit diagram (see [7]) is non-spherical. In the case of Ly and Suz ∘═══∘═══∘ this is also known from the fact that $\tilde{G}$ is a building (see [13]).

8. SOME CONCLUDING REMARKS

It is perhaps worth mentioning here that if G is a simply-connected rank 3 geometry, then it is of course homotopy equivalent to a bouquet of 2-spheres, the number of such spheres being $\chi - 1$ where χ is the Euler characteristic of G. Thus one knows the homotopy type of $\tilde{G}$ if G has rank 3. It would be interesting to know something of the homotopy type of $\tilde{G}$, or at least its homology groups, if G has rank at least 4 in the examples considered here. For the F_{24} geometry one can however say that $\tilde{G}$ is certainly not a bouquet of 5-spheres since $\chi(\tilde{G}) > 1$, and consequently one of the even dimensional homology groups of $\tilde{G}$ must be non-trivial.

It is also worth noticing that the top homology group of G, being free abelian, provides a Z-module for Aut G (see Curtis [2] for more information). For example, this representation for the 2-local M_{24} geometry has degree $2^{10} \cdot 3 \cdot 7$, splits into three irreducible constituents, and vanishes on elements of even order.

ACKNOWLEDGEMENT

This research was partially supported by NSF grant 800 2876.

BIBLIOGRAPHY

1. F. Buekenhout, "Diagrams for geometries and groups", *J. Combin. Theory Ser. A,* 27 (1979), 121-151.
2. C.W. Curtis, "Homology representations of finite groups", to appear.
3. B. Fischer, *Finite Groups Generated by 3-transpositions,* Lecture Notes (University of Warwick, 1979).
4. J. Folkman, "The homology groups of a lattice", *J. Math. Mech.* 15 (1966), 631-636.
5. D. Quillen, "Higher algebraic K-theory : I", *Algebraic K-theory I,* Lecture Notes in Math. 341 (Springer-Verlag, Berlin, 1973), pp. 83-147.
6. M.A. Ronan, "Coverings and automorphisms of chamber systems", *Europ. J. Combinatorics,* 1 (1980), 259-269.
7. M.A. Ronan, "On the second homotopy group of certain simplicial complexes and some combinatorial applications", *Quart. J. Math.* (Oxford), to appear.
8. M.A. Ronan and S.D. Smith, "2-local geometries for some sporadic groups", *Proc. A.M.S. Summer Inst. on Finite Groups,* Santa Cruz 1979 (Amer. Math. Soc.), to appear.
9. J.J. Seidel, "A survey of two-graphs", *Teorie Combinatorie,* Tomo I, Atti dei Convegni Lincei 17 (Acc. Naz. Lincei, Rome, 1976), pp. 481-511.
10. E. Shult and A. Yanushka, "Near n-gons and line systems", *Geom. Dedicata,* 9 (1980), 1-72.
11. B. Stellmacher, "Einfache Gruppen, die von einer Konjugiertenklasse von Elementen der Ordnung drei erzeugt werden", *J. Algebra,* 30 (1974), 320-354.

12. J. Tits, *Buildings of Spherical Type and Finite BN-pairs*, Lecture Notes in Math. 386 (Springer-Verlag, Berlin, 1974).
13. J. Tits, "Local characterizations of buildings", to appear.

Department of Mathematics
University of Illinois at Chicago Circle
Chicago
Illinois 60680
U.S.A.

ON CLASS-REGULAR PROJECTIVE HJELMSLEV PLANES

S.S. Sane

Regular projective Hjelslev planes (PH-planes), i.e. PH-planes admitting an automorphism group which is regular on the points and blocks (i.e. a Singer group) were introduced and studied by Jungnickel. In this paper, we study (t,r)-PH-planes admitting an automorphism group which is regular on each point and block class (neighbourhood). We prove that this notion is equivalent to the existence of a projective plane of order r and a (t,r)-Hjelmslev matrix in an abelian group of order t^2 as defined by Jungnickel. This enables us to construct many class-regular PH-planes which do not admit a Singer group. Further, all the results of Jungnickel (on regular PH-planes) also hold for class-regular PH-planes.

1. **REMARKS**: Projective Hjelmslev planes (H-planes) (more generally, projective Klingenberg planes or K-planes) are generalizations of projective planes in which two points (resp. two lines) are allowed to have more than one line passing through them (resp. more than one point of intersection) and which admit an epimorphism onto a projective plane. For a nice introduction to H-planes (and K-planes) and their automorphism groups, we refer to [3], [4] and [5]; [1] gives an extensive bibliography of the literature on H-planes and related structures.

2. **DEFINITION**: An automorphism group N of a K-plane (H-plane) π is said to be class-regular (on π) if and only if N is abelian and is regular on the elements of each point-class and each block-class (neighbourhoods) of π. In this case π will

be said to be class-regular.

Note that in the above definition, if π is a (t,r)-K-plane (H-plane), then $|N| = t^2$.

3. **DEFINITION**: ([3], definition 2.1). An automorphism group G of a K-plane (H-plane) π is called a Singer group if and only if $G = Z \oplus N$ where G is regular on the points and blocks of π and N is class-regular (on π). In this case, π will be called a Singer K-plane (H-plane).

4. **LEMMA**: *Let π be a K-plane and let π' be the underlying projective plane. Then every automorphism group G of π induces an automorphism group G' of π'.*

In particular, if G is class-regular (resp. Singer) then G is identity (resp. Singer).

5. **DEFINITION**: ([5], definition 1.2). Let N be an abelian group of order t^2 and let $A = (a_{ik})$, $i = 0,1,\ldots,r$, $k = 1,2,\ldots,t$ be a matrix with entries from N. A is called a (t,r)-K-matrix if and only if for each pair (i,j) with $i \neq j$ and $i,j = 0,1,\ldots,r$, $\{a_{ik} - a_{jm} : k, m = 1,2,\ldots,t\} = N$. A is called an H-matrix if and only if it furthermore satisfies: the $(r+1)t(r-1)$ differences $a_{ik} - a_{im}$, $i = 0,1,\ldots,r$, $k, m = 1,2,\ldots,t$ contain each $x \in N$ with $x \neq 0$ at least twice.

6. **THEOREM**: *Given a (t,r)-K-matrix (resp. H-matrix) and a projective plane π' of order r, there exists a class-regular (t,r)-K-plane (H-plane) π with π' as the underlying projective plane. Conversely, given a (t,r)-K-plane (H-plane) which is class-regular, there exists a (t,r)-K-matrix (H-matrix) in some group N.*

PROOF : Let the points of π' be denoted by lower case letters x, y, z etc. and lines by capital letters X, Y, Z etc.; $x\ I\ X$ means x is incident with X. Let F be the set of all the flags of π' i.e.

$$F = \{(x,X) : x\ I\ X,\ x, X \in \pi'\} .$$

For each point x, there are precisely $r+1$ X's such that (x,X) is a flag and dually. Hence, using König's lemma (see [6]), there exists a mapping $\phi : F \to \{0,1,\ldots,r\}$ such that ϕ satisfies the following: For two distinct flags (x,X) and (y,Y), $\phi(x,X) = \phi(y,Y)$ implies $X \neq Y$ and $x \neq y$. Let $A = (a_{ik})$, $i = 0,1,\ldots,r$, $k = 1,\ldots,t$ be the given (t,r)-matrix (H-matrix). Now define a new structure π whose points are ordered pairs (x,g) and whose lines are (X,h) where $x, X \in \pi'$, $g, h \in N$; (x,g) will be incident with (X,h) if and only if $h - g = a_{ik}$ for some k where $i = \phi(x,X)$.

We want to show that π is a K-plane (H-plane) with $\{(x,g) : g \in N\}$, as the point-classes and $\{(X,h) : h \in N\}$, as the block-classes. Let (x,g) and (y,g') be two points of π with $x \neq y$. There is a unique X in π passing through both x and y. Let $i = \phi(x,X)$ and $k = \phi(y,X)$. Then a block (X,h) of π passing through (x,g) and (y,g') must satisfy $h - g = a_{ij}$ for some $j = 1,\ldots,t$ and $h - g' = a_{k\ell}$ for some $\ell = 1,\ldots,t$. Hence, $g' - g = a_{k\ell} - a_{ij}$. We note that this uniquely determines k and ℓ, which follows from the definition of a K-matrix and hence h is uniquely determined. It is now easy to see that π is a K-plane with N as a class-regular automorphism group. Further, if A is an H-matrix, then for two distinct points (x,g) and (x,g') the difference $g' - g \neq 0$ arises from precisely one row (see [5], lemma 2.5), say i, of A. But then, by the definition of an H-matrix, $g' - g$ occurs twice as a difference from the i-th row. There is a unique X with $\phi(x,X) = i$. Hence, (x,g) and (x,g') occur on two lines (X,h) and (X,h') for some $h, h' \in N$. This proves that π is an H-plane.

For the converse, we start with any point-class (p) of π and label some point p of (p) as 0 where 0 is the identity of N. Now label other points of (p) from the elements of N as follows: if $q = pg$, then q is labelled g. Since N is regular on (p), this is clearly possible. Choose a set $\{L_0, L_1, \dots, L_r\}$ of $r+1$ pairwise non-neighbour lines passing through 0. Since N is regular on each block-class, the blocks in the neighbourhood of L_i can be denoted by $L_i + g$, $g \in N$, $i = 0,1,\dots,r$. Let $L_i = \{a_{ij}\}$, $j = 1,2,\dots,t$, $i = 0,1,\dots,r$ where a_{ij}'s are the points of (p) on L_i. Now set $A = (a_{ij})$, $i = 0,1,\dots,r$ and $j = 1,2,\dots,t$. It is easy to verify that A is a K-matrix (H-matrix).

7. COROLLARY : (Jungnickel, [5], theorem 1.3). *A Singer (t,r)-K-plane exists if and only if there exists a Singer projective plane of order r and a (t,r)-K-matrix in some group N.*

PROOF : Follows from theorem 6 and lemma 4.

8. COROLLARY : *Let A be a (t,r)-K-matrix (H-matrix). Then, for every choice π' of a projective plane of order r, there exists a class-regular (t,r)-K-plane (H-plane) π with π' as the underlying projective plane. Consequently, if π_1' and π_2' are non-isomorphic projective planes and if there exists a (t,r)-H-matrix, then there exist class-regular (t,r)-H-planes π_1 and π_2 with π_1' and π_2' as the underlying projective planes. Hence, π_1 and π_2 are non-isomorphic.*

9. REMARK : For non-isomorphic projective planes (which can be used in corollary 8) we refer to [2], Chapter IX. As an example, let $r = 9$. There are at least 4 distinct projective planes of order 9. Hence, for each t such that $(t,9)$ is a Lenz-pair (see [5], definition 1.12), there exist at least 4 non-isomorphic class-regular $(t,9)$-H-planes only one of which admits a Singer group.

10. REMARK : We observe that the results in [3], [4] and [5] are, in fact dependant only on the properties of (t,r)-K-matrices (H-matrices)(and not on the underlying projective plane). Hence, these results also hold with the term 'regular' appropriately replaced by 'class-regular'.

ACKNOWLEDGEMENTS : The author acknowledges the support of the British Council (via Commonwealth Scholarships Scheme). The author is also thankful to the Department of Mathematics, Westfield College, London for its hospitality during the time of this research.

The author is indebted to Professor Lenz for an important suggestion.

BIBLIOGRAPHY

1. B. Artmann, G. Dorn, D. Drake and G. Törner, "Hjelmslev'sche Inzidenzgeometrie und verwandte Gebiete-Literaturverzeichnis", *J. Geom.*, 7 (1976), 175-194.
2. D.R. Hughes and F.C. Piper, *Projective Planes* (Springer-Verlag, New York, 1972).
3. D. Jungnickel, "Regular Hjelmslev planes I", *J. Combin. Theory Ser. A*, 26 (1979), 20-37.
4. D. Jungnickel, "Regular Hjelmslev planes II", *Trans. Amer. Math. Soc.*, 241 (1978), 321-330.
5. D. Jungnickel, "On balanced regular Hjelmslev planes", *Geom. Dedicata*, 8 (1979), 445-462.
6. H.J. Ryser, *Combinatorial Mathematics* (John Wiley & Sons, New York, 1963).

Department of Mathematics
University of Florida
Gainsville
Florida 32611
U.S.A.

ON MULTIPLICITY-FREE PERMUTATION REPRESENTATIONS

Jan Saxl

1. INTRODUCTION

A permutation group H on the set of cosets of a subgroup G is said to be *multiplicity-free* if all the irreducible constituents of the permutation character 1_G^H are distinct; in other words, $\langle 1_G^H, \chi \rangle_H$ is 0 or 1 for all $\chi \in \mathrm{Irr}\, H$, where $\mathrm{Irr}\, H$ is the set of all irreducible characters of H. More generally, if $\Lambda \subseteq \mathrm{Irr}\, H$, we say that H is Λ-*multiplicity-free* on the set of cosets of G if $\langle 1_G^H, \chi \rangle_H$ is 0 or 1 for all χ in Λ. The aim of this note is to start an investigation of such permutation representations of some known groups. In particular, we shall show how to find all multiplicity-free representations of the symmetric groups and we shall give some interesting examples of Λ-multiplicity-free representations of the linear groups, where Λ is the set of unipotent characters (that is those irreducibles occurring in the permutation characters on the cosets of various parabolic subgroups - see below).

In recent years much work has been done on the classification of permutation representations of low rank of the known groups (see e.g. [1], [6],[5], [10]). Since the low rank representations are multiplicity-free (for instance if the rank is at most 5), our problem may be viewed as an attempt to generalize this. We should also recall that a permutation representation is multiplicity-free if and only if the corresponding centralizer ring is abelian (see [16, §29]); the abelian centralizer rings of parabolic representations of groups with a BN-pair were investigated in [4]. Finally, it is well known that if H is the automorphism group of a distance transitive graph then the permutation action of H on the vertices is multiplicity free.

2. IN THE SYMMETRIC GROUPS

Let H be the symmetric group S_n on a set Ω of n points. First we recall some facts about the irreducible characters of S_n. A partition of n is a string $\lambda = (\lambda_1, \ldots, \lambda_a)$ of positive integers with $\lambda_1 \geq \lambda_2 \geq \ldots \geq \lambda_a$ and $n = \lambda_1 + \lambda_2 + \ldots + \lambda_a$; we write $\lambda \vdash n$. The irreducible characters of S_n are in one-to-one correspondence with the partitions of n. Given $\lambda \vdash n$, we write π^λ for the permutation character of S_n on the set Ω^λ of cosets of the subgroup $S_{\lambda_1} \times S_{\lambda_2} \times \ldots \times S_{\lambda_a}$. There is a natural partial order on the set of partitions of n: we write $\lambda \trianglelefteq \mu$ if $\lambda_1 + \ldots + \lambda_i \leq \mu_1 + \ldots + \mu_i$ for all i. This has the nice property that there is a unique irreducible character χ^λ in π^λ which occurs in no π^μ with $\lambda \not\trianglelefteq \mu$ (see e.g. section 1 of [13] for more details of this approach). One other fact we shall need is the Branching rule of Schur (see [8, 9.2]):
If $\lambda \vdash n$ then

$$(\chi^\lambda)^{S_{n+1}} = \sum \chi^\nu ,$$

where the summation is over those partitions ν of $n+1$ which differ from λ in only one place, and similarly for $(\chi^\lambda)_{S_{n-1}}$.

EXAMPLE 2.1: (see [11], but known before). Let $k \leq \frac{1}{2}n$, let $\lambda = (n-k, k)$. Then $\pi^{(n-k,k)}$ is the permutation character of S_n on the set $\binom{\Omega}{k}$ of all k-subsets of Ω. It is well-known that

$$\pi^{(n-k,k)} = \sum_{i \leq k} \chi^{(n-i,i)} = \pi^{(n-k+1,k-1)} + \chi^{(n-k,k)} ;$$

this is easily seen inductively, using the fact that if $i \leq k$ then

$$\langle \pi^{(n-k,k)}, \pi^{(n-i,i)} \rangle_{S_n} = \langle 1, \pi^{(n-k,k)} \pi^{(n-i,i)} \rangle_{S_n}$$

$$= \# \operatorname{orb}(S_n, \binom{\Omega}{k} \times \binom{\Omega}{i}) = i+1,$$

the last equality being clear on listing the orbits of S_n on $\binom{\Omega}{k} \times \binom{\Omega}{i}$.

In particular, each $\pi^{(n-k,k)}$ is multiplicity free.

EXAMPLE 2.2 : (See [15], but probably known before). Let $n = 2k$ and set $\pi_{2,k} = (1_{S_2 \text{ wr } S_k})^{S_{2k}}$. Then $\pi_{2,k} = \sum_{\lambda \vdash k} \chi^{2\lambda}$, where, if $\lambda = (\lambda_1, \lambda_2, \ldots, \lambda_a) \vdash k$ then 2λ is the partition $(2\lambda_1, 2\lambda_2, \ldots, 2\lambda_a)$ of $2k$.

PROOF : (G.D. James and Jan Saxl) This is by induction on k. Observe first that

$$(\pi_{2,k})_{S_{n-1}} = ((1_{S_2 \text{ wr } S_k})^{S_n})_{S_{n-1}} = (1_{(S_2 \text{ wr } S_k) \cap S_{2k-1}})^{S_{2k-1}}$$

$$= (1_{S_2 \text{ wr } S_{k-1}})^{S_{2k-1}} = ((1_{S_2 \text{ wr } S_{k-1}})^{S_{2k-2}})^{S_{2k-1}}$$

$$= (\pi_{2,k-1})^{S_{n-1}},$$

the second equality being a special case of a Theorem of Mackey (see [13, Lemma 3.3]). Now $\pi_{2,k-1}$ is (by the induction hypothesis) the sum of characters of S_{n-2} corresponding to even partitions of $n-2$, each with multiplicity 1. Hence $(\pi_{2,k-1})^{S_{n-1}}$ is by the Schur branching rule just the sum of all the irreducible characters of S_{n-1} corresponding to partitions of $n-1$ with precisely one odd part. By the above observation

$$(\pi_{2,k-1})^{S_{n-1}} = (\pi_{2,k})_{S_{n-1}},$$

so $\pi_{2,k}$ is multiplicity free and, if χ^{λ} is in $\pi_{2,k}$ then, since the constituents of $(\chi^{\lambda})_{S_{n-1}}$ correspond to partitions with exactly one odd part, λ is either a two-part partition or has all parts even (the branching rule again). Next we check that even if λ is a two-part partition then the parts are even, observing that the principal character $\chi^{(2k)}$ is of course a constituent of $\pi_{2,k}$. Finally, if $\lambda = (\lambda_1, \lambda_2, \ldots, \lambda_a) \vdash k$ then $\chi^{2\lambda}$ is a constituent of $\pi_{2,k}$ since $\chi^{(2\lambda_1, 2\lambda_2, \ldots, 2\lambda_a - 1)}$ is a constituent of $(\pi_{2,k})_{S_{n-1}}$. This completes the proof.

EXAMPLE 2.3 : (see [3], but certainly known before). Let $n = 2k$ and set $\pi_{k,2} = (1_{S_k \text{ wr } S_2})^{S_n}$. Then

$$\pi_{k,2} = \sum_{i \le k/2} \chi^{(n-2i,2i)} .$$

PROOF of this is easy: first observe that $S_k \times S_k \le S_k \text{ wr } S_2$, so that $\pi_{k,2} \subseteq \pi^{(k,k)}$, and then calculate $\langle \pi_{k,2}, \pi^{(n-i,i)} \rangle_{S_n}$ for $i \le k$.

THEOREM : *Let* S_n *be multiplicity-free on the set of cosets of a subgroup* G. *Assume that* $n > 18$. *Then one of:*

(i) $A_{n-k} \times A_k \le G \le S_{n-k} \times S_k$ *for some* k *with* $0 \le k < n/2$;

(ii) $n = 2k$ *and* $A_k \times A_k < G \le S_k \text{ wr } S_2$;

(iii) $n = 2k$ *and* $G \le S_2 \text{ wr } S_k$ *of index at most* 4 ;

(iv) $n = 2k+1$ *and* G *fixes a point of* Ω *and is one of the groups in (ii) or (iii) on the rest of* Ω;

or (v) $A_{n-k} \times G_k \le G \le S_{n-k} \times G_k$ *where* k *is* 5, 6 *or* 9 *and* G_k *is Frobenius of order* 20, PGL(2,5) *or* PΓL(2,8) *respectively.*

REMARKS : (a) All groups in (i) and (ii) satisfy the assumptions but not all groups in the other cases do.

(b) Although the condition $n > 18$ is not strictly necessary, there are many "sporadic" examples for smaller n, and other possible candidates can only be ruled out by *ad hoc* arguments. There are no new examples for $n = 18$, but a computation is required to rule out the possibility of $G = P\Gamma L(2,8) \text{ wr } S_2$. For $n = 17$, there is just one new example, $G = M_{12} \times S_5$, but there are many more as n gets smaller. Some of these appear in [1].

PROOF OF THE THEOREM : The heart of the proof is in the following two observations:

(2a) *We have a considerable amount of information about* G *on* Ω : If π is the character of S_n on the set of cosets of G then $\langle \pi, \chi^\lambda \rangle_{S_n} \le 1$ for each $\lambda \vdash n$, so that $\langle 1_G, \chi^\lambda_G \rangle_G \le 1$ by the Frobenius reciprocity. Hence

$$\# \operatorname{orb}(G, \Omega^\lambda) = \langle 1_G, \pi^\lambda_G \rangle_G \le \# \text{irreducible consistuents in } \pi^\lambda .$$

In particular, G has at most 2 orbits on Ω and at most $k+1$ orbits on $\binom{\Omega}{k}$ (see 2.1).

REMARK : The last sentence of (2a) can be used to verify reasonably easily that if G is primitive on Ω then it is 2-transitive there.

(2b) *The order of* G *is large :*

$$| S_n : G | = \pi(1) \le \sum_{\lambda \vdash n} \chi^\lambda(1) = \# \{ g \in S_n : g^2 = 1 \} ,$$

the last equality by a theorem of Frobenius and Schur (see [7, 4.6]). It follows that

$$| G | \ge \frac{2}{n+2} \cdot \min | C_{S_n}(t) | ,$$

where the minimum is taken over all involutions t in S_n. Hence

$$|G| \geq \frac{2}{n+2} \cdot \min 2^a \cdot a!(n-2a)! ,$$

where a is an integer with $0 \leq a \leq n/2$.

We shall now complete the proof of the Theorem. Assume first that G is primitive on Ω. If $A_n \leq G$ then we are in case (i), otherwise $|G| < 4^n$ by [12]. Comparing with (2b) forces $n \leq 60$, and the small cases are easily handled (especially in view of the Remark after (2a)).

Assume next that G is transitive but not primitive on Ω. Considering the action of G on $\binom{\Omega}{j}$ for various small j and using (2a) we see quickly that (since $n > 15$) the blocks of imprimitivity have size 2 or $n/2$; hence n is even, say $n = 2k$. In the latter case, using (2a) with care, we see that $A_k \times A_k \leq G$ (and in fact $A_k \times A_k \neq G$ by the Littlewood-Richardson rule). In the case of k blocks of size 2, write $\mathcal{B}$ for the set of blocks and let B be the kernel of the action of G on $\mathcal{B}$. We see as before that $G^{\mathcal{B}} \geq A_k$. From (2b) we get $|B| > 2$. Then B can be viewed as a non-trivial submodule of the natural permutation module of A_k over $GF(2)$; hence $|B| \geq 2^{k-1}$ (or we can also see if using an elementary argument, considering elements of B fixing as much as possible). Thus G has index at most 4 in S_2 wr S_k.

Assume finally that G is not transitive on Ω. Then, by (2a), G has two orbits Γ and Δ. Let Γ be the smaller orbit, $|\Gamma| = k \leq n/2$. Using (2a) once more we see that G^Δ and G^Γ are t-homogeneous for each $t \leq k$. Hence G^Γ is known (cf. [11]). Since $G \leq G^\Gamma \times G^\Delta \leq S_k \times S_{n-k}$, we see that S_{n-k} is multiplicity-free on G^Δ. We can now proceed inductively and obtain conclusions (i), (iv) and (v) (some of the possibilities are eliminated by the Littlewood-Richardson rule - see [8, 16.4]).

This completes the proof of the Theorem.

REMARK: Similar techniques can be used to prove similar results about subgroups G of S_n for which $\langle 1_G^{S_n}, \chi^\lambda \rangle_{S_n} \leq c$ for each $\lambda \vdash n$, for some small constant c.

3. IN THE LINEAR GROUPS

Let now $H = GL(n,q)$ and let $V = V(n,q)$ be a vector space on which H acts naturally. We first recall some facts concerning certain characters of $GL(n,q)$ (these are due to Steinberg [14], but see also [4] for a more general approach).

Let $\lambda \vdash n$, $\lambda = (\lambda_1, \lambda_2, \ldots, \lambda_a)$, and let
$V^\lambda = \{(V_1, V_2, \ldots, V_a) : V = V_1 \supset V_2 \supset \ldots \supset V_a,\ \dim V_i = \lambda_i + \lambda_{i+1} + \ldots + \lambda_a\}$.
(In Section 2, to make the parallel clearer, we could re-define Ω^λ by
$\Omega^\lambda = \{(\Omega_1, \Omega_2, \ldots, \Omega_a) : \Omega = \Omega_1 \supset \Omega_2 \supset \ldots \supset \Omega_a,\ |\Omega_i| = \lambda_i + \lambda_{i+1} + \ldots + \lambda_a\}$;
indeed, we shall use this definition of Ω^λ from now on.) Now $GL(n,q)$ acts on V^λ; let τ^λ be the corresponding permutation character. As before in S_n, for each $\lambda \vdash n$ there is a unique irreducible constituent ψ^λ in τ^λ, which does not occur in any τ^μ with $\lambda \ntrianglelefteq \mu$. In fact, as is proved in [14],

$$\langle \pi^\lambda, \chi^\mu \rangle_{S_n} = \langle \tau^\lambda, \psi^\mu \rangle_{GL(n,q)}$$

for any $\lambda, \mu \vdash n$. We shall give a very slight extension of this remarkable fact below in 3.2.

In the rest of the paper, we shall write Λ for the set of characters obtained above:

$$\Lambda = \{ \psi^\lambda : \lambda \vdash n \}.$$

Given any character τ, we shall write $\Lambda(\tau)$ for its Λ-part, that is, $\Lambda(\tau)$ is the sum of the ψ^λ each occurring with the same multiplicity as in τ. We shall be interested in the Λ-multiplicity-free characters of $GL(n,q)$. In the rest of the paper, we shall give three families of examples.

The above observations together with 2.1 give immediately

EXAMPLE 3.1 : $\tau^{(n-k,k)} = \sum_{i\le k} \psi^{(n-i,i)}$ for each $k \le n/2$.

In parallel to 2.2, we have 3.2 and 3.3 :

PROPOSITION 3.2 : Let $n = 2k$. Then

$$\Lambda\left(1_{Sp(2k,q)}^{GL(2k,q)}\right) = \sum_{\lambda\vdash k} \psi^{2\lambda} .$$

REMARKS : (a) Using the notation of groups of Lie type, the Proposition says that the irreducibles in $\Lambda\left(1_{C_k(q)}^{A_{2k-1}(q)}\right)$ correspond to those in $1_{C_k}^{A_{2k-1}}$, where C_k and A_{2k-1} are the corresponding Weyl groups.

(b) I do not know when the character $1_{Sp(2k,q)}^{GL(2k,q)}$ is multiplicity free (rather than just Λ-multiplicity-free).

PROOF OF 3.2 : Let $\tau = 1_{Sp(2k,q)}^{GL(2k,q)}$, and let also $\pi = \pi_{2,k}$. By the remarks in the introduction to this section, it is enough to show that

$$\langle \tau, \tau^{\lambda} \rangle_{GL(2k,q)} = \langle \pi, \pi^{\lambda} \rangle_{S_n} \quad \text{for all } \lambda \vdash n,$$

that is,

$$\#\,\text{orb}\,(Sp(2k,q), V^{\lambda}) = \#\,\text{orb}\,(S_2 \text{ wr } S_k, \Omega^{\lambda}) \quad \text{for each } \lambda \vdash n.$$

If V_r is a subspace of V, let $R(V_r) = V_r \cap V_r^{\perp}$ (where $V_r^{\perp}$ is the orthogonal complement of V_r under the non-degenerate skew-symmetric form (,) on V associated with our $Sp(2k,q)$). Fix $\lambda \vdash n$. For each $V = V_1 \supset V_2 \supset \ldots \supset V_a$ in V^{λ} we define a triangular array $D = (d_{rs})$ with $1 \le r \le s \le a$ by $d_{rs} = \dim\,(R(V_r) \cap V_s)$. Let $B_1, B_2, \ldots, B_k$ be the blocks of S_2 wr S_k on Ω, with $B_r = \{\alpha_r, \beta_r\}$. For a subset Ω_r of Ω we write $R(\Omega_r)$ for the set of points in Ω_r whose block is not in Ω_r. For each $\Omega = \Omega_1 \supset \Omega_2 \supset \ldots \supset \Omega_a$ in Ω^{λ} we define an array $C = (c_{rs})$ with $1 \le r \le s \le a$ by $c_{rs} = |R(\Omega_r) \cap \Omega_s|$.

We shall prove the Proposition in the following three steps:

(1) Two elements of Ω^λ lie in the same S_2 wr S_k - orbit if and only if they have the same array C ;

(2) Two elements of V^λ lie in the same $Sp(2k,q)$ - orbit if and only if they have the same array D ;

(3) A given triangular array is the array of some element in Ω^λ if and only if it is the array of some element in V^λ .

STEP 1 : This is easy - it is similar to but easier than Step 2.

STEP 2 : First of all, if $g \in Sp(2k,q)$ then

$$(R(V_r) \cap V_s)g = R(V_r)g \cap V_s g = R(V_r g) \cap V_s g ,$$

which establishes one part of the assertion.

To prove the converse, we shall produce inductively certain "canonical" symplectic bases. Let $V = V_1 \supset V_2 \supset \ldots \supset V_a$ be an element of V^λ . Let b be maximal such that $V_a \not\subseteq R(V_b)$ (so b is maximal with $d_{ba} < \lambda_a$) . Choose $v_1 \in V_a$ so that $(v_1, w_1) = 1$ for some $w_1 \in V_b$ (note that $w_1 \in V_b - V_{b+1}$ if $b < a$) . Let Y be the hyperbolic plane $\langle v_1, w_1 \rangle$, and let $X = Y^\perp$. Let also $X_i = X \cap V_i$. Then we observe the following:

if $r \le b$ then $\dim X_r = \dim V_r - 2$ and $R(X_r) \cap X_s = R(V_r) \cap V_s$ for $s \ge r$,

if $r > b$ then $\dim X_r = \dim V_r - 1$ and $\dim(R(X_r) \cap X_s) = \dim(R(V_r) \cap V_s)$ for $s \ge r$.

Now continue inside X , getting a symplectic basis $v_1, w_1, \ldots, v_k, w_k$. This contains a basis for each V_r , hence also for each $R(V_r)$ and therefore also for each $R(V_r) \cap V_s$.

Let now $V = V'_1 \supset V'_2 \supset \ldots \supset V'_a$ be another element of V_λ with the same array D . Let $v'_1, w'_1, \ldots, v'_k, w'_k$ be a corresponding canonical symplectic basis. Then we see from the construction above that $v_r \in V_s$ if and only if $v'_r \in V'_s$, and $w_r \in V_s$ if and only if $w'_r \in V'_s$. Let $g \in Sp(2k,q)$ with

$v_r g = v'_r$ and $w_r g = w'_r$ for $r \le k$. Then $V_s g = V'_s$ for $s \le a$, and the assertion of Step 2 is proved.

STEP 3 : Let $V = V_1 \supset V_2 \supset \ldots \supset V_a$ be an element in V^λ. Let $v_1, w_1, \ldots, v_k, w_k$ be a corresponding canonical symplectic basis associated with it (so that it contains a basis for each V_s and each $R(V_s) \cap V_t$). Define $\Omega_s = \{\alpha_r : v_r \in V_s\} \cup \{\beta_r : w_r \in V_s\}$. Then $\Omega = \Omega_1 \supset \Omega_2 \supset \ldots \supset \Omega_a$ is an element of Ω^λ. Also,

$$R(\Omega_s) = \{\alpha_r : v_r \in V_s,\ w_r \notin V_s\} \cup \{\beta_r : v_r \notin V_s,\ w_r \in V_s\}$$
$$= \{\alpha_r : v_r \in R(V_s)\} \cup \{\beta_r : w_r \in R(V_s)\}, \text{ and so}$$
$$R(\Omega_s) \cap \Omega_t = \{\alpha_r : v_r \in R(V_s) \cap V_t\} \cup \{\beta_r : w_r \in R(V_s) \cap V_t\}.$$

Hence $|R(\Omega_s) \cap \Omega_t| = \dim R(V_s) \cap V_t$ for $s \le t \le a$.

Conversely, let $\Omega = \Omega_1 \supset \Omega_2 \supset \ldots \supset \Omega_a$ be an element of Ω^λ. Let $\mathcal{B} = \{v_1, w_1, \ldots, v_k, w_k\}$ be a symplectic basis for V. Let $V_s = \langle v_r : \alpha_r \in \Omega_s \rangle + \langle w_r : \beta_r \in \Omega_s \rangle$. Then $V = V_1 \supset V_2 \supset \ldots \supset V_a$ is an element of V^λ, and $\mathcal{B}$ contains a basis for each V_s (by construction) and hence also each $R(V_s) \cap V_t$. It follows that

$$R(V_s) \cap V_t = \langle v_r : \alpha_r \in R(\Omega_s) \cap \Omega_t \rangle + \langle w_r : \beta_r \in R(\Omega_s) \cap \Omega_t \rangle,$$

and so

$$\dim R(V_s) \cap V_t = |R(\Omega_s) \cap \Omega_t| \quad \text{for } s \le t \le a.$$

This completes the proof of the claim of Step 3 and hence also the proof of Proposition 3.2.

PROPOSITION 3.3 : Let $n = 2k$. Then

$$\Lambda\left(1_{GL(k,q^2)}^{GL(2k,q)}\right) = \sum_{\lambda \vdash k} \psi^{2\lambda}.$$

REMARK : Hence $GL(k,q^2)$ and $Sp(2k,q)$ have the same number of orbits in their action on the set of cosets of any parabolic subgroup of $GL(2k,q)$ (and the proof also works for $GL(k,F)$ and $Sp(2k,K)$ inside $GL(2k,K)$, where K and F are fields and $K \subset F$ with F degree 2 over K).

PROOF OF PROPOSITION 3.3 : Choose $i \in GF(q^2) - GF(q)$ so that $i^2 = i + j$ for some $j \in GF(q)$. Let $G = GL(k,q^2)$ acting naturally on $V' = V(k,q^2)$. Then V' can be viewed as a $2k$-dimensional vector space over $GF(q)$. Given $v \in V'$, $v = (z_1, z_2, \dots, z_k) = (x_1 + iy_1, x_2 + iy_2, \dots, x_k + iy_k)$ with $x_r, y_r \in GF(q)$, let $v\phi = (x_1, y_1, x_2, y_2, \dots, x_k, y_k) \in V = V(2k,q)$. If $\mathcal{B}' = \{e_1, e_2, \dots, e_k\}$ is the standard basis in V' then $\mathcal{B} = \{e_1\phi, ie_1\phi, e_2\phi, ie_2\phi, \dots, e_k\phi, ie_k\phi\}$ is the standard basis of V.

Let $H = GL(2k,q)$ acting naturally on V. If g is in G with matrix (z_{rs}) with respect to $\mathcal{B}'$, where $z_{rs} = x_{rs} + iy_{rs}$ with $x_{rs}, y_{rs} \in GF(q)$, let $g\theta$ be the element of H with matrix (a_{rs}) with respect to $\mathcal{B}$, where

$$\begin{aligned} a_{2r-1,2s-1} &= x_{rs}, \\ a_{2r-1,2s} &= y_{rs}, \\ a_{2r,2s-1} &= j\,y_{rs}, \\ a_{2r,2s} &= x_{rs} + y_{rs}. \end{aligned}$$

Then $(v\phi)(g\theta) = (vg)\phi$ and θ is an embedding of G into H.

For $v = (v_1, v_2, \dots, v_{2k}) \in V$ we define $iv \in V$ by $iv = (jv_2, v_1 + v_2, jv_4, v_3 + v_4, \dots, jv_{2k}, v_{2k-1} + v_{2k})$, so that $v \mapsto iv$ is a non-singular linear transformation on V and $(iv)\phi^{-1} = i(v\phi^{-1})$. Given $V_r \subseteq V$, we define

$$iV_r = \{iv : v \in V_r\}, \quad C(V_r) = V_r \cap iV_r \quad \text{and}$$

$$\overline{V}_r = V_r + iV_r.$$

Each of these is a subspace of V and, moreover, $C(V_r)\phi^{-1}$ and $\overline{V}_r\phi^{-1}$ are subspaces of V'. Given $V = V_1 \supset V_2 \supset \dots \supset V_a$ in V^λ we define a triangular array $E = (e_{rs})$ with $1 \le r \le s \le a$ by $e_{rs} = \dim C(V_r) \cap V_s$.

Correspondingly, given $\Omega_r \subseteq \Omega$, we write

$$i\Omega_r = \{\alpha_s \in \Omega : \beta_s \in \Omega_r\} \cup \{\beta_s \in \Omega : \alpha_s \in \Omega_r\},$$

$C(\Omega_r) = \Omega_r \cap i\Omega_r$ and $\overline{\Omega}_r = \Omega_r \cup i\Omega_r$ (where as before each $\{\alpha_s, \beta_s\}$ is a block for S_2 wr S_k on Ω). Also, given $\Omega = \Omega_1 \supset \Omega_2 \supset \ldots \supset \Omega_a$ we define a triangular array $B = (b_{rs})$ with $1 \le r \le s \le a$ by $b_{rs} = |C(\Omega_r) \cap \Omega_s|$.

We shall now prove Proposition 3.3 in three steps analogous to the three steps in the proof of Proposition 3.2.

STEP 1 : Two elements of Ω^λ are in the same S_2 wr S_k - orbit if and only if they have the same array B.

This is easy - in fact it follows from step 1 of Proposition 3.2 since two elements have the same array B if and only if they have the same array C, as the (r,s) - entry of $B + C$ is $|\Omega_s|$.

STEP 2 : Two elements of V^λ are in the same $G\theta$ - orbit if and only if they have the same array E.

Firstly, if $g \in G$ then

$$(C(V_r) \cap V_s)g\theta = C(V_r)g\theta \cap V_s g\theta = C(V_r g\theta) \cap V_s g\theta,$$

since $(iv)(g\theta) = i(v(g\theta))$ for $v \in V$, $g \in G$; this gives one part of the claim.

To prove the converse, we define a canonical basis for $V = V_1 \supset V_2 \supset \ldots \supset V_a$ in V^λ. This is done inductively, starting from $C(V_a)$ at step (a,a), taking each row from right to left and having finished with a row, proceeding upwards :

(a,a) : Choose v_r, iv_r with $r \le \frac{1}{2}e_{aa}$ linearly independent in $C(V_a)$;

(s,t) with $s < t$: Add v_r from $C(V_s) \cap V_t$ up to $r = e_{st} - e_{s+1,t} + \frac{1}{2}e_{s+1,s+1}$ so that the set $\{v_r, iv_r : r \le e_{st} - e_{s+1,t} + \frac{1}{2}e_{s+1,s+1}\}$ is a linearly independent subset of $C(V_s)$;

(s,s) with $s < a$: Add v_r and iv_r up to $r = \frac{1}{2}e_{ss}$ so that $\{v_r, iv_r : r \le \frac{1}{2}e_{ss}\}$ is a basis in $C(V_s)$.

Thus at step (s,t) we are producing a basis for $C(V_{s+1}) + (C(V_s) \cap V_t)$ whose ϕ^{-1} - image contains a basis for

$(C(V_{s+1}) + \overline{C(V_s) \cap V_t})\phi^{-1}$ consisting of the images of the v_r.

This construction is possible:

Certainly the (a,a)-step is possible, since $C(V_a)$ is a subspace of V of dimension e_{aa} such that $C(V_a)\phi^{-1}$ is a subspace of V' of dimension $\frac{1}{2}e_{aa}$; hence we can find v_r in $C(V_a)$ for $v \leq \frac{1}{2}e_{aa}$ such that the $v_r\phi^{-1}$ form a basis of $C(V_a)\phi^{-1}$, and so $\{v_r, iv_r : r \leq \frac{1}{2}e_{aa}\}$ is a basis of $C(V_a)$.

The (s,s)-step is possible: We already have a basis for $C(V_s) \cap V_{s+1}$ containing the v_r with $r \leq e_{s,s+1} - \frac{1}{2}e_{s+1,s+1}$ such that $\{v_r\phi^{-1} : r \leq e_{s,s+1} - \frac{1}{2}e_{s+1,s+1}\}$ is a basis of $\overline{C(V_s) \cap V_{s+}}\,\phi^{-1}$; since $\overline{C(V_s) \cap V_{s+1}} \subseteq C(V_s)$, we can add v_r from $C(V_s)$ up to $r = \frac{1}{2}e_{ss}$ to get a basis $\{v_r\phi^{-1} : r \leq \frac{1}{2}e_{ss}\}$ for $C(V_s)\phi^{-1}$; then $\{v_r, iv_r : r \leq \frac{1}{2}e_{ss}\}$ is a basis for $C(V_s)$.

The (s,t)-step with $s < t$ is also possible: We already have a suitable basis for $C(V_{s+1})$ containing a basis for $C(V_{s+1}) \cap V_t$. Since $C(V_{s+1}) \cap (C(V_s) \cap V_t) = C(V_{s+1}) \cap V_t$, we obtain a basis for $C(V_{s+1}) + (C(V_s) \cap V_t)$ by adding to our basis for $C(V_{s+1})$ any $e_{st} - e_{s+1,t}$ vectors of $C(V_s) \cap V_t$ which are independent modulo $C(V_{s+1}) \cap V_t$. In particular, if $t < a$, we choose these to include all the vectors we already have for $C(V_s) \cap V_{t+1}$ (using the modular law on the way). Thus we obtain a basis $\{v_r : r \leq e_{st} - e_{s+1,t} + \frac{1}{2}e_{s+1,s+1}\} \cup \{iv_r : r \leq \frac{1}{2}e_{s+1,s+1}\}$ for $C(V_{s+1}) + (C(V_s) \cap V_t)$. We finally claim that the set $\{v_r, iv_r : r \leq e_{st} - e_{s+1,t} + \frac{1}{2}e_{s+1,s+1}\}$ is a basis for $C(V_{s+1}) + \overline{C(V_s) \cap V_t}$ and hence a linearly independent subset of $C(V_s)$: certainly it is a spanning set. Now $C(V_{s+1}) \cap \overline{C(V_s) \cap V_t} = \overline{C(V_{s+1}) \cap V_t}$, as one can check. Hence

$$\begin{aligned}\dim(C(V_{s+1}) + \overline{C(V_s) \cap V_t}) &= \dim C(V_{s+1}) + \dim \overline{C(V_s) \cap V_t} - \\ &\quad - \dim \overline{C(V_{s+1}) \cap V_t} = e_{s+1,s+1} + (2e_{st} - e_{tt}) - (2e_{s+1,t} - e_{tt}) \\ &= 2(e_{st} - e_{s+1,t} + \tfrac{1}{2}e_{s+1,s+1}),\end{aligned}$$

so it is a basis.

Hence the construction indeed is possible, and canonical bases do exist.

We claim next that if $\{v_1, v_2, \ldots, v_k\}$ and $\{w_1, w_2, \ldots, w_k\}$ are such that $\{v_r, iv_r : r \le k\}$ and $\{w_r, iw_r : r \le k\}$ are bases for V then there is an element g in G such that $v_r(g\theta) = w_r$ and $(iv_r)(g\theta) = iw_r$ for $r \le k$. This is true since $\{v_r\phi^{-1} : r \le k\}$ and $\{w_r\phi^{-1} : r \le k\}$ are bases in V', so there is a g in G with $(v_r\phi^{-1})g = w_r\phi^{-1}$ for $r \le k$; then $v_r(g\theta) = w_r$ and $(iv_r)(g\theta) = iw_r$ for $r \le k$.

Let now $V = V_1 \supset V_2 \supset \ldots \supset V_a$ and $V = W_1 \supset W_2 \supset \ldots \supset W_a$ be two elements of V^λ corresponding to the same array E. Let $\{v_r, iv_r : r \le k\}$ and $\{w_r, iw_r : r \le k\}$ be corresponding canonical bases. Then there is a g in G with $v_r(g\theta) = w_r$ and $(iv_r)(g\theta) = iw_r$ for $r \le k$, and it follows (by the construction of canonical bases) that $(C(V_s) \cap V_t)(g\theta) = C(W_s) \cap W_t$ for $1 \le s \le t \le a$. In particular, $V_s(g\theta) = W_s$ for $s \le a$.

STEP 3 : If a triangular array is the array of an element in Ω^λ then it is also the array of some element in V^λ, and vice versa.

Firstly, let $V = V_1 \supset V_2 \supset \ldots \supset V_a$ be an element of V^λ. Let $\{v_r, iv_r : r \le k\}$ be a corresponding canonical basis; this contains a basis for each V_s and for each $C(V_s) \cap V_t$. Define $\Omega_s = \{\alpha_r : v_r \in V_s\} \cup \{\beta_r : iv_r \in V_s\}$. Then $\Omega = \Omega_1 \supset \Omega_2 \supset \ldots \supset \Omega_a$ is in Ω^λ, and

$$C(\Omega_s) \cap \Omega_t = \{\alpha_r, \beta_r : v_r, iv_r \in V_s\} \cap (\{\alpha_r : v_r \in V_t\} \cup$$

$$\{\beta_r : iv_r \in V_t\}) = \{\alpha_r : v_r \in C(V_s) \cap V_t\} \cup \{\beta_r : iv_r \in C(V_s) \cap V_t\},$$

so that $\quad |C(\Omega_s) \cap \Omega_t| = \dim C(V_s) \cap V_t$

(since the canonical basis contains a basis for $C(V_s)$ and a basis for V_t).

Conversely, let $\Omega = \Omega_1 \supset \Omega_2 \supset \ldots \supset \Omega_a$ be an element of Ω^λ. Let $\mathcal{B} = \{ v_r, i v_r : r \le k \}$ be a basis for V. Let $V_s = \langle v_r : \alpha_r \in \Omega_s \rangle + \langle i v_r : \beta_r \in \Omega_s \rangle$. Then $V = V_1 \supset V_2 \supset \ldots \supset V_a$ is in V^λ, and $\mathcal{B}$ contains a basis for each V_s. We claim next that $C(V_s) = \langle v_r, i v_r : \alpha_r \in C(\Omega_s) \rangle$, so that $\mathcal{B}$ also contains a basis of each $C(V_s)$: certainly $\alpha_r \in C(\Omega_s) \Longrightarrow v_r, i v_r \in C(V_s)$, since

$$\alpha_r \in C(\Omega_s) \Longrightarrow \alpha_r, \beta_r \in \Omega_s \Longrightarrow v_r, i v_r \in V_s \Longrightarrow v_r, i v_r \in C(V_s).$$

Now

$$\begin{aligned} \overline{V}_s &= V_s + i V_s \\ &= \langle v_r : \alpha_r \in \Omega_s \rangle + \langle i v_r : \beta_r \in \Omega_s \rangle + \langle i v_r : \alpha_r \in \Omega_s \rangle + \\ &\qquad + \langle i^2 v_r : \beta_r \in \Omega_s \rangle \\ &= \langle v_r, i v_r : \alpha_r \in \Omega_s \rangle + \langle v_r, i v_r : \beta_r \in \Omega_s \rangle \\ &= \langle v_r, i v_r : \alpha_r \in \overline{\Omega}_s \rangle , \end{aligned}$$

so that $\dim \overline{V}_s = 2 \, |\Omega_s| - |C(\Omega_s)|$; but on the other hand,

$$\dim \overline{V}_s = \dim V_s + \dim i V_s - \dim C(V_s) = 2|\Omega_s| - \dim C(V_s).$$

Hence $\dim C(V_s) = |C(\Omega_s)|$, and so $C(V_s) = \langle v_r, i v_r : \alpha_r \in C(\Omega_s) \rangle$. It now follows that

$$C(V_s) \cap V_t = \langle v_r : \alpha_r \in C(\Omega_s) \cap \Omega_t \rangle + \langle i v_r : \beta_r \in C(\Omega_s) \cap \Omega_t \rangle ,$$

and so $\dim C(V_s) \cap V_t = |C(\Omega_s) \cap \Omega_t|$ for $1 \le s \le t \le a$.

This completes the Proof of Proposition 3.3.

REMARK: As far as I know, the classification problem for Λ-multiplicity-free permutation representations of $GL(n,q)$ remains open. The observation (2a) carries over unchanged, but the Remark following it of course not. However, this should still be useful in view of [2].

The observation (2b) gets somewhat weakened to

(3b) *The order of* G *is larger than* $\frac{1}{c}|GL(n,q) : B|$, *where* $c < n!$ *and* B *is the Borel subgroup.* (This looks hopeful in view of [9].)

For, if the character of $GL(n,q)$ on the cosets of G is τ and if c is the number of orbits of G on $V^{(1^n)}$ then

$$c = \langle 1_G, \tau_G^{(1^n)} \rangle_G = \langle \tau, \tau^{(1^n)} \rangle_{GL(n,q)} \le \sum_{\lambda \vdash n} \langle \psi^\lambda, \tau^{(1^n)} \rangle_{GL(n,q)}$$

$$= \langle \chi^\lambda, \pi^{(1^n)} \rangle_{S_n} ,$$

the second equality by Frobenius reciprocity and the last by the result of Steinberg at the beginning of this section. But $\pi^{(1^n)}$ is the regular character of S_n, so

$$\sum_{\lambda \vdash n} \langle \chi^\lambda, \pi^{(1^n)} \rangle_{S_n} = \sum_{\lambda \vdash n} \chi^\lambda(1) ,$$

and so, as in (2b),

$$c \le \# \{ g \in S_n : g^2 = 1 \} < n! .$$

Hence

$$|G| \ge \frac{1}{c} |V^{(1^n)}| = \frac{1}{c} | GL(n,q) : B |$$

and (3b) holds.

BIBLIOGRAPHY

1. E. Bannai, "Maximal subgroups of low rank of finite symmetric and alternating groups", *J. Fac. Sci. Univ. Tokyo,* 18 (1972), 475-486.
2. P.J. Cameron and W.M. Kantor, "2-Transitive and antiflag transitive collineation groups of finite projective spaces", *J. Algebra,* 60 (1979), 384-422.
3. P.J. Cameron, P.M. Neumann and J. Saxl, "An interchange property in finite permutation groups", *Bull. London Math. Soc.,* 11 (1979), 161-169.
4. C.W. Curtis, N. Iwahori and R. Kilmoyer, "Hecke algebras and characters of parabolic type of finite groups with BN-pairs", *Inst. Hautes Etudes Sci. Publ. Math.,* 40 (1972), 81-116.
5. C.W. Curtis, W.M. Kantor and G.M. Seitz, "The 2-transitive permutation representations of the finite Chevalley groups", *Trans. Amer. Math. Soc.,* 218 (1976), 1-57.

6. R.B. Howlett, "On the degrees of Steinberg characters of Chevalley groups", *Math. Z.*, 135 (1974), 125-135.
7. I.M. Isaacs, *Character Theory of Finite Groups* (Academic Press, New York, 1976).
8. G.D. James, *The Representation Theory of the Symmetric Group* (Springer-Verlag, Berlin, 1978).
9. W.M. Kantor, "Permutation representations of the finite classical groups of small degree or rank", *J. Algebra*, 60 (1979), 158-168.
10. W.M. Kantor and R.A. Liebler, "The rank 3 permutation representations of the finite classical groups", to appear.
11. D. Livingstone and A. Wagner, "Transitivity of finite permutation groups on unordered sets", *Math. Z.*, 90 (1965), 393-403.
12. C.E. Praeger and J. Saxl, "On the orders of primitive permutation groups", *Bull. London Math. Soc.*, 12 (1980), 303-307.
13. J. Saxl, "Characters of multiply transitive permutation groups", *J. Algebra*, 34 (1975), 528-539.
14. R. Steinberg, "A geometric approach to the representations of the full linear group over a Galois field", *Trans. Amer. Math. Soc.*, 71 (1951), 274-282.
15. R.M. Thrall, "On symmetrized Kronecker powers and the structure of the free Lie ring", *Amer. J. Math.*, 64 (1942), 371-388.
16. H. Wielandt, *Finite Permutation Groups* (Academic Press, New York, 1964).

Department of Pure Mathematics & Mathematical Statistics
16 Mill Lane
Cambridge CB2 1SB
U.K.

ON A CHARACTERIZATION OF THE GRASSMANN MANIFOLD REPRESENTING THE LINES IN A PROJECTIVE SPACE

Giuseppe Tallini

The pair (S, R), where S is a set of elements, *points*, and R is a family of subsets of S, *lines*, will be called a *partial line space* if

(i) R is a covering of S;

(ii) any member of R contains at least two points;

(iii) given any two distinct points in S, there exists at most one line through them. (If such a line exists, we say the points are *collinear*.)

A *subspace* of (S, R) is a subset T of S such that any two distinct points in T are collinear and the line joining them belongs to T; thus T is a line space contained in (S, R).

We shall assume (S, R) is a *proper* partial line space; that is there are at least two non-collinear points in S.

A *maximal* subspace in (S, R) is a subspace which is not properly contained in any subspace. By Zorn's Lemma, any subspace in (S, R) is contained in a maximal subspace.

Now suppose that (S, R) satisfies the following axioms:

(A1) Given three pairwise collinear points, there exists a subspace containing them.

(A2) No line is a maximal subspace. Furthermore, there are two families, say Σ and P, of maximal subspaces in (S, R) such that any maximal subspace belongs either to Σ or to P and the following hold :

(I) any two subspaces in Σ meet in exactly one point;

(II) if $T \in \Sigma$ and $\pi \in P$, then $T \cap \pi$ is either empty or is a line;

(III) for every line r, there exist a unique T in Σ and a unique π in P such that $r \subseteq T \cap \pi$.

We study $(S, \mathcal{R})$ in order to characterize the Grassmann manifold representing the lines in a projective space.

Since $\mathcal{R}$ covers S, for each p in S, there exists a line r in $\mathcal{R}$ through p. Then, by (III), there exists π in $\mathcal{P}$ through r and so through p, and $\pi \neq r$ (r not being a maximal subspace); therefore, there are two distinct lines through p in π; thus, by (III), there are two members of Σ through p (and they are distinct because of (II)). Hence, there are at least two distinct elements of Σ through each $p \in S$. We set

$$\ell_p = \{ T \in \Sigma \ : \ T \ni p \} ;$$

then $|\ell_p| \geq 2$, and

$$\mathcal{L} = \{ \ell_p \ : \ p \in S \} \tag{1}$$

is a proper family of subsets of Σ (recall (III)). By (I), $(\Sigma, \mathcal{L})$ is a line space.

Now we prove that

LEMMA 1 : *Each π in $\mathcal{P}$ is a projective plane; that is, any two lines in π meet at a point.*

PROOF : Suppose the contrary; then in π there are two lines r_1 and r_2 such that $r_1 \cap r_2 = \emptyset$. Let T_1 and T_2 be the elements in Σ through r_1 and r_2 respectively (they exist because of (III)). From (I) it follows that T_1 and T_2 meet at a point p which does not belong to π (otherwise $r_1 \cap r_2 = \{p\}$). Consider p_1 in r_1 and p_2 in r_2; the points p_1, p_2, p are pairwise collinear (since $p_1, p_2 \in \pi$; $p_1, p \in T_1$; $p_2, p \in T_2$); thus, by (A1), there exists a subspace containing p_1, p_2, p. Let τ be a maximal subspace containing it; τ must belong either to Σ or to $\mathcal{P}$ by (A2). But τ cannot belong to Σ; for τ and T_1 $(\in \Sigma)$ contain the line p_1p, by (I), and $\tau \in \Sigma \Longrightarrow \tau = T_1 \Longrightarrow p_2 \in T_1 \Longrightarrow p_2 \in r_1 \cap r_2$, a contradiction. Thus $\tau \in \mathcal{P}$. It follows that there are two distinct members of $\mathcal{P}$ (π and τ) through the line p_1p_2, a contradiction. $\square$

Now, let T_1, T_2, T_3 be any three distinct and independent elements of (Σ, L), that is, they pairwise meet at distinct points: $p_1 = T_2 \cap T_3$, $p_2 = T_3 \cap T_1$, $p_3 = T_1 \cap T_2$. The points p_1 and p_2 are collinear; so (S, R) contains the line p_1p_2. Similarly, the lines p_2p_3 and p_3p_1 exist. From (A1) it follows that p_1, p_2, p_3 belong to a subspace in (S, R) and thus to a maximal subspace τ. Also, τ does not belong to Σ. In fact, $p_3 \in \tau$ and $p_3 \notin T_3$ implies $\tau \neq T_3$. Since $\tau \cap T_3 \supseteq p_1p_2$, $\tau \in \Sigma$ would contradict (I). Hence $\tau \in P$ and τ is a projective plane, by Lemma 1.

Next, consider the set

$$\alpha = \{T \in \Sigma \ : \ T \cap \tau \ \text{ is a line}\}.$$

The set α contains T_1, T_2, T_3. Furthermore, α is a projective plane in (Σ, L). In fact, by (II), two distinct elements T, T' in α belong to a unique ℓ_p (in L), completely contained in α, where $p = (T \cap \tau) \cap (T' \cap \tau)$; moreover, if $\ell_p, \ell_{p'} \subset \alpha$ are distinct elements in L, then p and $p' \in \tau$ so that, by (III), there is exactly one element in α belonging both to ℓ_p and $\ell_{p'}$ through the line pp' in τ. It follows that in (Σ, L) the span of T_1, T_2, T_3 is a projective plane, namely α. Hence (Σ, L) is a projective space.

The mapping $p \in S \longrightarrow \ell_p \in L$ is bijective. Let

$$\rho \ : \ \ell_p \in L \longrightarrow p \in S$$

be its inverse mapping. Clearly, ρ maps each line through T in Σ in the space (Σ, L) onto a point in the space T in (S, R), this mapping being bijective for all T in Σ.

LEMMA 2 : *ρ maps the lines in a ruled plane in (Σ, L) onto the points in a plane π in P, defining an isomorphism between them. Conversely, for any π in P, $\rho^{-1}(\pi)$ is a ruled plane in (Σ, L).*

PROOF : Let α be a plane in (Σ, L) and let T_1, T_2, T_3 be any three independent points in α. Then T_1, T_2, T_3 are three subspaces in (S, R) belonging to Σ which pairwise meet at distinct and independent points $p_1 = T_2 \cap T_3$, $p_2 = T_3 \cap T_1$, $p_3 = T_1 \cap T_2$. Since the points p_1, p_2, p_3 are pairwise collinear, there exists a maximal subspace π in (S, R) through them by (A1); by (I), $\pi \in P$. The set of all subspaces T in Σ meeting π in a line turns out to be, in (Σ, L), a plane containing T_1, T_2, T_3, namely the plane α. Therefore each line ℓ_p in α consists of all elements in Σ meeting π in lines of the pencil through p in π. Thus ρ maps the line ℓ_p in α onto the point p in π, and conversely. □

Since ρ maps the lines through a point (star of lines) onto a subspace T in Σ and a ruled plane onto a subspace π in P, we have

LEMMA 3 : *ρ maps a pencil of lines in (Σ, L) onto a line in (S, R). Conversely, for any r in R, $\rho^{-1}(r)$ is a pencil of lines in (Σ, L).* □

It follows that ρ induces an isomorphism between the star of lines through T in Σ in (Σ, L), which is a projective space taking as lines the pencils of lines, and the subspace T in (S, R) : thus these subspaces are projective spaces.

We note that if (S, R) is irreducible (i.e. all its lines have at least three points), then the same is true for (Σ, L), and conversely.

The previous results are summarized in the following propositions.

PROPOSITION I : *Let (S, R) be a proper partial line space satisfying axioms (A1) and (A2). Then there exist a projective space (Σ, L) and a bijection $\rho : L \longrightarrow S$ mapping pencils of lines in (Σ, L) onto lines in (S, R); also $\rho^{-1} : S \to L$ maps lines onto pencils of lines. Hence ρ maps stars of lines and ruled planes* *in* (Σ, L) *onto subspaces* T *in* Σ *and* π *in* P, *inducing an isomorphism between them; further,* $\rho^{-1} : S \to L$ *maps* T *in* Σ *and* π *in* P *onto stars of lines and ruled planes, respectively. Finally,* (Σ, L) *is irreducible iff* (S, R) *is irreducible. It follows that if* (Σ, L) *is irreducible, finitely generated and Pascalian, then* (S, R) *is isomorphic to the Grassmann manifold representing the lines in* (Σ, L). □

Since any finite irreducible projective space is finitely generated and Pascalian, from Proposition I we get the following characterization of the Grassmann manifold representing the lines in a Galois space.

PROPOSITION II : *Any irreducible, finite, proper partial line space satisfying axioms* (A1) *and* (A2) *is isomorphic to the Grassmann manifold representing the lines in a Galois space.* □

BIBLIOGRAPHY

1. B. Segre, *Lectures on Modern Geometry* (Cremonese, Rome, 1961).
2. G. Tallini, "Problemi e risultati sulle geometrie di Galois", Relazione 30 (Istituto Matematico, Università di Napoli, 1973).
3. G. Tallini, "Spazi parziali di rette, spazi polari. Geometrie subimmerse", Seminario di Geometrie Combinatorie 14 (Università di Roma, 1979).

Istituto Matematico
Università di Roma
00185 Roma
Italy.

AFFINE SUBPLANES OF PROJECTIVE PLANES

K. Vedder

If a finite projective plane of order n contains an affine subplane of order $m < n$, such that every point is incident with at least one line of the affine subplane, then $m = 3$ and $n = 4$ or $m^2 - \frac{1}{2}m - 1 < n \leq m^2 - W$, where W depends on the configurations formed by the parallel classes of the subplane. We have $W = 2$ for $m = 3$ and $W > 2$ for $m > 3$.

INTRODUCTION

Let π be a finite projective plane of order n which contains an affine subplane π' of order $m < n$. If we denote the number of lines of π' through a point P of π by $w(P)$, then a theorem of Ostrom and Sherk says that

(i) if $w(P) \geq 1$ for all points P in π then $m = 3$ and $n = 4$ or $m^2 - m + 1 \leq n \leq m^2 - 1$, or

(ii) if $w(P) = 0$ for some point P in π then $n \geq m^2 - 1$.

To prove (ii) one considers a point P in π with $w(P) = 0$. Now, for each point Q of π' the line through P and Q does not contain any other point of π'. Since there are m^2 points in π' it follows that $n + 1 \geq m^2$.

In the same paper, Ostrom and Sherk show that a Desarguesian plane of order n contains an affine subplane of order 3 if and only if $n = 3$ or $n \equiv 1 \pmod 3$. This means that the plane of order 8 does not contain an affine subplane of order 3, and every example satisfying $w(P) = 0$, for some point P, and $n = m^2 - 1$ must involve a plane of non-prime power order.

Apart from the case $m = 3$ and $n = 4$ there are only two other examples known which satisfy (i); these are the projective planes of order 3 and 7 which contain affine subplanes of order 2 and 3 respectively.

The author would like to point out that Proposition 1 was found independently, using different counting arguments, by P.J. Cameron.

PROOF OF THEOREM

In the following, let π be a projective plane of order n, π' be an affine subplane of π of order $m < n$, and let $\bar{\pi}$ denote all those points of π which are not contained in π'.

To construct bounds for the order of π in terms of the order of the affine subplane, we consider the number of points of $\bar{\pi}$ which can be covered by a parallel class of π'. This number is maximal if the m lines of the parallel class are concurrent in π. In this case the parallel class is incident with $m(n-m)+1$ points of $\bar{\pi}$. The number of points of $\bar{\pi}$ which are incident with a parallel class of π' is minimal if each of these points lies on at most 2 lines of the parallel class. The number of "covered" points is equal to $\sum_{i=0}^{m-1} (n+1-m-i)$. This gives us the following.

LEMMA 1: *A parallel class of an affine subplane of order* m *of a projective plane of order* n, $m < n$, *covers*

at least $nm - \frac{3}{2}m^2 + \frac{3}{2}m$ *and*

at most $nm - m^2 + 1$

points outside the affine subplane. □

The following proposition gives a lower bound for the order of π, and shows that not every parallel class can cover a minimal number of points if $m > 3$.

PROPOSITION 1: *Let* π *be a projective plane of order* n *containing an affine subplane of order* m. *If* $w(P) \geq 1$ *for all points* P *in* π, *then*

(i) $m^2 - \frac{1}{2}m - 1 < n$,

or (ii) $m = 3$ and $n = 4$.

PROOF : The number of points of $\bar{\pi}$ is $n^2 + n + 1 - m^2$. By Lemma 1, the lines of π' can cover all the points of $\bar{\pi}$ only if

$$(m+1)(nm - \tfrac{3}{2}m^2 + \tfrac{3}{2}m) \le n^2 + n + 1 - m^2 . \qquad (1)$$

The two sides are equal if and only if every parallel class covers a minimal number of points. Solving (1) for n , we get

$$0 \le n^2 + n(-m^2 - m + 1) + (\tfrac{3}{2}m^3 - m^2 - \tfrac{3}{2}m + 1). \qquad (2)$$

The integer solutions n of (2) have to satisfy $n \ge \frac{1}{2}(m^2 + m - 1 + \sqrt{\Delta})$ or $n \le \frac{1}{2}(m^2 + m - 1 - \sqrt{\Delta})$, where $\Delta = m^4 - 4m^3 + 3m^2 + 4m - 3$. We note that $\Delta = (m^2 - 2m)^2 - (m-1)(m-3)$.

If $m = 2$, then $\Delta = 1$. Thus $n \ge 3$ and, therefore, $n > m^2 - \frac{1}{2}m + 1$ or $n \le 2$, which contradicts $m < n$.

If $m = 3$, then $\Delta = 9$. Either $n > 7$ and n and m satisfy (i), or $n \le 4$. If $n \le 4$, then $n = 4$ since $m < n$; this is the only exception.

If $m \ge 4$ then Δ lies between two consecutive squares; so $(m^2 - 2m)^2 > \Delta > (m^2 - 2m - 1)^2$. This means we do not have equality in (2), and thus not a minimal cover for $m \ge 4$. Using $\sqrt{\Delta} > m^2 - 2m - 1$, we get $n > \frac{1}{2}(m^2 + m - 1 + m^2 - 2m - 1)$ or $n < \frac{3}{2}m$. A point P with $w(P) < m$ lies on $w(P) + (m - w(P))m \ge 2m - 1$ lines which intersect π'. Thus $n \ge 2m - 2 \ge \frac{3}{2}m$, contradicting $n < \frac{3}{2}m$. □

In the proof of the above proposition, it may be noted that there cannot be a minimal cover of the points of $\bar{\pi}$ if $m \ge 4$. Clearly, for $m = 2$ and $n = 3$ we have both a minimal and a maximal cover. For $m = 3$, a minimal cover can exist only if $n = 4$ or $n = 7$. By Theorem 2 of Ostrom and Sherk (1964), both the projective plane of order 7 and the plane of order 4 contain affine subplanes of order 3 . In the case $n = 4$, the set $\bar{\pi}$ consists of 12 points and each of the four parallel classes of π' covers exactly 3 of them. Since every parallel class consists of 3 lines, and each line contains 2 points of $\overline{\pi}$, each point of $\overline{\pi}$ lies on exactly 2 lines of π'. This is the only case where each point of $\overline{\pi}$ lies on at least 2 lines of π' as we show in the following; see also [2], Theorem 1, (iv).

LEMMA 2 : *Let* π *be a projective plane of order* n *containing an affine subplane of order* m. *Then*

(i) *every point of* $\overline{\pi}$ *is incident with at most* 2 *lines of* π' *if and only if* $m = 2$ *and* $n = 3$ *or* $m = 3$ *and* $n = 4$ *or* 7;

(ii) *every point of* $\overline{\pi}$ *is incident with at least* 2 *lines of* π' *if and only if* $m = 3$ *and* $n = 4$.

PROOF : It remains to show that if every point P of $\overline{\pi}$ satisfies $w(P) \geq 2$ then $n = 4$ and $m = 3$.

Let b be a line of π'. Since each of the $n + 1 - m$ points on b which lie in $\overline{\pi}$ is incident with at least one other line of the same parallel class of π', we have $m-1 \geq n+1-m$. Thus $2m-2 \geq n$ which, using Proposition 1, implies $n = 4$ and $m = 3$. □

Using the same type of argument as in Proposition 1 with a maximal number of points covered by each parallel class, we would get $n \leq m^2 - 1$. We will show, however, that n takes on the upper bound only in the case $m = 2$ and that for $m > 2$ at most one parallel class of π' forms part of a pencil in π.

LEMMA 3 : *Let* π *be a finite projective plane of order* n *containing an affine subplane of order* $m < n$. *If* $w(P) \geq 1$ *for all points* P *in* π, *then*

(i) $w(P) = m$ *for at most one point* P,

or (ii) $m = 2$ *and* $n = 3$.

PROOF : Suppose there exist two distinct points P and Q with $w(P) = w(Q) = m$. The line b through P and Q does not intersect π', and each of its points lies on at least one line of π'.

Since $2m$ lines of the affine subplane intersect b in P or Q, it follows that the remaining $n-1$ points on b are covered by the remaining m^2-m lines of π'. Hence

$$n-1 \leq m^2-m . \qquad (3)$$

If $m=3$ then $n \leq 7$. It follows from Proposition 1 that $n=4$ or $n=7$. In both these cases $W(R) \leq 2$ for all points R in $\overline{\pi}$. We may therefore assume that $m \neq 3$ and combine inequality (3) with inequality (i) of Proposition 1. This yields $m<4$. Thus $m=2$, and substitution of $m=2$ in (3) gives the required result. □

Let W denote the maximal number amongst all $w(R)$, with R in $\overline{\pi}$ and $w(R) \neq m$.

PROPOSITION 2: *Let* π *be a finite projective plane of order* n *containing an affine subplane of order* $m<n$. *If* $w(P) \geq 1$ *for all points* P *in* π, *then*

(i) $n \leq m^2-W$, *or*

(ii) $m=2$ *and* $n=3$.

PROOF: If $m=2$ then it follows from Lemma 3 that $n=3$. So let $m>2$, and let Q be a point of $\overline{\pi}$ with $w(Q)=W$. By Lemma 3, such a point exists and $2 \leq W < m$.

Since $w(Q)<m$ there exists a point S in π' such that the line b through Q and S is not a line of π'. Now each of the remaining $n-1$ points on b lies on at least one line of π'. This implies $n-1 \leq m^2+m-w(S)-W$. The result follows with $w(S)=m+1$. □

The two inequalities (i) of Propositions 1 and 2 combined read for $m>3$, $m^2-\frac{1}{2}m-1<n \leq m^2-W$. This implies $W<\frac{1}{2}m+1$ if $m>3$. Since this holds for $m=3$ as well, $w(P)<\frac{1}{2}m+1$ $(m \neq 2)$ for all points P in $\overline{\pi}$, with the possible exception of at most one point. If $m>3$ then we cannot have a minimal cover, and there exists at least one point P in $\overline{\pi}$ with $w(P)>2$. Using Lemma 1, one can show that it is not possible that $w(P)=m$

for exactly one point P of $\overline{\pi}$ and that $w(Q) \le 2$ for all other points Q of $\overline{\pi}$. Hence $\frac{1}{2}m + 1 > W > 2$ for all $m > 3$.

This means in particular that $m \neq 4$. For $m = 5$, we get $W = 3$ and $n = 22$. By the Bruck-Ryser Theorem, however, there does not exist a plane of order 22. Since there is no plane of order 6 either we have proved the following

THEOREM: *If a finite projective plane of order* n *contains an affine subplane of order* m, $m < n$, *such that every point is incident with at least one line of the affine subplane, then*

(i) $m = 2$ *and* $n = 3$,

or (ii) $m = 3$ *and* $n = 4$ *or* 7,

or (iii) $m > 6$ *and* $m^2 - \frac{1}{2}m - 1 < n < m^2 - 2$. □

BIBLIOGRAPHY

1. D.R. Hughes and F.C. Piper, *Projective Planes* (Springer-Verlag, New York, 1973).
2. T.G. Ostrom and F.A. Sherk, "Finite projective planes with affine subplanes", *Canad. Math. Bull.*, 7 (1964), 549-559.

NOTE ADDED IN PROOF:

The author is indebted to Professor G. Pickert for pointing out that the results of this note are contained in J.F. Rigby's paper "Affine subplanes of finite projective planes", *Canad. J. Math.*, 17 (1965), 977-1009.

Mathematisches Institut
Justus-Liebig-Universität
Arndtstrasse 2
6300 Giessen
Federal Republic of Germany

POINT STABLE DESIGNS

Karl Erich Wolff

1. INTRODUCTION

The characterization of connected regular graphs by Hoffman [3] and of partial geometric designs by Bose, Bridges, Shrikhande [1] led the author [6] to the introduction of point stable designs, which are defined by the equation $NJ = JN$, where $N = AA^T$, A is the incidence matrix of the design and J the all-one matrix. This equation is equivalent to $NJ = \alpha J$ for some $\alpha \in \mathbb{N}$, which means geometrically that

$$\sum_{pIB} [B] = \alpha \quad \text{for all points } p,$$

where $[B]$ denotes the number of points on B. Clearly any $1-(v, k, r)$-design is point stable with $\alpha = rk$. Hoffman's characterization of connected regular graphs may be generalized to

THEOREM 1 : [6] *A design is connected and point stable if and only if there is a polynomial* $f(X) \in \mathbb{R}[X]$ *with* $f(N) = J$. *In this case the unique polynomial of minimal degree with this property is the Hoffman polynomial*

$$h(X) = v \prod_{\rho \in \mathrm{spec}N\setminus\{\alpha\}} \frac{X-\rho}{\alpha-\rho},$$

where v *is the number of points,* $\mathrm{spec}N$ *the spectrum of* N *and* $NJ = \alpha J$.

Besides the Hoffman equation

$$h(N) = J \qquad (1)$$

two similar equations play an important role. A *partial geometric design* (introduced by Bose, Shrikhande, Singhi [2] and Neumaier

[5]) is a 1-design satisfying

$$(N - \rho E) A = tJ \quad \text{for some } \rho, \ t \in \mathbb{N}\setminus\{0\}. \tag{2}$$

As a generalization of (2) the author [6] defined a *semi partial geometric design* by

$$(N - \rho E) A = J_{v,1} Y_{1,b} \quad \text{for some } \rho \in \mathbb{R}, \ Y_{1,b} \in \mathbb{N}^b. \tag{3}$$

Motivated by these equations, we shall define four ranks of a design and use them to classify point stable designs.

2. RANK AND COLUMN RANK

Following Neumaier [4] we define the *rank* $R(D)$ of a design D as the minimal degree of a monic polynomial $f(X) \in \mathbb{R}[X]$ with

$$f(N)A = tJ \quad \text{for some } t \in \mathbb{R}.$$

In the following let

$$s = |\mathrm{spec} N\setminus\{0\}| \quad (= |\mathrm{spec} A^T A\setminus\{0\}|)$$

denote the number of nonzero eigenvalues of D. To exclude trivial cases we assume

$$s \geq 2 \quad \text{and} \quad 2 \leq [B] < v \quad \text{and} \quad 2 \leq [p] < b$$

for all blocks B and points p.

THEOREM 2 : [7] *If* D *is point stable with* s *nonzero eigenvalues, then*

(a) $R(D) \in \{s-1, s\}$;

(b) $R(D) = s-1$ *if and only if* D *is a connected 1-design.*

For the proof of Theorem 2 use the minimal polynomial of N, the Hoffman equation and the formula

$$(N^n A)_{p,B} = \text{the number of p-B-paths of length } n.$$

Motivated by (3), we define the *column rank* $C(D)$ of a design D as the minimal degree of a monic polynomial $f(X) \in \mathbb{R}[X]$ with $f(N)A = J_{v,1} Y_{1,b}$ for some real $(1\times b)$-matrix

$Y_{1,b}$.

THEOREM 3 : [7] *If* D *is point stable with* s *nonzero eigenvalues, then*

(a) $C(D) \in \{s-1, s\}$;

(b) $C(D) = s-1$ *if and only if* D *is connected.*

The proof of Theorem 3 requires similar arguments to the proof of Theorem 2.

Theorems 2 and 3 give us the following "outer" classification of the point stable designs with s nonzero eigenvalues into three classes:

$R(D) = s-1, \quad C(D) = s-1$;

$R(D) = s, \quad C(D) = s-1$;

$R(D) = s, \quad C(D) = s$.

Now we partition each of these three classes into four subclasses according to the values of two other ranks.

3. POINT RANK AND BLOCK RANK

Motivated by the Hoffman equation $h(N) = J$, we define (following Neumaier) the *point rank* $Pr(D)$ of a design D as the minimal degree of a polynomial $f(X) \in R[X]$ with $f(N) = tJ$ for some $t \in R$. The *block rank* $Br(D)$ is defined as the point rank $Pr(D')$ of the dual D' of D. Using similar arguments as in the preceeding proofs, we obtain

THEOREM 4 : [7] *If* D *is point stable with* s *nonzero eigenvalues,* $M = A^T A$, *then*

(a) $Pr(D) \in \{s-1, s, s+1\}$;

(b) *if* $C(D) = s-1$, *then* $Pr(D) = s-1 \Leftrightarrow \det N > 0$,
$Pr(D) = s \Leftrightarrow \det N = 0$;

(c) *if* $C(D) = s$, *then* $Pr(D) = s \Leftrightarrow \det N > 0$,
$Pr(D) = s+1 \Leftrightarrow \det N = 0$;

(d) $Br(D) \in \{s-1, s, s+1\}$;

(e) *if* $R(D) = s-1$, *then* $Br(D) = s-1 \Leftrightarrow \det M > 0$,
$Br(D) = s \Leftrightarrow \det M = 0$;

(f) *if* $R(D) = s$, *then* $Br(D) = s \Leftrightarrow \det M > 0$,
$Br(D) = s+1 \Leftrightarrow \det M = 0$.

REMARK : In [8] the author proves that, for each of the four ranks of a point stable design D, there is exactly one real monic polynomial of minimal degree satisfying the corresponding rank equation.

4. CHARACTERIZATIONS

We characterize semi partial geometric designs, partial geometric designs, (r,λ)-designs and 2-designs by their ranks and obtain a generalization of the Bose-Bridges-Shrikhande-characterization of partial geometric designs.

From the preceeding theorems we easily obtain

THEOREM 5 : [7]

(a) $C(D) = 1$ *if and only if* D *is semi partial geometric;*

(b) $Pr(D) = 1$ *if and only if* D *is an* (r,λ)*-design;*

(c) $R(D) = Pr(D) = 1$ *if and only if* D *is a* 2*-design;*

(d) $R(D) = Br(D) = 1$ *if and only if* D' *is a* 2*-design;*

(e) $R(D) = Pr(D) = Br(D) = 1$ *if and only if* D *is a square* $(v=b)$ 2*-design;*

(f) *If* D *is point stable, then* $R(D) = 1$ *if and only if* D *is partial geometric.*

QUESTION : May we omit the hypothesis "D is point stable" in (f) ?

From Theorems 3 and 5 we obtain

COROLLARY 1 : *If* D *is connected and point stable with* s *non-zero eigenvalues, then* D *is semi partial geometric if and only if* $s = 2$.

An immediate consequence of Corollary 1 is

COROLLARY 2 : (Bose, Bridges, Shrikhande [1]) *If* D *is a connected* 1*-design with* s *nonzero eigenvalues, then* D *is partial geometric if and only if* $s = 2$.

BIBLIOGRAPHY

1. R.C. Bose, W.G. Bridges and M.S. Shrikhande, "A characterization of partial geometric designs", *Discrete Math.*, 16 (1976), 1-7.
2. R.C. Bose, S.S. Shrikhande and N.M. Singhi, "Edge regular multigraphs and partial geometric designs with an application to the embedding of quasi-residual designs", *Teorie Combinatorie*, Tomo I, Atti dei Convegni Lincei 17, (Accad. Naz. Lincei, Rome, 1976), pp.49-81.
3. A.J. Hoffman, "On the polynomial of a graph", *Amer. Math. Monthly*, 70 (1963), 30-36.
4. A. Neumaier, "Classification of 1-designs", personal communication.
5. A. Neumaier, "t ½-designs", *J. Combin. Theory Ser. A*, 28 (1980), 226-248.
6. K.E. Wolff, "Punkt-stabile und semi-partial-geometrische Inzidenzstrukturen", *Mitt. Math. Sem. Giessen*, 135 (1978), 1-96.
7. K.E. Wolff, "Rank classification of point stable designs", *Europ. J. Combinatorics*, to appear.
8. K.E. Wolff, "Uniqueness of the rank polynomials of point stable designs", *Math. Z.*, to appear.

Mathematisches Institut
Justus-Liebig-Universität
Arndstrasse 2
6300 Giessen
Federal Republic of Germany

OTHER TALKS

L. Babai : Primitive coherent configurations and permutation groups.

L.M. Batten : Quadrics and the Higman-Sims geometry.

J.-C. Bermond : On G-designs.

A. Beutelspacher : A characterization of finite projective spaces.

A. Brouwer : Construction of sets of mutually orthogonal Latin squares.

H.S.M. Coxeter : My graph.

F. De Clerck : Quadrics and partial geometries.

M. Dehon : Constructions of designs $S_\lambda(2, 3, v)$ without repeated blocks.

J. Doyen : Weiss graphs.

W.L. Edge : The tritangent planes of Bring's curve.

D. Glynn : What is a matrix?

R.L. Graham : Some recent results on null-designs.

X. Hubaut : Regular arcs.

D.R. Hughes : A failed attempt to construct a symmetric (78, 22, 6)-design.

J. Kahn : Varieties of combinatorial geometries.

E.S. Lander : Coding theory and a new multiplier theorem.

C. Mitchell : Block intersections in divisible designs.

S.E. Payne : Generalized quadrangles and the Higman-Sims technique.

L. Teirlinck : Embeddings in projective spaces.

R. Weiss : Moufang graphs.

P. Wild : Semisymmetric designs.

M. Willems : Restricted D i-spaces.

B.J. Wilson : Incompleteness of $(2^{2h-1} + 2^{h-1} - 2^h - 2, 2^{h-1})$-arcs in $PG(2, 2^h)$.

PARTICIPANTS

[illegible]. Andreassian	Central College Iowa
E.F. Assmus	Lehigh
L. Babai	Budapest
L.M. Batten	Winnipeg
J. -C. Bermond	Orsay
T. Beth	Erlangen
A. Beutelspacher	Mainz
A. Bichara	Rome
N. Biggs	London (Royal Holloway)
A. Brouwer	Amsterdam
A.A. Bruen	Western Ontario
P.J. Cameron	Oxford
P.V. Ceccherini	Rome
A.M. Cohen	Amsterdam
H.S.M. Coxeter	Toronto
R.M. Damerell	London (Royal Holloway)
F. De Clerck	Ghent
M. de Finis	Rome
M. Dehon	Brussels
M.J. de Resmini	Rome
M. Deza	Paris
J. Doyen	Brussels
W.L. Edge	Edinburgh
D.A. Foulser	Chicago Circle
M.J. Ganley	Glasgow
D. Glynn	Bologna
R. Graham	Bell Labs
W. Haemers	Eindhoven
M. Hall	Pasadena
C. Hering	Tübingen
D.G. Higman	Michigan
R. Hill	Salford
J.W.P. Hirschfeld	Sussex
X. Hubaut	Brussels
D.R. Hughes	London (Westfield)
D. Jungnickel	Giessen
J. Kahn	M.I.T.
M.J. Kallaher	Washington State
A.D. Keedwell	Surrey
G. Kelly	Waterloo
J. Key	Birmingham
E.S. Lander	Oxford
C. Lefèvre-Percsy	Brussels
H. Lenz	Berlin
R. Liebler	Colorado State
R.J. List	Birmingham
D. Livingstone	Birmingham
M. Marchi	Milan
C. Mitchell	Racal, Salisbury
A. Neuen	London (Westfield)
A. Neumaier	Freiburg
M. Oraee Yazdi	Sussex
U. Ott	Braunschweig
S.E. Payne	Miami, Ohio
N. Percsy	Mons
M. Ronan	Chicago Circle
A.R. Sadeh	Sussex
S.S. Sane	Florida
J. Saxl	Cambridge
E. Shult	Kansas State
D. Smit Ghinelli	London (Westfield)
G. Tallini	Rome
L. Teirlinck	Brussels
J.A. Thas	Ghent
J.H. van Lint	Eindhoven
K. Vedder	Giessen
A. Wagner	Birmingham
R. Weiss	Tufts
P. Wild	Ohio State
M. Willems	Antwerp
B.J. Wilson	London (Chelsea)
K.E. Wolff	Giessen